Membrane Distillation and Other Membrane-Related Applications for Water Cleaning and Desalination

Membrane Distillation and Other Membrane-Related Applications for Water Cleaning and Desalination

Guest Editor

Simona Renda

MDPI

Basel • Beijing • Wuhan • Barcelona • Belgrade • Novi Sad • Cluj • Manchester

Guest Editor
Simona Renda
Aragon Institute for
Engineering Research
University of Zaragoza
Zaragoza
Spain

Editorial Office
MDPI AG
Grosspeteranlage 5
4052 Basel, Switzerland

This is a reprint of the Special Issue, published open access by the journal *Membranes* (ISSN 2077-0375), freely accessible at: https://www.mdpi.com/journal/membranes/special_issues/G77J570JZQ.

For citation purposes, cite each article independently as indicated on the article page online and as indicated below:

Lastname, A.A.; Lastname, B.B. Article Title. *Journal Name* **Year**, *Volume Number*, Page Range.

ISBN 978-3-7258-7232-9 (Hbk)
ISBN 978-3-7258-7233-6 (PDF)
https://doi.org/10.3390/books978-3-7258-7233-6

Contents

About the Editor . . . . **vii**

Simona Renda
Membrane Distillation and Other Membrane-Related Applications for Water Cleaning and Desalination
Reprinted from: *Membranes* **2026**, *16*, 81, https://doi.org/10.3390/membranes16030081 . . . . . **1**

Nunzio Cancilla, Andrea Cipollina, Luigi Gurreri and Michele Ciofalo
Direct Contact Membrane Distillation: A Critical Review of Transmembrane Heat and Mass Transfer Models
Reprinted from: *Membranes* **2026**, *16*, 64, https://doi.org/10.3390/membranes16020064 . . . . . **4**

Qiaoying Wang, Bingbing Zhang, Jicheng Sun, Wenjia Zheng, Jie Zhang and Zhichao Wu
Comprehensive Investigation of Iron Salt Effects on Membrane Bioreactor from Perspective of Controlling Iron Leakage
Reprinted from: *Membranes* **2025**, *15*, 297, https://doi.org/10.3390/membranes15100297 . . . . . **41**

Lei Ni, Qiancheng Cui, Zhe Wang, Xueting Zhang, Jun Ma, Wenjuan Zhang and Caihong Liu
Zn^{2+}-Mediated Co-Deposition of Dopamine/Tannic Acid/ZIF-8 on PVDF Hollow Fiber Membranes for Enhanced Antifouling Performance and Protein Separation
Reprinted from: *Membranes* **2025**, *15*, 277, https://doi.org/10.3390/membranes15090277 . . . . . **58**

Saeedeh Saremi, Leonie Marie Scheer, Gerhard Braun, Marcus Koch, Markus Gallei and Matthias Faust
Investigations on the Particle Fouling and Backwash Efficiency During Microplastic Microfiltration–Particle Size Aspects
Reprinted from: *Membranes* **2025**, *15*, 272, https://doi.org/10.3390/membranes15090272 . . . . . **76**

Ester Komačková, Lukáš Sedlák, Ivan Červeňanský and Jozef Markoš
Exploration of Methods for In Situ Scale Removal During Magnesium Hydroxide Membrane Crystallization
Reprinted from: *Membranes* **2025**, *15*, 267, https://doi.org/10.3390/membranes15090267 . . . . . **97**

Tshifhiwa T. Tshauambea, Soraya P. Malinga and Patrick G. Ndungu
Treatment of Industrial Brine Using a Poly (Vinylidene Fluoride) Membrane Modified with Carbon Nanotubes
Reprinted from: *Membranes* **2025**, *15*, 220, https://doi.org/10.3390/membranes15080220 . . . . . **116**

Junliang Dong, Qianzhi Sun, Xiaolin Feng and Ruijun Zhang
High-Performance Thin Film Composite Nanofiltration (NF) Membrane Constructed on Modified Polyvinylidene Fluoride (PVDF) Substrate
Reprinted from: *Membranes* **2025**, *15*, 216, https://doi.org/10.3390/membranes15070216 . . . . . **141**

Saqr A. A. Al-Muraisy, Jiamin Wu, Mingliang Chen, Begüm Tanis, Sebastiaan G. J. Heijman, Shahrul bin Ismail, et al.
Oil and Water Recovery from Palm Oil Mill Effluent: A Comparative Study of PVDF and α-Al_2O_3 Ultrafiltration Membranes
Reprinted from: *Membranes* **2025**, *15*, 176, https://doi.org/10.3390/membranes15060176 . . . . . **155**

Qiang Zhou, Meiling Chen, Yushan Jiang, Linnan Zhang and Yanhong Wang
Development and Application of a Novel Ultrafiltration Membrane for Efficient Removal of Dibutyl Phthalate from Wastewater
Reprinted from: *Membranes* **2025**, *15*, 142, https://doi.org/10.3390/membranes15050142 . . . . . **178**

Samraa R. Khaleel, Salah S. Ibrahim, Alessandra Criscuoli, Alberto Figoli, Dahiru U. Lawal and Qusay F. Alsalhy
Influence of Silane Treatment on CNM/PAC/PVDF Properties and Performance for Water Desalination by VMD
Reprinted from: *Membranes* **2025**, *15*, 104, https://doi.org/10.3390/membranes15040104 . . . . . **193**

About the Editor

Simona Renda

Simona Renda is a postdoctoral researcher, currently under the Juan de la Cierva competitive program. She has been a member of the consolidated Catalysis and Reactor Engineering Group of the University of Zaragoza (Spain) since 2023 and a member of the Aragón Institute for Engineering Research (I3A) since 2025. She holds a BSc and MSc in Chemical Engineering and a PhD in Chemical Engineering from the University of Salerno (Italy), awarded with honors. Her research focuses on catalysis and reactor engineering for environmental applications, including, but not limited to, carbon capture and utilization, Claus tail gas treatment, and water desalination via membrane distillation. She has experience in both experimental work and CFD modelling, and her projects combine academic research with industrial collaboration. She has participated in more than 10 research and technological transfer projects, including two European H2020 projects, and has published more than 30 scientific papers in SCI-indexed journals, with more than 800 citations and an h-index of 14 (based on Scopus data, March 2026).

membranes

MDPI

Editorial

Membrane Distillation and Other Membrane-Related Applications for Water Cleaning and Desalination

Simona Renda

Catalysis and Reactor Engineering Group (CREG), Aragon Institute of Engineering Research (I3A), Universidad de Zaragoza, 50018 Zaragoza, Spain; srenda@unizar.es

1. Introduction

The increasing demand for clean and safe water across the globe represents one of the most pressing challenges of modern society. Population growth, industrialization, agricultural activities, and climate change are placing severe stress on conventional freshwater resources, accelerating the need for advanced and sustainable water purification technologies. Against this background, membrane distillation (MD) has undergone a significant "research boom" over the past decade, evolving from a niche seawater desalination method into a pivotal technology in various complex industrial wastewater treatments.

2. Overview of Special Issue Contributions

Membrane distillation is a thermally driven membrane process that exploits a hydrophobic, microporous membrane to separate a warm feed stream from a cooler permeate side [1]. The resulting vapor pressure difference enables the selective transport of water vapor through the membrane's pores while effectively retaining non-volatile species such as salts, organic compounds, and pathogens. Unlike pressure-driven membrane processes, MD operates at low hydraulic pressure and moderate temperatures, making it well suited for coupling with low-grade waste heat and renewable energy sources, including solar and geothermal energy [2].

A key advantage of membrane distillation is its ability to achieve a theoretically complete rejection of dissolved contaminants while maintaining stable performance, even at very high salinities [3]. This makes MD particularly attractive for challenging applications such as seawater desalination [4], hypersaline brine concentration [5], industrial wastewater treatment [6], and the removal of emerging contaminants [7] and microplastics [8]. In addition, MD systems' modularity and operational flexibility support their potential deployment in decentralized and small-scale water treatment configurations [9].

However, despite its promising characteristics, progress in the application of MD has been hindered by a lack of specifically designed membranes, recurrent pore wetting, high energy consumption compared to reverse osmosis, and limited consensus on mathematical modeling for heat and mass transfer. For this reason, the technique is still under active development, with research efforts focused on addressing limitations related to membrane wetting, fouling, thermal efficiency, and long-term operational stability. Recent developments in membrane material design, surface functionalization, module configuration, and process integration has led to significant performance improvements, reinforcing MD's position as an innovative and versatile technology within the broader membrane science landscape.

3. Conclusions

This Special Issue, "Membrane Distillation and Other Membrane-Related Applications for Water Cleaning and Desalination", aims to capture recent advances in membrane distillation and related water purification technologies. It comprises one comprehensive review article, which critically examines the state of the art in MD, alongside nine original research papers, addressing critical knowledge gaps through a multidisciplinary collection of research focused on advanced materials, process modeling, and practical industrial implementation.

Together, these contributions provide a coherent overview of current research trends and highlight emerging directions that are expected to drive future developments in the field.

Despite these advancements, future research must prioritize the scaling up of these processes into continuous industrial operations and the development of multi-fiber modules to maximize effective surface area. Priority should be given to the design of nano-engineered or 3D-printed modules that allow for unprecedented control over temperature polarization and membrane architecture. Moreover, future efforts should integrate molecular-simulation-driven precision to target emerging contaminants such as PFAS and microplastics, ensuring that membrane technologies remain a sustainable and economically viable cornerstone of global water security.

Acknowledgments: I would like to express my sincere gratitude to Miguel Menéndez for encouraging me to take on the role of Guest Editor for this Special Issue. I am also deeply thankful to all the authors who chose to contribute their work, as well as to the Editorial Board of *Membranes* for their invaluable support throughout the process.

Conflicts of Interest: The author declares no conflicts of interest.

List of Contributions:

1. Cancilla, N.; Cipollina, A.; Gurreri, L.; Ciofalo, M. Direct Contact Membrane Distillation: A Critical Review of Transmembrane Heat and Mass Transfer Models. *Membranes* **2026**, *16*, 64. https://doi.org/10.3390/membranes16020064.
2. Wang, Q.; Zhang, B.; Sun, J.; Zheng, W.; Zhang, J.; Wu, Z. Comprehensive Investigation of Iron Salt Effects on Membrane Bioreactor from Perspective of Controlling Iron Leakage. *Membranes* **2025**, *15*, 297. https://doi.org/10.3390/membranes15100297.
3. Ni, L.; Cui, Q.; Wang, Z.; Zhang, X.; Ma, J.; Zhang, W.; Liu, C. Zn2+-Mediated Co-Deposition of Dopamine/Tannic Acid/ZIF-8 on PVDF Hollow Fiber Membranes for Enhanced Antifouling Performance and Protein Separation. *Membranes* **2025**, *15*, 277. https://doi.org/10.3390/membranes15090277.
4. Komačková, E.; Sedlák, L.; Červeňanský, I.; Markoš, J. Exploration of Methods for In Situ Scale Removal During Magnesium Hydroxide Membrane Crystallization. *Membranes* **2025**, *15*, 267. https://doi.org/10.3390/membranes15090267.
5. Dong, J.; Sun, Q.; Feng, X.; Zhang, R. High-Performance Thin Film Composite Nanofiltration (NF) Membrane Constructed on Modified Polyvinylidene Fluoride (PVDF) Substrate. *Membranes* **2025**, *15*, 216. https://doi.org/10.3390/membranes15070216.
6. Tshauambea, T.T.; Malinga, S.P.; Ndungu, P.G. Treatment of Industrial Brine Using a Poly (Vinylidene Fluoride) Membrane Modified with Carbon Nanotubes. *Membranes* **2025**, *15*, 220. https://doi.org/10.3390/membranes15080220.
7. Al-Muraisy, S.A.A.; Wu, J.; Chen, M.; Tanis, B.; Heijman, S.G.J.; bin Ismail, S.; van Lier, J.B.; Lindeboom, R.E.F. Oil and Water Recovery from Palm Oil Mill Effluent: A Comparative Study of PVDF and α-Al_2O_3 Ultrafiltration Membranes. *Membranes* **2025**, *15*, 176. https://doi.org/10.3390/membranes15060176.

8. Zhou, Q.; Chen, M.; Jiang, Y.; Zhang, L.; Wang, Y. Development and Application of a Novel Ultrafiltration Membrane for Efficient Removal of Dibutyl Phthalate from Wastewater. *Membranes* **2025**, *15*, 142. https://doi.org/10.3390/membranes15050142.
9. Khaleel, S.R.; Ibrahim, S.S.; Criscuoli, A.; Figoli, A.; Lawal, D.U.; Alsalhy, Q.F. Influence of Silane Treatment on CNM/PAC/PVDF Properties and Performance for Water Desalination by VMD. *Membranes* **2025**, *15*, 104. https://doi.org/10.3390/membranes15040104.

References

1. Francis, L.; Ahmed, F.E.; Hilal, N. Advances in Membrane Distillation Module Configurations. *Membranes* **2022**, *12*, 81. [CrossRef] [PubMed]
2. Zhang, Z.; Atia, A.A.; Andrés-Mañas, J.A.; Zaragoza, G.; Fthenakis, V. Comparative Techno-Economic Assessment of Osmotically-Assisted Reverse Osmosis and Batch-Operated Vacuum-Air-Gap Membrane Distillation for High-Salinity Water Desalination. *Desalination* **2022**, *532*, 115737. [CrossRef]
3. Bush, J.A.; Vanneste, J.; Cath, T.Y. Membrane Distillation for Concentration of Hypersaline Brines from the Great Salt Lake: Effects of Scaling and Fouling on Performance, Efficiency, and Salt Rejection. *Sep. Purif. Technol.* **2016**, *170*, 78–91. [CrossRef]
4. Hsu, S.T.; Cheng, K.T.; Chiou, J.S. Seawater Desalination by Direct Contact Membrane Distillation. *Desalination* **2002**, *143*, 279–287. [CrossRef]
5. Hickenbottom, K.L.; Cath, T.Y. Sustainable Operation of Membrane Distillation for Enhancement of Mineral Recovery from Hypersaline Solutions. *J. Membr. Sci.* **2014**, *454*, 426–435. [CrossRef]
6. Ugarte, P.; Renda, S.; Cano, M.; Pérez, J.; Peña, J.Á.; Menéndez, M. Air-Gap Membrane Distillation of Industrial Brine: Effect of Brine Concentration and Temperature. *Ind. Eng. Chem. Res.* **2024**, *63*, 1546–1553. [CrossRef]
7. Costa, F.C.R.; dos Santos, C.R.; Amaral, M.C.S. Trace Organic Contaminants Removal by Membrane Distillation: A Review on Mechanisms, Performance, Applications, and Challenges. *Chem. Eng. J.* **2023**, *464*, 142461. [CrossRef]
8. Saremi, S.; Scheer, L.M.; Braun, G.; Koch, M.; Gallei, M.; Faust, M. Investigations on the Particle Fouling and Backwash Efficiency During Microplastic Microfiltration–Particle Size Aspects. *Membranes* **2025**, *15*, 272. [CrossRef] [PubMed]
9. Schwantes, R.; Bauer, L.; Chavan, K.; Dücker, D.; Felsmann, C.; Pfafferott, J. Air Gap Membrane Distillation for Hypersaline Brine Concentration: Operational Analysis of a Full-Scale Module–New Strategies for Wetting Mitigation. *Desalination* **2018**, *444*, 13–25. [CrossRef]

https://doi.org/10.3390/membranes16030081

membranes

MDPI

Review

Direct Contact Membrane Distillation: A Critical Review of Transmembrane Heat and Mass Transfer Models

Nunzio Cancilla [1], Andrea Cipollina [1,*], Luigi Gurreri [2] and Michele Ciofalo [1]

[1] Dipartimento di Ingegneria, Università degli Studi di Palermo, Viale delle Scienze Ed. 6, 90128 Palermo, Italy; nunzio.cancilla@unipa.it (N.C.); michele.ciofalo@unipa.it (M.C.)
[2] Dipartimento di Ingegneria Elettrica, Elettronica e Informatica, Università di Catania, Viale Andrea Doria 6 Ed. 3, 95125 Catania, Italy; luigi.gurreri@unict.it
* Correspondence: andrea.cipollina@unipa.it

Abstract

The present review summarizes a vast body of literature on the subject of Membrane Distillation (MD), with a special emphasis on the existing results and correlations for the transmembrane transport of heat and mass. The issue of saltwater physical properties is also discussed in depth, whereas the advective transport of heat and salt concentration in the feed and permeate compartments is only briefly mentioned but is beyond the scope of this review. The paper does not aim to provide a complete treatment of the subject of MD, which can be found in other publications. Rather, it suggests the data and correlations most suitable for the range of operating conditions typically expected in actual units implementing Direct Contact Membrane Distillation (DCMD), including hollow fiber designs, with a view to assist model development. The focus is on MD for water desalination, although some results may apply well to other fields.

Keywords: direct contact membrane distillation; heat and mass transfer; transmembrane transport model; desalination; hollow-fiber membrane; seawater

1. Introduction

To face the growing global concern of water scarcity, desalination has expanded rapidly, with a total installed capacity of ~35 billion m^3/year [1,2]. However, conventional desalination technologies like reverse osmosis (RO), multi-stage flash distillation (MSF), and multi-effect distillation (MED) are energy-intensive and often costly. Each cubic meter of produced fresh water consumes on the order of 2.6–8.5 kWh and emits significant CO_2 [2–5].

Membrane Distillation (MD) has emerged as a promising alternative desalination method [4,6] that could alleviate these energy and sustainability concerns [3,7]. MD is a thermally driven membrane process that operates at ambient or low pressures with a hydrophobic microporous membrane, which mediates evaporation and condensation [8–14]. In an MD system, a heated feed solution is brought into contact with one side of the membrane, while a cooler stream (distillate or coolant) is in contact with the opposite side (Figure 1a). The hydrophobic membrane prevents liquid penetration but allows water vapor to diffuse through its pores. A vapor–pressure gradient, maintained by the temperature difference between the hot feed and the cold permeate sides, drives water to evaporate from the feed, traverse the membrane as vapor, and condense into purified distillate on the cold side. This process yields a high-purity water product because

 https://doi.org/10.3390/membranes16020064

only volatile molecules (e.g., water) pass through the membrane, leaving non-volatile solutes behind.

Notably, producing water via MD typically requires much less electrical energy (e.g., ~0.6–1.8 kWh/m^3) since the primary energy input is thermal, and can be supplied by renewable or waste heat [3,7,9,12,15,16]. Indeed, since MD can operate at relatively low temperatures (40–80 °C), it can utilize inexpensive heat sources such as solar thermal collectors, geothermal heat, or industrial waste heat. By decoupling desalination from high electricity demands and leveraging abundant low-temperature heat, MD offers a route to reducing the carbon footprint of water production [3]. Moreover, MD offers other benefits [17]; for example, compared to RO, it requires less pre-treatment, exhibits energy consumptions which are less sensitive to the feed salinity, and has the potential for a far greater water recovery [18].

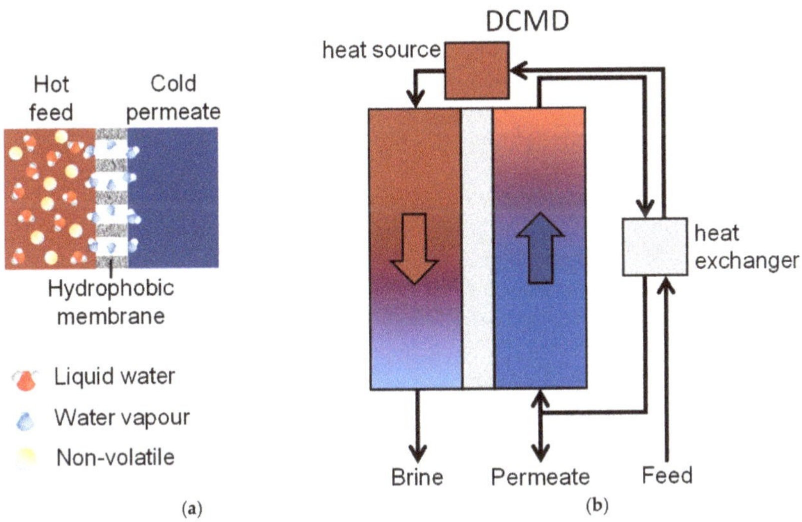

Figure 1. MD basic principles (**a**) and schematic of Direct Contact Membrane Distillation (DCMD) (**b**). Adapted from [17].

Despite its potential advantages, MD is still considered an emerging technology, with only a few pilot and full-scale implementations reported worldwide [3]. In particular, the cost-effectiveness of MD is hindered by several drawbacks [8,12,15,18,19], such as the high sensitivity to temperature polarization, risk of pore wetting, membrane fouling and scaling issues and, ultimately, uncertain energy consumption and costs, which span wide ranges [17].

MD can be configured in several modes [8,9,11,12,19,20]. In all MD configurations, evaporation at the hot side and condensation at the cold side are spatially separated by the hydrophobic membrane, but they differ in how the vapor is collected and condensed. The four classical configurations are as follows:

- Direct Contact MD (DCMD), where both the feed and permeate liquids are in direct contact with the membrane (Figure 1b);
- Air Gap MD (AGMD), where an air layer between the membrane and condensation surface serves as a barrier;
- Sweep Gas MD (SGMD), which uses a carrier gas to sweep vapor out of the membrane module to an external condenser;
- Vacuum MD (VMD), where a vacuum on the permeate side draws vapor, which then condenses externally.

https://doi.org/10.3390/membranes16020064

DCMD is the simplest of these: in this configuration, the vapor condenses directly into the cold stream inside the module. This simplicity means that DCMD requires no specialized condensers, vacuum pumps, or sweeping gas circuits, making it easy to implement with a compact layout. Indeed, the majority of bench-scale MD studies have employed the DCMD configuration due to its straightforward design and operation [20,21]. DCMD also tends to produce higher instantaneous flux than configurations with an air gap, since the vapor path is short and the membrane is in direct contact with the cooling medium. These benefits of DCMD, together with the general advantages of MD such as low operating pressure, moderate temperature requirements, and flexibility in handling a wide range of feeds, make it attractive for diverse applications [21]. It has been successfully tested on seawater, brines from RO plants, industrial effluents, and even in the food and beverage sector for concentrating liquids like fruit juices and dairy streams.

A crucial aspect of DCMD (and MD in general) is the module design, which strongly influences heat and mass transfer efficiency. The membranes used in MD are typically available in flat-sheet or tubular formats. Flat-sheet membranes are commonly assembled into plate-and-frame or spiral-wound modules, while tubular membranes, including capillary and hollow fiber types, are integrated into shell-and-tube configurations. Plate-and-frame modules are widely adopted in laboratory-scale research due to their simple assembly, compatibility with all MD configurations, and relatively low fouling propensity. However, their packing density is limited (surface/volume ratios of 100–800 m^2/m^3) [9,10,22]. Spiral-wound modules offer higher compactness (surface/volume ratios of 100–1200 m^2/m^3) [10,22], making them more suitable for commercial use, but they tend to be more prone to fouling, and their maintenance is more problematic. Tubular, capillary, and hollow fiber modules have exhibited through the years a progressive trend towards a reduction in membrane diameter and a corresponding increase in packing density, with surface/volume ratios ranging today from 20 to 300 m^2/m^3 for tubular, 600 to 1200 m^2/m^3 for capillary, and 2000 to 5000 m^2/m^3 for hollow fiber modules [9,10,22]. However, this increase in packing density also correlates with a higher susceptibility to fouling, especially in hollow fiber modules, which often require appropriate pre-treatment of the feed stream.

Over the years, the MD research community has developed a variety of innovative module configurations, including hollow fiber membranes, multi-effect modules, and solar-assisted systems. The goal of new designs is to enhance the thermal energy efficiency of MD by minimizing temperature polarization, increasing mass transfer across the membrane, and improving heat recovery from condensed water vapor. A recent review by Ali et al. [23] reports advances in the design of hollow fiber and flat sheet membrane modules for MD applications, discussing current research directions in module development and highlighting potential future trends. The authors suggest that, in the coming years, the choice of module configuration will increasingly depend on the specific needs of each application. For example, in desalination and brine valorization processes, larger and more efficient spiral-wound units may become the dominant option. Conversely, hollow fiber modules featuring design modifications that reduce polarization effects (e.g., undulated fibers) may be better suited for recovering minerals and nutrients from concentrated brines and industrial effluents. Meanwhile, plate-and-frame systems equipped with flat sheet membranes are likely to be preferred in vacuum multi-effect configurations for desalination and brine concentration, as well as in AGMD setups for wastewater treatment or niche applications where ease of maintenance is a critical requirement.

The dominance of heat and mass transfer phenomena in MD performance means that rigorous modeling is indispensable for design and scale-up. Physically, DCMD involves a complex interplay of simultaneous heat and mass transport through membrane and fluid

 https://doi.org/10.3390/membranes16020064

channels [8,11,12]. On the feed side, heat is consumed by evaporation at the membrane surface, cooling the adjacent liquid; on the permeate side, heat is released by condensation, warming the permeate stream. Within the membrane, heat is conducted across the polymer matrix and stagnant air in the pores, while vapor diffuses from feed to permeate driven by the vapor partial pressure gradient. Thus, mass transfer in DCMD is typically governed by a combination of molecular diffusion and Knudsen diffusion of water vapor through the pores (their relative importance depending on pore size and mean free path), whereas heat transfer encompasses convective heat transfer in the feed and permeate streams, conductive heat leak through the membrane, and the latent heat carried by the vapor [20].

These mechanisms are tightly coupled: as vapor is transported, it carries away latent heat, which in turn reduces the local temperature difference across the membrane, a critical feedback effect that can limit the driving force, referred to as temperature polarization [24–26]. Likewise, if the feed contains non-volatile solutes (salt, organic compounds, etc.), evaporation concentrates these solutes at the membrane interface, potentially reducing the vapor pressure. This phenomenon is known as concentration polarization and plays a minor role [27–31]. Because of these coupled effects, the accurate modeling of DCMD must simultaneously solve heat and mass balances in the system [20]. Typical models involve energy and mass conservation equations for the feed and permeate flow channels and within the membrane, coupled via boundary conditions at the membrane surfaces where evaporation/condensation occurs. Solving these equations can be complex, and researchers have employed a range of approaches from analytical solutions with simplifying assumptions, to numerical methods and computational fluid dynamics (CFD) [17,20,32]. Regardless of the approach, reliable models are needed to predict how operating conditions (feed temperature, flow rates, salt concentration), membrane properties (thickness, porosity, tortuosity, thermal conductivity), and module design (channel geometry, flow regime) influence the water flux and energy efficiency of DCMD. Such predictive capability is essential for process optimization and scale-up.

The present paper arises from a critical review of the literature on DCMD transmembrane models, which revealed that there is no universally adopted set of equations or standards. Different studies often employ different assumptions and correlations to describe the same underlying transport phenomena. Therefore, in the authors' opinion, there is a need for a critical appraisal of current heat and mass transfer models for DCMD, and there is room to establish a common basis for future modeling efforts. In the present review, the authors did their best to systematically analyze recent and classical models, highlighting inconsistencies, overlaps, and gaps. This comprehensive appraisal reveals that, while numerous models exist, many can be reconciled by recognizing equivalent formulations, excluding non-justified or even non-physically based equations, and acknowledging a few errors prevalent in past studies. The paper emphasizes the need for a harmonized modeling framework based on well-validated assumptions and a consistent terminology. This paper does not aim to provide a complete treatment of the subject of MD, which can be found in other publications [8,10,20,28]. Rather, it suggests the data and correlations most suitable for the range of operating conditions typically expected in actual units implementing DCMD, including hollow fiber designs, with a view to assist model development. The focus is on MD for water desalination, although some results may apply well to other fields.

2. Thermophysical Properties of Salt Water

A crucial step in the prediction of MD performance is the accurate expression of the thermophysical properties of salt water (assumed here to be a simple NaCl solution) as functions of temperature and salt concentration. Several quantities are used to measure the

https://doi.org/10.3390/membranes16020064

concentration of salt in water. It is important to define them rigorously and to provide the relevant conversion formulae because different definitions are used in different correlations for physical properties and transport phenomena.

- Salinity S, or mass fraction, is defined as the ratio of mass of salt to mass of solution and, of course, is dimensionless and insensitive to the mass units adopted, provided they are the same for salt and the solution (although it is often measured in g/kg, i.e., as $1000 \cdot S$, in percent, i.e., as $100 \cdot S$, or in ppm, or parts per million, i.e., as 10^6 S).
- Molar fraction x is defined as the ratio of moles of salt to moles of solution (water + salt) and thus, like the salinity, it is dimensionless. The quantity x is related to S as follows:

$$x = \frac{S \cdot MW_w}{S \cdot MW_w + (1 - S)MW_s} \tag{1}$$

in which MW_w and MW_s are the molecular weights of water and salt, expressed, as is usual, in g/mol (18 and 58.44, respectively).

- Molality m is defined as the ratio of moles of salt to the mass of water, expressed in mol/kg. It is related to the above-defined salinity S as follows:

$$m = \frac{1000}{MW_s} \frac{S}{1 - S} \tag{2}$$

- Molarity M is defined as the ratio of moles of salt to volume of solution. If expressed, as is usual, in mol/L, it is related to the above-defined salinity S as follows:

$$M = \frac{S \cdot \rho(S)}{MW_s} \tag{3}$$

in which $\rho(S)$ is the density of the solution of salinity S, expressed in kg/m^3, as discussed below.

- Finally, for computational purposes, it may be preferable to express the molar concentration in moles of salt per cubic meter of solution rather than per liter, using the molar concentration C related to molarity M as follows:

$$C = 1000 \cdot M \tag{4}$$

For seawater Membrane Distillation, of particular importance are the case $C = 0$ (representative of the permeate, i.e., practically pure water) and the range C = 500–1000 mol/m^3, representative of the feed (seawater) and the resulting retentate (brine). Lower ranges become of interest for the MD of brackish water, while higher ranges are pertinent applications of MD to desalinate hypersaline solutions (e.g., RO brine). In regard to temperature, the range of interest is typically 10 to 90 °C, the lower end corresponding to the minimum expected temperature of the permeate and the upper end to the maximum expected temperature of the feed.

An extensive and recent review of the physical properties of seawater is provided by Sharqawy et al. [33]. It includes correlations, tables and charts for density ρ, dynamic viscosity μ, specific heat c_p, thermal conductivity k, vapor saturation pressure p_{sat}, latent heat of vaporization H_{fg} (present notation) and other properties. Another less recent but very extensive source of thermodynamics data, tables and correlations can be found in "Appendix A" of a textbook by El-Dessouky and Ettouney [34], who, however, omit to acknowledge the original sources of the various results presented and the expected accuracy of the various correlations. Other proposed data and correlations are scattered through the literature.

 https://doi.org/10.3390/membranes16020064

2.1. Density

The most accurate and recent correlation, valid to within ±0.1% in the range t = 0–180 °C, S = 0–0.16 kg/kg, is that proposed by Sharqawy et al. [33], which can be written as follows:

$$\rho(t,S) = \rho_0(t) + \Delta\rho(t,S) \tag{5}$$

in which

$$\rho_0(t) = 9.999 \cdot 10^2 + 2.034 \cdot 10^{-2} t - 6.162 \cdot 10^{-3} t^2 + 2.261 \cdot 10^{-5} t^3 - 4.657 \cdot 10^{-8} t^4 \tag{6}$$

is the density of pure water, while

$$\Delta\rho(t,S) = 8.020 \cdot 10^2 S - 2.001 S\, t + 1.677 \cdot 10^{-2} S\, t^2 - 3.060 \cdot 10^{-5} S\, t^3 - 1.613 \cdot 10^{-5} S^2 t^2 \tag{7}$$

is an additive term accounting for the influence of salinity. Similar results in the same range of temperature and salinity are provided by the more complex correlation reported by El-Dessouki and Ettouney [34], in their Equation (A.1), which, however, does not indicate its expected accuracy.

The behavior of ρ as a function of temperature t in the range 0–100 °C for different values of the salinity S from 0 to 0.1 kg/kg is shown in Figure 2. Note that density is affected to a similar extent by temperature and salinity; seawater (S = 0.03 kg/kg) at 70° has about the same density as freshwater at 10 °C.

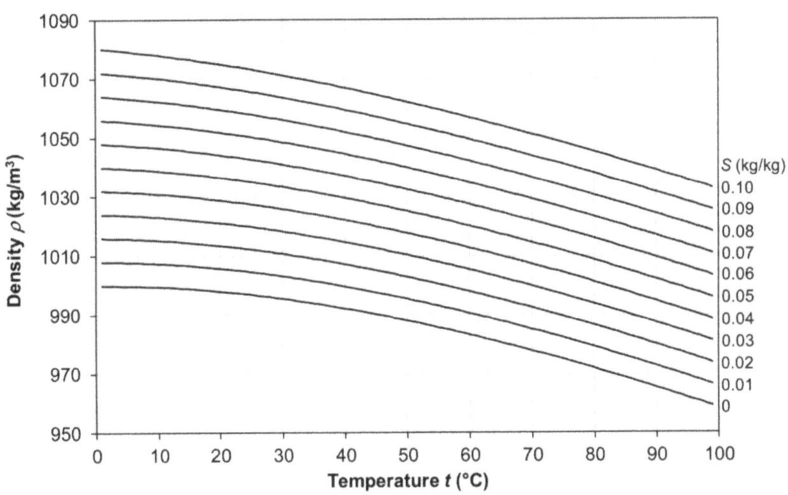

Figure 2. Density ρ of a saline solution as a function of temperature t for different values of the salinity S, based on Equations (5)–(7).

If one wishes to express the solution density, or any other of the physical properties described in the following, as a function of the molarity M rather than of the salinity S, simply inverting Equation (3) to obtain S as a function of M yields the following:

$$S = \frac{M \cdot MW_s}{\rho(t,S)} \tag{8}$$

which, however, contains the unknown density as a function of S. To resolve this issue, for practical purposes, ρ can be approximated as an explicit function of t and M as follows:

$$\rho(t,M) = \rho_0(t) + 41.6M \tag{9}$$

https://doi.org/10.3390/membranes16020064

in which $\rho_0(t)$ is given by Equation (6). The obtained value can then be substituted for the denominator in Equation (8) to obtain in explicit form the salinity S as a function of t and M:

$$S \approx \frac{M \cdot MW_s}{\rho_0(t) + 41.6M} \tag{10}$$

2.2. Dynamic Viscosity

A correlation offering a good compromise between accuracy and range of validity is that proposed by Isdale et al. [35], declared to be valid to within ±1% in the range t = 10–180 °C, S = 0–0.15 kg/kg:

$$\mu(t, S) = \mu_0(t)\left(1 + A_1 S + B_1 S^2\right) \tag{11}$$

in which $\mu_0(t)$ is the dynamic viscosity of pure water, for which the authors used the correlation proposed by Korosi and Fabuss [36]:

$$\mu_0(t) = \exp\left(-10.7019 + \frac{604.129}{139.18 + t}\right) \tag{12}$$

while the terms A_1 and B_1 are the following:

$$A_1 = 1.474 \cdot 10^{-3} + 1.5 \cdot 10^{-5} t - 3.927 \cdot 10^{-8} t^2 \tag{13}$$

$$B_1 = 1.073 \cdot 10^{-5} - 8.5 \cdot 10^{-8} t + 2.23 \cdot 10^{-10} t^2 \tag{14}$$

The same correlation, in a slightly different formulation, is provided by El-Dessouki and Ettouney [34], in their Equation (A.3), which is declared valid in the range t = 10–180 °C, S = 0–0.13 kg/kg (however, they omit a crucial factor 10^{-3} in their Table A.3).

The behavior of μ as a function of temperature t in the range 10–100 °C for different values of the salinity S from 0 to 0.1 kg/kg, as predicted by Equations (11)–(14), is shown in Figure 3.

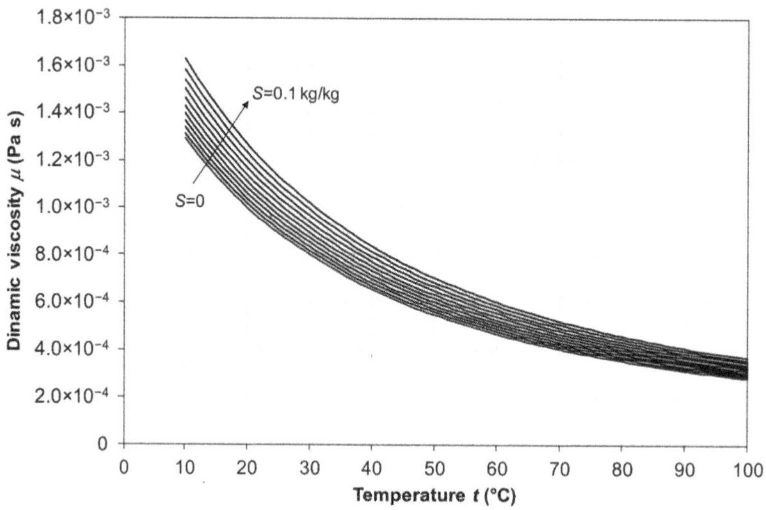

Figure 3. Dynamic viscosity μ of a saline solution as a function of temperature t for different values of the salinity S, based on Equations (11)–(14).

 https://doi.org/10.3390/membranes16020064

2.3. Specific Heat Capacity

A correlation declared to be valid to within ±0.28% in the range t = 0–180 °C, S = 0–0.18 kg/kg, is that proposed by Jamieson et al. [37]:

$$c_p(T,S) = A_2 + B_2T + C_2T^2 + D_2T^3 \tag{15}$$

in which T is the absolute temperature ($T = t + 273.15$) and the terms A, B, C and D are as follows:

$$A_2 = 5.328 - 9.76 \cdot 10^{-2}S + 4.04 \cdot 10^{-4}S^2 \tag{16}$$

$$B_2 = -6.913 \cdot 10^{-3} + 7.351 \cdot 10^{-4}S - 3.15 \cdot 10^{-6}S^2 \tag{17}$$

$$C_2 = 9.6 \cdot 10^{-6} - 1.927 \cdot 10^{-6}S + 8.23 \cdot 10^{-9}S^2 \tag{18}$$

$$D_2 = 2.5 \cdot 10^{-9} + 1.666 \cdot 10^{-9}S - 7.125 \cdot 10^{-12}S^2 \tag{19}$$

Almost identical results, despite the different coefficients and units, are provided by the correlation reported in El-Dessouki and Ettouney [34], in their Equation (A.2), which is declared valid in the range t = 20–180 °C, S = 0.02–0.16 kg/kg.

The behavior of c_p as a function of temperature t in the range 0–100 °C for different values of the salinity S from 0 to 0.1 kg/kg, as provided by Equations (15)–(19), is shown in Figure 4. Note the comparable influence of salinity and temperature on specific heat capacity.

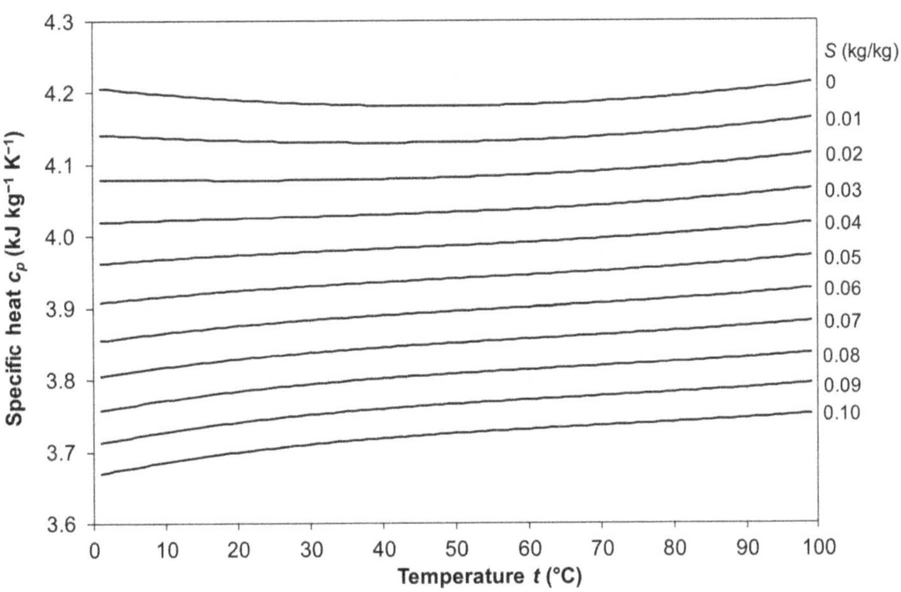

Figure 4. Specific heat capacity c_p of a saline solution as a function of temperature t for different values of the salinity S, based on Equations (15)–(19).

2.4. Thermal Conductivity

Among the experimental studies presented in the literature, the most extensive (t = 0–180 °C, S = 0–0.16 kg/kg) is probably that by Jamieson and Tudhope [38], who

https://doi.org/10.3390/membranes16020064

declared their results to be approximate to within ±3% in this range of validity. Unfortunately, the correlation proposed by these authors, in their Equation (2), is obviously affected by errors and does not match the results presented by the same authors in tabular form, in their Table VI. A modified form of Jamieson and Tudhope's correlation which matches the original experimental data was presented by El Dessouki and Ettouney [34]

$$\log_{10}(k) = \log_{10}\left(240 + 2 \cdot 10^{-4}S\right) + 0.434\left(2.3 - \frac{343.5 + 0.037S}{T}\right)\left(1 - \frac{T}{647.3 + 0.03S}\right)^{\frac{1}{3}} \tag{20}$$

This was also reported, almost identically, by Sharqawy et al. [33] (who, however, attributed it to Jamieson and Tudhope).

The behavior of k as a function of temperature t in the range 0–100 °C for different values of the salinity S from 0 to 0.1 kg/kg as predicted by Equation (20) is shown in Figure 5. Only a few curves (corresponding to S = 0, 0.05 and 0.1 kg/kg) are shown for clarity purposes. Note that the influence of salinity on k is negative and small; letting S increase from 0 to 0.1 kg/kg causes a decrease in k of only ~1%.

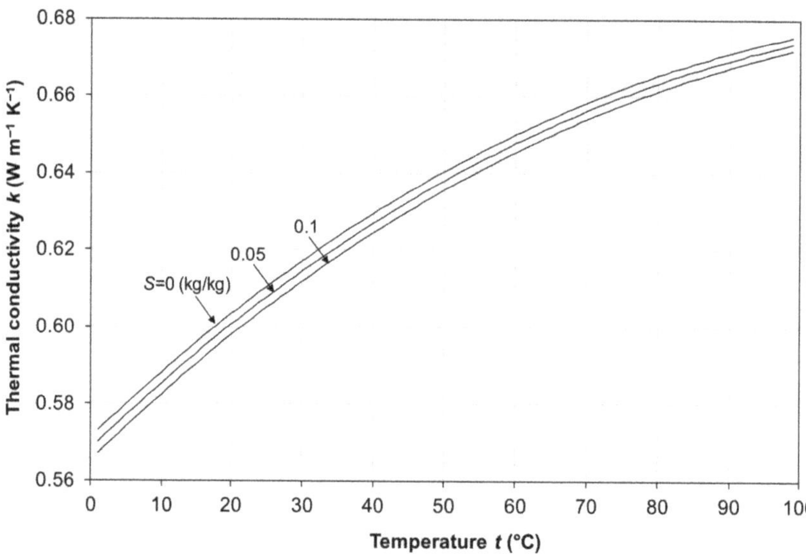

Figure 5. Thermal conductivity k of a saline solution as a function of temperature t for different values of the salinity S, based on Equation (20).

Two properties of special interest, in that heat and mass transfer across a membrane are heavily influenced by them, are the vapor saturation pressure p_{sat} and the latent heat of vaporization H_{fg}.

2.5. Vapor Saturation Pressure

The vapor saturation pressure of salt water is commonly expressed by several authors, e.g., Khayet and Matsuura [28] and Alkhudhiri et al. [8], as follows:

$$p_{sat}(T, S) = p_{sat,0}(T)a_w(S) \tag{21}$$

in which $p_{sat,0}$ is the saturation pressure of pure water and a_w is the water activity in a saline solution, expressed as a function of the salinity S or of any other equivalent quantity.

In its turn, $p_{sat,0}$ is commonly expressed using Antoine's law in one of its many different forms. In all of the following expressions, (22)–(26), the saturation pressure $p_{sat,0}$

https://doi.org/10.3390/membranes16020064

is in Pa and T is the absolute temperature in K. The following formulation of Antoine's equation is adopted by Khayet and Matsuura [28], Yadav et al. [39], Khalifa et al. [40] and Momeni et al. [41] (in this last paper, in their Equation (12), there is a wrong "plus" sign instead of "minus" in the denominator):

$$p_{sat,0}(T) = \exp\left(23.1964 - \frac{3816.44}{T - 46.13}\right) \tag{22}$$

(ranges of validity and accuracy are not provided).

Formulae possessing the same structure as Equation (22), but with slightly different numerical coefficients, are adopted by Ansari et al. [42] and Song et al. [43] and yield values of $p_{sat,0}$ practically identical to Equation (22). Other authors, e.g., Bouguecha et al. [44], Eykens et al. [45], Kuang et al. [46], Lai and Zhang [47] and Kariman et al. [48], adopt yet another set of coefficients, which, however, yields values of $p_{sat,0}$ ~1% larger over most of the range 0–100 °C.

Finally, a different formulation is reported by the ASHRAE Fundamentals Handbook [49]:

$$p_{sat,0}(T) = \exp\left[\begin{array}{l} \frac{-5800.2206}{T} + 1.3914993 - 4.8640239 \cdot 10^{-2}T + \\ +4.1764768 \cdot 10^{-5}T^2 - 1.4452093 \cdot 10^{-8}\, T^3 + 6.5459673 \cdot \ln(T) \end{array}\right] \tag{23}$$

and is claimed to be valid to within ±0.1% in the range 0–200 °C. Note that some authors, e.g., Wang et al. [50], correctly report Equation (23), while the version of the same equation reported by Sharqawy et al. [33] contains an apparently minor typo in the quadratic coefficient, but actually yields values ~5–6% too large. Equations (22) and (23) yield very similar values, the discrepancy being less than 100 Pa for t = 0 to 100 °C.

In regard to the dimensionless factor a_w (water activity in the solution), it is generally regarded as independent of temperature. Some authors, e.g., Bouguecha et al. [44], assume a_w = 1, thus neglecting altogether the influence of salinity. Several authors, e.g., Song et al. [43], Khayet and Matsuura [28], Chang et al. [51], Yadav et al. [39], Lai and Zhang [47] and Kariman et al. [48], write the activity as follows (using the present notation):

$$a_w = x_w\gamma_w = (1 - x)\gamma_w \tag{24}$$

$x_w = 1 - x$ being the molar fraction of water and γ_w an activity coefficient. Alkhudhiri et al. [8] suggest assuming, at least as a first approximation, γ_w = 1, i.e., using Raoult's law:

$$a_w \approx 1 - x \tag{25}$$

Kuang et al. [46], Khalifa et al. [40] and Kariman et al. [48] adopt the following more accurate formulation:

$$a_w = (1 - x)\left(1 - 0.5x - 10x^2\right) \tag{26}$$

As an alternative to Equation (24), Ansari et al. [42] suggest for a_w the following expression:

$$a_w = 1 - 0.03112m - 0.001482m^2 \tag{27}$$

in which m is declared to be the "molality" but is attributed the (wrong) units of mol/L, appropriate for "molarity".

The saturation pressure values for salt water can also be deduced from experimental Boiling Point Elevation (BPE) data as reported by various authors, e.g., Bromley et al. [52]. A comparison between Equation (25) (Raoult's law), Equation (26) and Equation (27) (with m = molality) is conducted in Figure 6 by reporting the decrement in p_{sat} with respect to that of pure water as a function of the temperature for a salinity S = 0.03 kg/kg, a value typical of seawater. Values deduced from BPE measurements [52] are also reported as symbols for some temperatures. Equation (27) is that which best approximates Bromley's

https://doi.org/10.3390/membranes16020064

data, with Equation (26) yielding only slightly lower values and the simplest Equation (25) (Raoult's law) underpredicting the p_{sat} decrement rather severely.

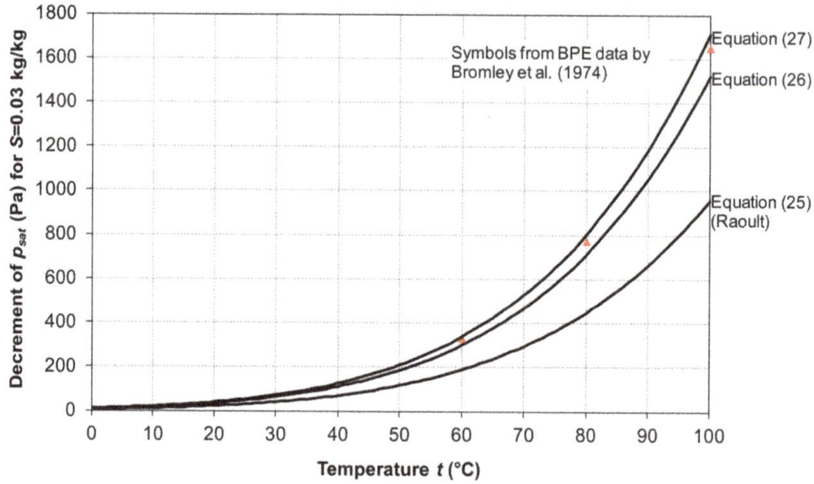

Figure 6. Decrement of the vapor saturation pressure p_{sat} predicted by using alternative formulations of the water activity a_w as a function of temperature t for a salinity S of 0.03 kg/kg. Data deduced from BPE measurements of Bromley et al. [52] are reported as symbols.

The saturation pressure p_{sat}, computed by using Equations (21), (22) and (26), is represented in Figure 7 as a function of temperature t for different values of the salinity S (related to the molar fraction x by Equation (1)). As in Figure 5, only a few values of S are reported for clarity purposes. The figure shows that the presence of salt up to concentrations of 0.1 kg/kg reduces only marginally the saturation pressure of a solution.

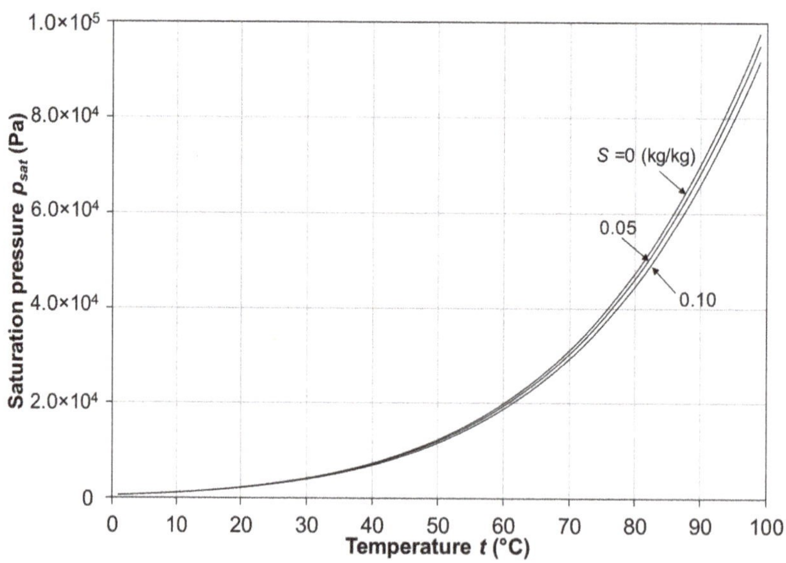

Figure 7. Vapor saturation pressure p_{sat} of a saline solution as a function of temperature t for different values of the salinity S, based on Equations (21), (22) and (26).

https://doi.org/10.3390/membranes16020064

2.6. Latent Heat of Vaporization

According to Sharqawy et al. [33], the latent heat of vaporization H_{fg} of a saline solution can be assumed equal to that of the liquid water mass fraction $(1 - S)$ and thus can be evaluated as follows:

$$H_{fg} = H_{fg,0}(1 - S) \tag{28}$$

in which $H_{fg,0}$ is the latent heat of pure water in J/kg and S is the salinity in kg/kg. In regard to $H_{fg,0}$, Sharkawy et al. [33] propose a 4th-order polynomial fit, declared to be valid to within ±0.01% in the range t = 0–200 °C:

$$H_{fg,0} = 2.501 \cdot 10^6 - 2.369 \cdot 10^3 t + 2.678 \cdot 10^{-1} t^2 - 8.103 \cdot 10^{-3} t^3 - 2.079 \cdot 10^{-5} t^4 \tag{29}$$

A simpler expression which departs from Equation (29) only ~1 kJ/kg in the present range of interest (t = 10–90 °C) is the linear fit reported by Ansari et al. [42]:

$$H_{fg,0} = 2.5028 \cdot 10^6 - 2.43818 \cdot 10^3 t \tag{30}$$

The formula reported by Khayet and Matsuura [28], Khalifa et al. [40] and other authors, i.e., $H_{fg,0} = 2.0243 \cdot 10^6 + 1753.5 \cdot T$, is obviously wrong, since it predicts $H_{fg,0}$ increasing with T.

The behavior of the latent heat of vaporization H_{fg} as a function of temperature t in the range 0–100 °C for different values of the salinity S from 0 to 0.1 kg/kg, as computed using Equations (28) and (29), is reported in Figure 8. Note the comparable influence of temperature and salinity on the latent heat.

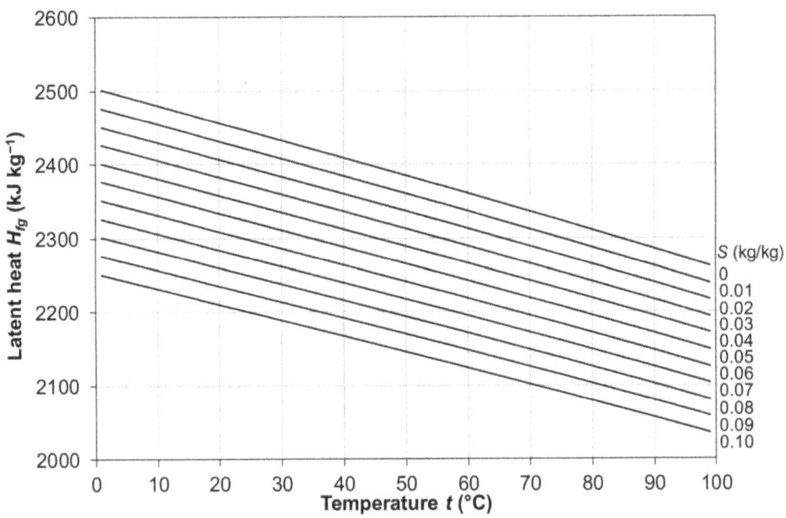

Figure 8. Latent heat of vaporization H_{fg} of a saline solution as a function of temperature t for different values of the salinity S, based on Equations (28) and (29).

3. Membrane Morphology and Properties

In Membrane Distillation, non-wetting, hydrophobic microporous membranes are commonly employed. Typical polymeric materials used for the fabrication of these membranes are polyvinylidene fluoride (PVDF), polytetrafluoroethylene (PTFE), polypropylene (PP) and polyethersulfone (PES). Ideally, optimal membranes adopted in MD systems should facilitate efficient (i.e., low) resistance to mass transfer while exhibiting minimal thermal conductivity to minimize heat dissipation across the membrane. Furthermore, they

 https://doi.org/10.3390/membranes16020064

ought to maintain structural integrity at elevated temperatures and demonstrate robust tolerance to corrosive substances, including acids and alkalis. There are several membrane properties, such as thermal conductivity, porosity, pore size distribution and tortuosity, the mean free path of water vapor molecules within the pores and the wetting pressure, for which correlations and equations can be found in the literature to allow for their estimation. These will be presented in the following subsections.

3.1. Membrane Thermal Conductivity

According to various authors, such as Alkhudhiri et al. [8], Khayet and Matsuura [28], Zhang et al. [53] and Olatunji and Camacho [20], most models for the effective thermal conductivity k_m of the membrane as a whole can be reconducted to either of the following expressions:

$$k_m = (1-\varepsilon)k_{sm} + \varepsilon k_g \tag{31}$$

$$\frac{1}{k_m} = \frac{(1-\varepsilon)}{k_{sm}} + \frac{\varepsilon}{k_g} \tag{32}$$

where k_{sm} is the thermal conductivity of the solid matrix (polymer), k_g is the thermal conductivity of filling gas (air and/or water vapor), and ε is the porosity of the membrane. Equation (31) (called the "parallel", "Voigt" or "isostrain" model by analogy with the elastic modulus of composites) is preferred by Chang et al. [51], Yadav et al. [39], Schofield et al. [54], Momeni et al. [41] and several other authors, while Equation (32) ("series", "Reuss" or "isostress" model) is preferred by Phattaranawik et al. [55].

García-Payo and Izquierdo-Gil [56] compared nine different models for the prediction of the membrane effective thermal conductivity against experimental data for two PVDF, two unsupported PTFE and two supported PTFE membranes. They concluded that the parallel model (31) largely overestimates the thermal conductivity, whereas the series model (32) slightly underestimates it. The best fit for all tested membranes, especially at large porosities (>60%), was found to be obtained by the following formula, recommended also by Hitsov et al. [32] and Ansari et al. [42]:

$$k_m = k_g \frac{1+2\zeta(1-\varepsilon)}{1-\zeta(1-\varepsilon)} \tag{33}$$

in which the coefficient ζ is

$$\zeta = \frac{k_{sm}-k_g}{k_{sm}+2k_g} \tag{34}$$

Figure 9 compares the values of the membrane effective thermal conductivity k_m (in W m^{-1} K^{-1}), predicted as a function of the porosity ε by using Equation (31) (parallel model), Equation (32) (series model), and Equations (33) and (34) (combined model) for k_{sm} = 0.14 W m^{-1} K^{-1} (representative of PP at 25 °C) and k_g = 0.026 W m^{-1} K^{-1} (representative of air at the same temperature). The experimental value measured by Schofield et al. [54] for dry air-filled Enka PP membranes with a thickness of 0.1 mm, porosity of 0.75, and nominal pore size of 0.1 μm (k_m = 0.046 ± 0.005 W m^{-1} K^{-1}) is also reported with error bars.

All three models correctly converge to k_{sm} in the limiting case ε = 0 and to k_g in the limiting case ε = 1, but the values predicted at intermediate porosities vary broadly. The experimental result of Schofield et al. [54] is overpredicted by ~13% by the parallel model, which yields k_m = 0.052 W/(m K); underpredicted by ~15% by the combined model, which yields k_m = 0.039 W/(m K); and grossly underpredicted (by ~30%) by the series model, which yields k_m = 0.0325 W/(m K). Note that these results partially contradict the remarks by Garcıa-Payo and Izquierdo-Gil [56] mentioned above. According

 https://doi.org/10.3390/membranes16020064

to Schofield et al. [54], it would be expected that most commercial microfiltration membranes would have a thermal conductivity between 0.04 and 0.06 W/(m K), increasing with decreasing porosity.

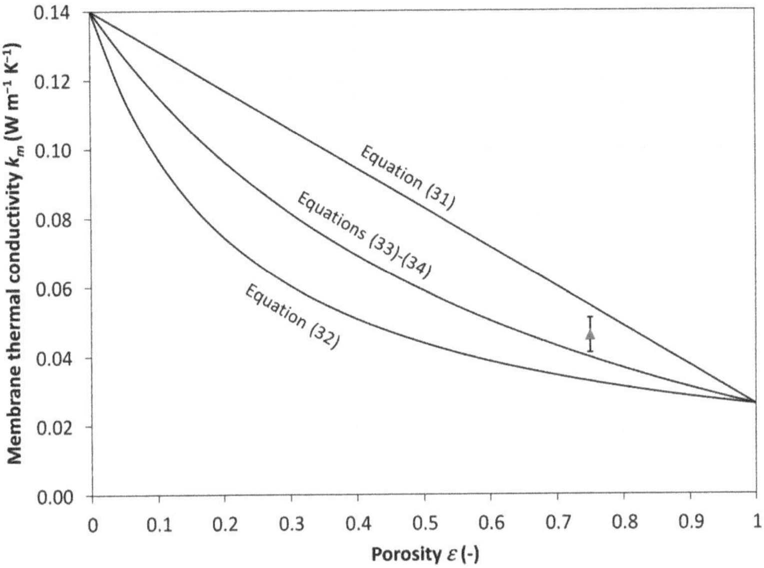

Figure 9. Effective thermal conductivity km of the membrane predicted by three different models (see text) for k_{sm} = 0.14 W m^{-1} K^{-1} and k_g = 0.026 W m^{-1} K^{-1}. The experimental result of Schofield et al. [54] is also reported with error bars.

The application of any of the above formulae requires a knowledge of the thermal conductivity of the pore-filling gas and of the polymeric matrix, both preferably evaluated at the membrane mean temperature $\langle t_m \rangle = (t_{f,m} + t_{p,m})/2$ (where $t_{f,m}$ and $t_{p,m}$ are the feed–membrane and membrane–permeate interface temperatures, respectively). In regard to the gas phase conductivity k_g, the membrane pores can be assumed to be filled by a mixture of air and water vapor. For air, accurate thermal conductivity values are reported by Stephan and Laesecke [57]. In the temperature range of interest (t = 10–90 °C) and at a pressure p of 1 bar, they can be approximated with excellent accuracy by the following quadratic formula:

$$k_{air} = -4.107 \cdot 10^{-8}t^2 + 7.631 \cdot 10^{-5}t + 2.417 \cdot 10^{-2} \quad (35)$$

Practically identical results are provided in the range of interest by the simpler (linear) formula adopted by Hitsov et al. [32] and Ansari et al. [42]:

$$k_{air} = 7.77 \cdot 10^{-5}t + 2.394 \cdot 10^{-2} \quad (36)$$

Alkhudhiri et al. [8] report the following even simpler formula:

$$k_{air} = 1.5 \cdot 10^{-3}\sqrt{T} \quad (37)$$

which, however, approximates Equations (35) and (36) only at low temperatures (~300 K) but departs from them at higher temperatures and thus is not recommended. Also, the following formula is reported by Olatunji and Camacho [20]:

$$k_{air} = 1.36 \cdot 10^{-3} + 3.885 \cdot 10^{-5}T + 1.66 \cdot 10^{-3}\sqrt{T} \quad (38)$$

This is unacceptable, since it predicts excessively high values of k_{air} at all temperatures.

https://doi.org/10.3390/membranes16020064

The influence of pressure is very small, and thus the above values, Equation (35) or Equation (36), can be used with confidence for the air filling the membrane pores with its partial pressure (possibly slightly less than atmospheric pressure).

For water vapor, properties (including thermal conductivity k_{vap}) should be evaluated along the saturation line rather than at a constant pressure. The NIST data [58] can be used, which can be fitted in the range of interest using the following simple quadratic formula:

$$k_{vap}= 1.6185 \cdot 10^{-7}t^2 + 6.1488 \cdot 10^{-5}t + 1.6779 \cdot 10^{-2} \quad (39)$$

The thermal conductivities of air at p = 1 bar (k_{air}) and water vapor along its saturation line (k_{vap}) are reported as solid lines in Figure 10 as functions of temperature t in the range 0–100 °C, computed by using Equations (35) and (39), respectively.

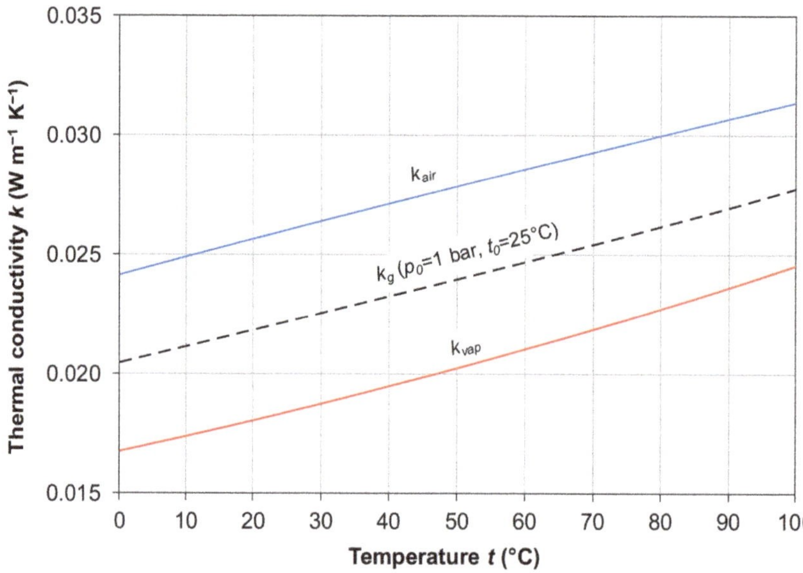

Figure 10. Solid lines: thermal conductivities in W/(mK) of air at p = 1 bar (k_{air}) and water vapor along its saturation line (k_{vap}), computed by using Equations (35) and (39), respectively, as functions of temperature t in °C. Broken line: thermal conductivity of the gas mixture filling the membrane pores according to Equations (40) and (41) for t_0 = 25 °C, p_0 = 1 bar.

In regard to the thermal conductivity of the mixture of air–water vapor (pore filling gas), calculating it as a simple weighted average of k_{air} and k_{vap}, using the molar fractions (or relative partial pressures) as weights, can be inaccurate. According to Udoetok [59], for binary mixtures, better results are obtained using the following formula (present notation):

$$k_g= 0.5 \cdot \frac{k_{air}k_{vap}}{x_{air}k_{air} + x_{vap}k_{vap}} + 0.5 \cdot \left(x_{air}k_{air} + x_{vap}k_{vap}\right) \quad (40)$$

which expresses the conductivity k_g of the gas mixture as an average between the values that would be obtained by assuming the two fractions as thermal conductances in series or in parallel. The molar fractions x of air and water vapor in the membrane pores can be estimated under the assumptions that, before operation, the pores are filled by air only at an absolute temperature T_0 and pressure p_0 and that neither the pore volume nor their air content change during operation. Therefore, at a mean membrane absolute temperature T, the partial pressure of air is $p_{air} = p_0T/T_0$, while the partial pressure of water vapor can be

https://doi.org/10.3390/membranes16020064

identified with $p_{sat}(T)$. The molar fractions to be used in Equation (40) are then given by the following:

$$x_{vap} = \frac{p_{sat}(T)}{p_{sat}(T) + p_0 T/T_0}; \qquad x_{air} = \frac{p_0 T/T_0}{p_{sat}(T) + p_0 T/T_0} \tag{41}$$

For example, assuming p_0 = 1 bar (10^5 Pa) and t_0 = 25 °C (T_0 = 298.15 K), and using Equations (35) and (39) for k_{air} and k_{vap}, Equations (40) and (41) yield for k_g the behavior shown in Figure 10 as a broken line. As expected, k_g is an intermediate between k_{air} and k_{vap}; it can be verified that moderate changes in p_0 and t_0 affect the computed values of k_g only negligibly.

In regard to the conductivity of the solid polymeric matrix, k_{sm} (the subscript "*sm*" is for "solid matrix"), according to Alkhudhiri et al. [8] and Khayet and Matsuura [28], the typical values of k_{sm} at 23 °C (~296 K) are in the range 0.17–0.19 W m^{-1} K^{-1} for PVDF, 0.25–0.27 W m^{-1} K^{-1} for PTFE and 0.11–0.16 W m^{-1} K^{-1} for PP. In addition, Khayet and Matsuura [28] report thermal conductivities at 348 K (~75 °C) of 0.21 W m^{-1} K^{-1} for PVDF, 0.29 W m^{-1} K^{-1} for PTFE and 0.2 W m^{-1} K^{-1} for PP.

Hitsov et al. [32], based mainly on data from Phattaranawik et al. [55], account for the temperature variation in k_{sm} through the following linear fit:

$$k_{sm} = \alpha \cdot 10^{-4} T + \beta \cdot 10^{-2} \tag{42}$$

where α and β are reported in Table 1 for different membrane materials. The same correlations are used by Ansari et al. [42] and Olatunji and Camacho [20].

Table 1. Thermal conductivity of common hydrophobic commercial membrane materials at different values of the mean membrane temperature $\langle t_m \rangle$. From data in Hitsov et al. [32].

Polymeric Material	Constants	k_{sm} at 20 °C (W m^{-1} K^{-1})	k_{sm} at 40 °C (W m^{-1} K^{-1})	k_{sm} at 60 °C (W m^{-1} K^{-1})	k_{sm} at 80 °C (W m^{-1} K^{-1})	k_{sm} at 100 °C (W m^{-1} K^{-1})
Polypropylene (PP)	α = 12.5 β = −23.5	0.13	0.16	0.18	0.21	0.23
Polyethersulfone (PES)	α = 4.17 β = 1.45	0.14	0.15	0.15	0.16	0.17
Polyvinylidene fluoride (PVDF)	α = 5.77 β = 0.914	0.18	0.19	0.20	0.21	0.22
Polytetrafluoroethylene (PTFE)	α = 5.77 β = 8.914	0.26	0.27	0.28	0.29	0.30

The values of k_{sm} for the same materials are represented as functions of temperature t in the range 20–100 °C in Figure 11.

For PTFE, Alkhudhiri et al. [8] report the following correlation:

$$k_{sm} = 4.86 \cdot 10^{-4} t + 0.253 \tag{43}$$

which correlates k_{sm} with the temperature t in Celsius rather than with the absolute temperature T. Equation (43) yields values similar to Equation (42), written for PTFE.

As a word of caution, it must be kept in mind that the conductivity data for polypropylene are highly scattered and depend on various factors such as crystallinity, the presence of additives and the kind of treatment performed. The data reported above lie in the lower range of experimental values from the literature.

https://doi.org/10.3390/membranes16020064

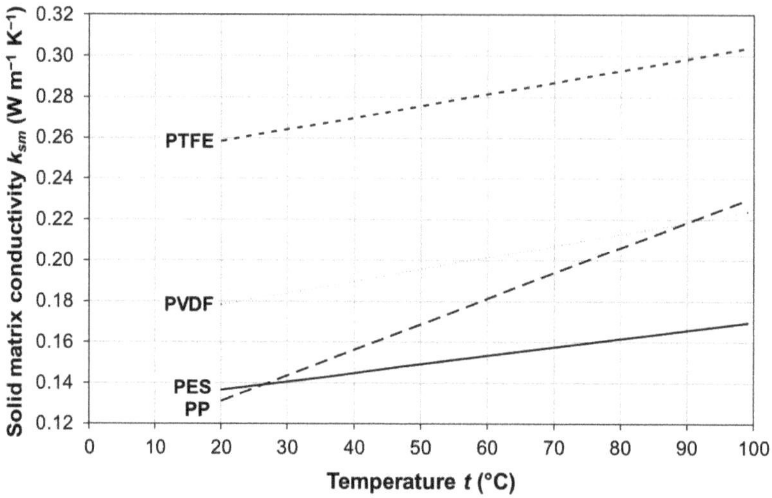

Figure 11. Thermal conductivity of the solid matrix for polymeric materials commonly adopted in membranes for MD, k_{sm}, as a function of temperature t in °C. Based on literature correlation in Equation (42).

3.2. Pore Size and Its Distribution

El-Bourawi et al. [60] and Lawson and Lloyd [61] report that, in MD systems, the pore size usually ranges between 100 nm and 1 μm. In general, the larger the pore size, the higher the permeate flux; on the other hand, the smaller the pore size, the lower the liquid penetration. For every feed solution and operating condition, the optimal pore size should be established. Moreover, the mass transfer mechanism varies with the pore size, as will be discussed in Section 4.

The pore diameter in an MD membrane is not uniform but follows a statistical distribution. A uniform pore size distribution is typically desirable in MD systems to enhance overall efficiency. A narrow distribution can also be advantageous, as it helps prevent the wetting of the largest pores or a subset of pores, which would otherwise reduce the effective membrane surface area and diminish membrane performance. Uniform pores lead to steadier fluxes and more reliable operation, particularly during extended use [28], whereas membranes with a large and non-uniform pore distribution are susceptible to wetting, which compromises the salt rejection efficiency [62].

Phattaranawik and Jiraratananon [63] evaluated the pore size distribution in PVDF and PTFE membranes by using scanning electron microscopy (Hitachi model S-900, Tokyo, Japan) and image processing software (AnalySIS provided by Soft Imaging System). They found that pore size is a broadly scattered variable, and even the mean pore diameter does not always comply with the nominal data provided by the manufacturer. For example, in PTFE Sartorious membranes with a nominal pore diameter of 0.2 μm, they measured an average pore diameter of 0.253 μm, with more than 10% of the pores above a diameter of 0.4 μm and ~5% below 0.1 μm. They suggested the log-normal law:

$$n(d_p) = \frac{1}{SD_{\log} d_p \sqrt{2\pi}} \exp\left\{ -\frac{1}{2} \left[\frac{\ln(d_p / \langle d_p \rangle)}{SD_{\log}} \right]^2 \right\} \tag{44}$$

where d_p and $\langle d_p \rangle$ are the generic and mean pore diameter, respectively, and $SD_{\log}$ is the standard deviation of the dimensionless variable $d_p/\langle d_p \rangle$. The dimensionless quantity $\exp(SD_{\log})$ is called the geometric standard deviation of the distribution.

 https://doi.org/10.3390/membranes16020064

More recently, Fan et al. [64] measured the pore size distribution in PVDF membranes using a gas–liquid displacement method. A membrane sample was first soaked in n-butanol, and then the wetting liquid was displaced by ultra-high purity N_2 gas with a stepwise increase of pressure, so that the largest pores were voided first while increasingly smaller pores required increasingly higher pressures, as dictated by the Young–Laplace equation. As remarked by Dong et al. [65], most models of transmembrane mass transfer make use of a mean pore size and could incur significant error due to the distribution of the pore size. In particular, Khayet et al. [66], using their theoretical model, found that the predicted permeance increases as the geometric standard deviation of the pore size distribution increases. The increment in membrane permeance is small in commercial membranes, which generally exhibit a narrow pore size distribution, whereas larger discrepancies may arise when laboratory-made membranes with broader pore size distributions are used. Woods et al. [67] used a numerical model to predict the water vapor flux through all pore sizes and estimate the effect of considering the pore size distribution on the estimates MD fluxes. They state that the uncertainties in modeling and experimental design tend to overshadow the influence of pore size distribution in DCMD (but not in VMD). The inaccuracy in the predicted water vapor flux due to neglecting the pore size distribution is closely related to the width of the distribution. For DCMD, where molecular diffusion usually dominates, the authors report that the discrepancy is limited to ~3.5% for membranes with a mean pore size of at least 100 nm and a geometric standard deviation of the pore size of 1.2.

3.3. Membrane Porosity and Pore Tortuosity

Membrane porosity ε is defined as the ratio of the volume of pores to the total volume of the membrane. In general, the larger the porosity, the larger the permeate flux across the membrane and the lower the conductive heat dissipation. In Membrane Distillation, ε ranges between 30 and 85% [60]. The membrane porosity ε can be determined, as suggested by Smolder and Franken [68], from density measurements using the following equation [68,69]:

$$\varepsilon = 1 - \frac{\rho_m}{\rho_{pol}} \tag{45}$$

in which ρ_m and ρ_{pol} are the densities of the membrane and polymeric material, respectively, and the mass of the filling gas is neglected. More recently, Fan et al. [64] estimated the porosity of PVDF membranes by means of gravimetric measurements, using the following equation:

$$\varepsilon = 1 - \frac{w/\rho_{pol}}{A\,\delta} \tag{46}$$

where w, A and δ are the mass, membrane area and thickness of the membrane sample, respectively, and ρ_{pol} is the density of the bulk polymer. Note that Equation (46) can be traced back to the more general Equation (45). Zou et al. [70] measured the porosity ε of a PVDF hollow fiber membrane using a kerosene immersion method, in which ε was evaluated by the following equation:

$$\varepsilon = \frac{4(w_2 - w_1)}{\pi(OD^2 - ID^2)L\rho_k} \tag{47}$$

in which w_2 and w_1 are the masses of the membrane after and before the immersion in kerosene, respectively; OD and ID are the outer and inner diameters of the hollow fibers, respectively; L is the length of the hollow fibers; and ρ_k is the density of kerosene.

Pore tortuosity τ is defined as the ratio of pore length l to membrane thickness δ. As remarked by Field et al. [71], it can play an important role in determining the membrane permeance to water vapor (see following Section 4). In general, the larger the pore tortuosity,

https://doi.org/10.3390/membranes16020064

the lower the permeate flux across the membrane. Pore tortuosity can be experimentally measured using gas permeation methods [72], but only under a considerable amount of assumptions, and the results often depend on the specific experimental technique adopted. Often, τ is simply assumed to be equal to 2. Khayet and Matsuura [28] observe that "...*in various theoretical models the tortuosity factor was used as an adjusting parameter to obtain better agreement between the simulated and the experimental permeate fluxes. In fact, it is difficult to measure the pore tortuosity.*"

Chang et al. [51], Khalifa et al. [40] and Yadav et al. [39] assume τ to be the inverse of the porosity ε:

$$\tau = \frac{1}{\varepsilon} \tag{48}$$

Other authors, including Srisurichan et al. [73], Alkhudhiri et al. [8], Lai and Zhang [47] and Kariman et al. [48], correlate the tortuosity τ to the porosity ε using the correlation suggested by Macki and Meares [74]:

$$\tau = \frac{(2-\varepsilon)^2}{\varepsilon} \tag{49}$$

For example, for the case of Liqui-Cel® hollow fiber membrane modules, Bouguecha et al. [44] report a tortuosity τ = 2.6 with a porosity ε = 0.4. Calculating the pore tortuosity via Equation (49) would yield the broadly different value τ = 6.4, while computing τ as the inverse of porosity, Equation (48), yields the more correct value of 2.5.

3.4. Liquid Entry Pressure

Liquid Entry Pressure (*LEP*) is a key membrane characteristic in MD. It is the threshold transmembrane pressure $p_f - p_p$ (in which the subscripts "*f*" and "*p*" are for feed and permeate, respectively), above which the feed liquid begins to pass across the pores of the hydrophobic membrane. It depends on the membrane hydrophobicity and pore size. To calculate the *LEP* value, Alkhudhiri et al. [8], Zhang et al. [53], Ciofalo et al. [75] and many other authors report an expression proposed by Franken et al. [76] and derived from the Young–Laplace equation:

$$LEP = -\frac{2B\Gamma\cos\theta}{r_{\max}} \tag{50}$$

in which B is a geometric pore coefficient (B = 1 for cylindrical pores), Γ is the liquid surface tension, θ is the contact angle between the solution and the membrane surface (see Figure 12) and r_{max} is the largest pore radius. Lopez et al. [77] provide for *LEP* an equation identical to Equation (50) apart from the nomenclature.

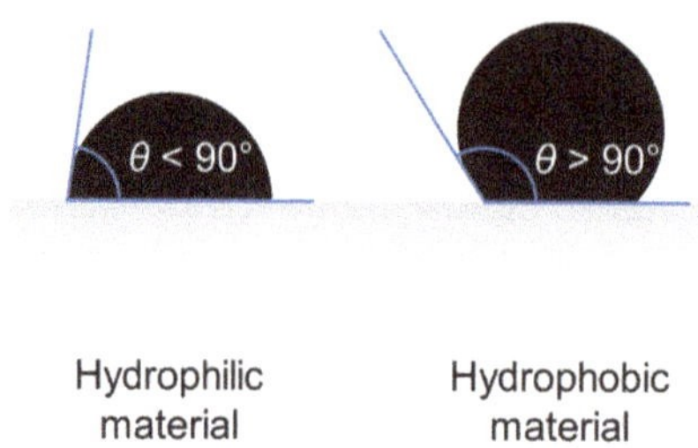

Figure 12. Schematic representation of the contact angle between a droplet of liquid and the membrane surface made by a hydrophilic or a hydrophobic material.

For the contact between a single water droplet and a polymeric surface, for PTFE, Alkhudhiri et al. [8] report a value of θ in the range 108–115°; for PVDF, both Curcio and Drioli [10] and Tomaszewska [78] suggest a value of θ = 107°; for PP, Curcio and Drioli [10]

https://doi.org/10.3390/membranes16020064

give $\theta = 120°$. In regard to the surface tension Γ, as reported by Sharkawy et al. [33], data for a saline solution are difficult to obtain and are scarce and rather scattered in the literature above a temperature of 35–40 °C. Zhang et al. [53], later followed also by Alkhudhiri et al. [8], express the influence of the molarity M on the surface tension as follows:

$$\Gamma = \Gamma_0 + \frac{\Delta\Gamma}{\Delta M}M \tag{51}$$

in which M is in mol/L, Γ_0 is the surface tension of pure water (72 mN/m at 25 °C) and the value of $\Delta\Gamma/\Delta M$ = 1.467 (mN/m)/(mol/L) was experimentally measured for NaCl solution by Sghaier et al. [79]. The influence of the dissolved salt is tiny up to seawater concentration; e.g., for M = 0.6 mol/L, one has Γ = 72.88 mN/m.

Rácz et al. [80] reported experimental LEP values for a variety of commercially available membranes including PTFE and PVDF flat-sheet and PP hollow fiber configurations, with mean pore diameters between 0.2 and 1.0 µm. Measured values of LEP ranged between 0.48 and 4.6 bar, with no correlation with material and geometry and only a loose correlation with the mean pore size.

4. Transmembrane Mass Transfer

With reference to Figure 13 below, the molar water vapor flux J_{vap} through the membrane, expressed in mol m^{-2} s^{-1}, can be modeled as follows:

$$J_{vap} = C_m\left(p_{sat,f} - p_{sat,p}\right) \tag{52}$$

where $p_{sat,f}$ and $p_{sat,p}$ are the vapor saturation pressures corresponding to the temperatures and salt concentrations on the feed–membrane and permeate–membrane interfaces, respectively, expressed in Pa, and C_m is the membrane permeance, expressed, consistently, in mol m^{-2} s^{-1} Pa^{-1}.

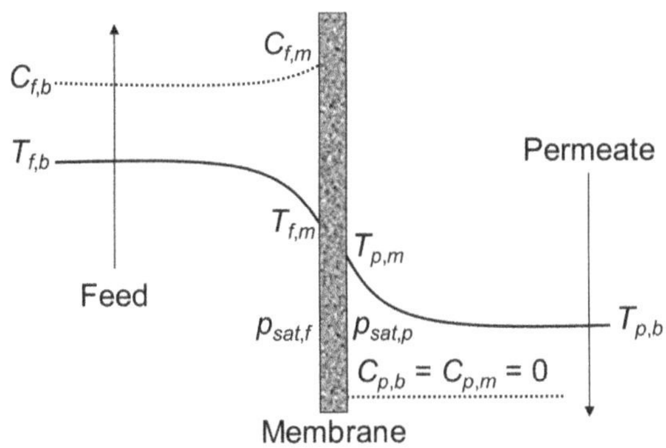

Figure 13. Schematic representation of the profiles of temperature (solid line) and concentration (broken line) on the two sides of the membrane.

Permeance is closely related to the mean free path of the water vapor molecules. The mean free path λ (expressed in m) of a molecule in a one-component gas, as given by Khayet and Matsuura [28], Alkhudhiri et al. [8] and several other authors, is as follows:

$$\lambda = \frac{K_B T}{\sqrt{2}\pi p \sigma^2} \tag{53}$$

where K_B is the Boltzmann constant (1.380649×10^{-23} J K^{-1}), T is the absolute temperature, p is the mean pressure in a pore (in Pa) and σ is the collision diameter of the molecule. Cussler [81], Mills [82] and Momeni et al. [41] report $\sigma_{vap} = 2.641 \cdot 10^{-10}$ m and $\sigma_{air} = 3.711 \cdot 10^{-10}$ m. A slightly lower value is reported by Alkhudhiri et al. [8] for

https://doi.org/10.3390/membranes16020064

σ_{air} ($3.66 \cdot 10^{-10}$ m). For example, for p = 1 bar (10^5 Pa) and T = 323.15 K (50 °C), Equation (53) yields for λ a value of $1.44 \cdot 10^{-7}$ m, or 0.144 µm, for water vapor and $0.73 \cdot 10^{-7}$ m, or 0.073 µm, for air.

For a binary gas mixture (in particular, air–water vapor), several authors, including Phattaranawik and Jiraratanon [63], Khayet and Matsuura [28], Ghaleni et al. [83], Dong et al. [65], Momeni et al. [41] and Kariman et al. [48], use the following formula:

$$\lambda_g = \frac{K_B T}{\pi p \left(\frac{\sigma_{vap}+\sigma_{air}}{2}\right)^2} \frac{1}{\sqrt{1+\frac{MW_{vap}}{MW_{air}}}} \tag{54}$$

For example, for a binary 50–50% mixture of air and water vapor under the same conditions as above (p = 1 bar, T = 323.15 K), Equation (54) yields $\lambda_g = 1.11 \cdot 10^{-7}$ m, a value intermediate, as expected, between those computed for the two separate components.

The membrane permeance C_m has different expressions depending on the value of the Knudsen number Kn = λ / d_p, the ratio of the mean free path λ to the mean pore diameter d_p. Two extreme cases are sketched in Figure 14.

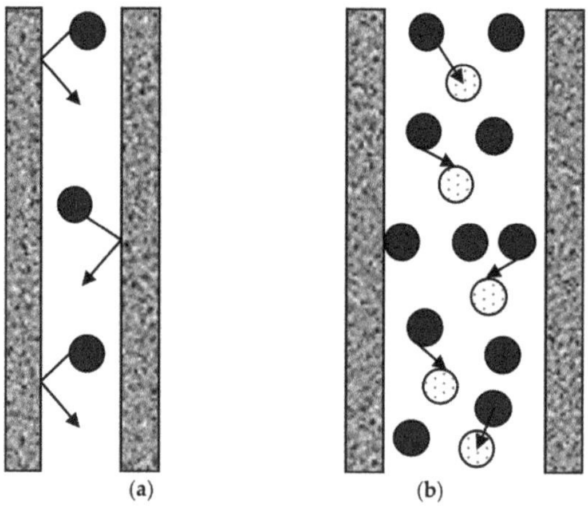

Figure 14. Transport mechanism through a pore of a membrane: (**a**) Knudsen regime and (**b**) diffusion regime.

Under conditions representative of MD, the mean free path of the water vapor molecules in a pore is of the order of 0.1 µm. As remarked by Dong et al. [65], since pore sizes typically range between 0.1 and 1 µm, the Knudsen number typically ranges between 0.1 and 1, thus falling in the intermediate, or transitional, region between the two above limiting cases. Similar conclusions are reached by Momeni et al. [41].

Thus, three cases are considered in the literature. They will be discussed in Section 4.1 through Section 4.3 here below.

4.1. Case I (Small Pores, Knudsen Regime): $d_p < \lambda$ or $Kn > 1$

According to most authors, e.g., Hitsov et al. [32] and Dong et al. [65], in this case (Knudsen regime), molecule–pore wall collisions dominate with respect to molecule–molecule collisions. Some authors, e.g., Khayet and Matsuura [28], apply a more stringent criterion considering the Knudsen regime to be dominant only if Kn > 10, i.e., $d_p / \lambda < 0.1$. Phattaranawik et al. [55] express the membrane permeance C_m, defined by Equation (52) but written as ${C_m}^{Kn}$ for "Knudsen", in terms of a diffusivity D^{Kn}:

$$C_m^{Kn} = \frac{D^{Kn}}{R_g T \delta} \tag{55}$$

 https://doi.org/10.3390/membranes16020064

In its turn, D^{Kn} (m^2 s^{-1}) is expressed as follows:

$$D^{Kn} = \frac{4}{3}\frac{\varepsilon d_p}{\tau}\sqrt{\frac{R_g T}{2\pi MW_{vap}}} \tag{56}$$

where ε is the membrane porosity, τ is the pore tortuosity, T is the absolute temperature, R_g is the universal gas constant (8.314 J mol^{-1}·K^{-1}) and MW_{vap} is the molecular weight of water (18·10^{-3} kg/mol). In regard to d_p, in [55], it is defined as the mean diameter of those pores that lie in the Knudsen portion of the pore size distribution (i.e., whose diameter is $<\lambda$); however, for simplicity, it can be identified with the mean pore diameter (implicitly assuming that all pores fall in the Knudsen regime).

From Equations (55) and (56) it follows that the membrane permeance $C_m{}^{Kn}$, expressed in mol m^{-2} s^{-1} Pa^{-1}, can be explicitly written as follows:

$$C_m^{Kn} = \frac{1}{3}\frac{\varepsilon d_p}{\tau\delta}\sqrt{\frac{8}{\pi R_g T\, MW_{vap}}} \tag{57}$$

Dong et al. [65], in their Equation (11.a), Olatunji and Camacho [20], in their Equation (36), Chang et al. [51], in their Equation (2), and Yadav et al. [39], in their Equation (6a), all give expressions for the membrane permeance in the Knudsen regime in terms of mass flux (i.e., in kg m^{-2} s^{-1} Pa^{-1}), which, once converted into molar permeance (i.e., divided by MW_{vap}) and allowing for changes in notation, become identical to Equation (57). Khayet and Matsuura [28] give an expression, their Equation (10.17), for $B_m{}^{Kn}$ (defined as the molar permeance in the Knudsen regime *per single pore*); this must be multiplied by the number of pores per unit area, equal to $4\varepsilon/(\pi\, d_p{}^2)$, to obtain the molar permeance per unit membrane area, $C_m{}^{Kn}$. The result, allowing for changes in notation, becomes again identical to Equation (57) above.

4.2. Case II (Large Pores, Diffusion Regime): $d_p > 100\lambda$ or $Kn < 0.01$

In this case (molecular diffusion regime), molecule–molecule collisions dominate with respect to molecule–pore wall collisions, and water vapor transport occurs through the concentration-driven diffusion of vapor molecules through the filling gas (air + water vapor). Phattaranawik and Jiraratananon [63], assuming the log-normal distribution of pore diameters in Equation (44), observe that, for a mean free path $\lambda \approx 0.11$ μm, this case can typically be ignored because it interests only a small fraction of the pores (those with $d_p > 11$ μm). However, Phattaranawik et al. [55] presented a detailed analysis of this case. By considering, for generality, opposite fluxes of air and water vapor through the pores, they obtained, in the present notation, the following expression for C_m (mol m^{-2} s^{-1} Pa^{-1}), called $C_m{}^D$ for "diffusion":

$$C_m^D = \frac{\varepsilon D_g}{\tau\delta R_g T}\left[1-\left(\frac{p_{vap}}{p}\right)\left(1-\frac{J_{air}}{J_{vap}}\right)\right]^{-1} \tag{58}$$

in which J_{vap} and J_{air} are the molar fluxes of water vapor and air, respectively, p_{vap} is the partial pressure of water vapor (which, as stated in Section 3.1, can be identified with $p_{sat}(T)$), $p = p_{vap} + p_{air}$ is the total pressure of the gas filling the pores and D_g is the gas diffusivity. Under the common assumption that air is trapped in the pores, one has $J_{air} = 0$, and Equation (58) becomes the following:

$$C_m^D = \frac{D_g}{R_g T}\frac{p}{p_{air}}\frac{\varepsilon}{\tau\delta} \tag{59}$$

Dong et al. [65], in their Equation (11.b), and Olatunji and Camacho [20], in their Equation (37), give expressions for the membrane permeance in the diffusive regime in terms of mass flux (i.e., in kg m^{-2} s^{-1} Pa^{-1}), which, once converted into molar perme-

 https://doi.org/10.3390/membranes16020064

ance (i.e., divided by MW_{vap}) and allowing for changes in notation, become identical to Equation (59). Chang et al. [51] report a slightly different expression, see their Equation (2), in which p_{air}/p is replaced by the equivalent quantity $\mid Y_m \mid_{ln}$ (log mean mole fraction of air). As for the previous case of the Knudsen regime, Khayet and Matsuura [28] give an expression, their Equation (10.18), for $B_m{}^D$ (defined as the permeance in the diffusive regime *per single pore*); this must be multiplied by the number of pores per unit area, $4\varepsilon/(\pi d_p{}^2)$, to obtain the permeance per unit membrane area, $C_m{}^D$. The result, allowing for changes in notation, becomes again identical to Equation (59).

For the diffusivity D_g, the Fuller equation for binary gas diffusion [81,84] can be used:

$$D_g = 1 \cdot 10^{-7} \frac{T^{1.75}}{p} \frac{\sqrt{\frac{1}{MW_{vap}} + \frac{1}{MW_{air}}}}{\left(V_{vap}^{1/3} + V_{air}^{1/3}\right)^2} \quad (60)$$

in which D_g is in m^2/s, T is in K, p is in atm, MW is in g/mol and V is the so called "diffusion volume" in cm^3/mol. Alkhudhiri et al. [8] report that V = 12.7 cm^3/mol for water vapor and 20.1 cm^3/mol for air; Poling et al. [85] report the similar values V = 13.1 cm^3/mol for water vapor and 19.7 cm^3/mol for air. For example, for air–water vapor at $p = 10^5$ Pa (1 bar) and T = 300 K, Equation (60) yields $D_g = 2.54 \cdot 10^{-5}$ m^2/s.

As an alternative, the diffusivity of water vapor through the stagnant air in the pores can be estimated by the simpler formula reported by many authors, including Phattaranawik et al. [55], Khayet and Matsuura [28], Olatunji and Camacho [20], Dong et al. [65] and Kariman et al. [48]:

$$D_g = 1.895 \cdot 10^{-5} \frac{T^{2.072}}{p} \quad (61)$$

in which D_g is in m^2/s, T is in K and p is in Pa. For T = 300 K and p = 1 atm., this yields $D_g = 2.57 \cdot 10^{-5}$ m^2/s, almost identical to the outcome of the more complex Equation (60).

Yet another correlation is proposed by Cussler [81] and Ansari et al. [42]:

$$D_g = 4.32 \cdot 10^{-4} \frac{T^{1.5}}{p} \quad (62)$$

in which, as in Equation (61), D_g is in m^2/s, T is in K and p is in Pa. For example, for $p = 10^5$ Pa (1 bar) and T = 300 K, Equation (62) yields $D_g = 2.24 \cdot 10^{-5}$ m^2/s, farther than Equation (61) from the more complete Equation (60).

4.3. Case III (Intermediate Pore Size, Transitional Regime): $\lambda < d_p < 100\lambda$ or $0.01 \leq Kn \leq 1$

In this case (transitional regime), both molecule–molecule collisions and molecule–pore wall collisions are important. Based on the Bosanquet theory for the overall self-diffusivity of a molecule under simultaneous Knudsen and molecular diffusion [86–88], and according to the majority of authors, e.g., Alkhudhiri et al. [8], Dong et al. [65], Chang et al. [51], Tsai et al. [89] and Kariman et al. [48], in this case, the resistances associated with molecular diffusion and Knudsen transport must be treated as resistances in series (Figure 15).

 https://doi.org/10.3390/membranes16020064

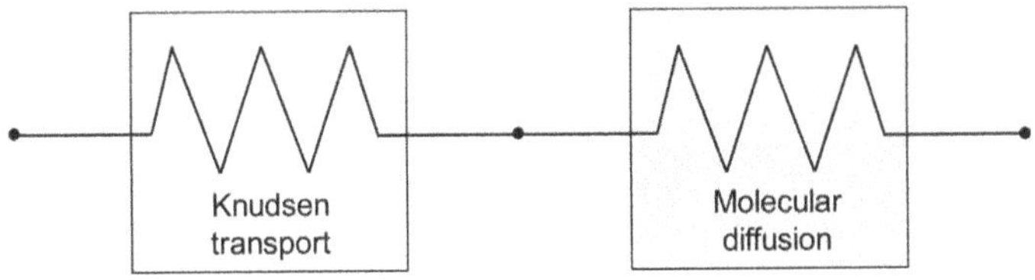

Figure 15. Transitional mass transfer regime with Knudsen transport and molecular diffusion in series.

Consistently, the membrane molar permeance C_m (now called $C_m{}^C$ for "combined") can be expressed as follows:

$$\frac{1}{C_m^C} = \frac{1}{C_m^{Kn}} + \frac{1}{C_m^D} \quad (63)$$

Kuang et al. [46] report an explicit form of Equation (63) in which $C_m{}^{Kn}$ is expressed using Equation (57), while $C_m{}^D$ is expressed using Equation (59). The resulting expression of $C_m{}^C$ (in mol m^{-2} s^{-1} Pa^{-1}) is as follows:

$$C_m^C = \frac{\varepsilon}{\tau\delta}\left[\frac{3}{d_p}\sqrt{\frac{\pi R_g T\, MW_{vap}}{8}} + \frac{p_{air}}{p}\frac{R_g T}{D_g}\right]^{-1} \quad (64)$$

Note that $C_m{}^C$ depends on several quantities for which a more or less severe uncertainty exists. In particular, it is inversely proportional to the pore tortuosity τ, a quantity for which different formulae yield broadly different results (see Section 3.3). $C_m{}^C$ is also directly proportional to the membrane porosity, which is also affected, albeit to a lesser extent, by uncertainty. Also, the choice of a single pore diameter d_p is an oversimplification with respect to the actual pore size distribution (see remarks at the end of Section 3.2). Finally, also for the diffusivity D_g and the ratio p_{air}/p, different values are provided according to the model used.

Equation (64) must be multiplied by MW_{vap} (in kg/mol) if one wishes to express $C_m{}^C$ as *mass* permeance, i.e., as mass flux per unit pressure difference (in units of kg m^{-2} s^{-1} Pa^{-1}), as, for example, in Olatunji and Camacho [20]. Similarly, Equation (64) has to be multiplied by the molar volume of liquid water MW_w/ρ_w (in m^3/mol) if one wishes to express $C_m{}^C$ as *volume* permeance, i.e., as liquid volume flux per unit pressure difference (i.e., in m s^{-1} Pa^{-1} units), as, for example, in Bouguecha et al. [44], who report for LiquiCel® hollow fiber membranes a volume permeance of $3.27{\cdot}10^{-10}$ m s^{-1} Pa^{-1}.

4.4. Contribution of Poiseuille Flow

As a further transport mechanism, Poiseuille flow might be considered, see, e.g., Khayet and Matsuura [28]. However, this mechanism is in contradiction with the assumption that, at least in DCMD, air is trapped in the membrane pores; it is expected to be important only if this assumption is removed, and, even so, only in the presence of a significant transmembrane total pressure difference and very large pores, otherwise being negligible. As remarked by Field et al. [71], *"it is necessary to emphasize that the presence of Knudsen diffusion necessarily excludes the presence of Poiseuille flow"*. Also, according to Dong et al. [65], as seen in their Table 1, in the presence of a gas mixture (air–water vapor) in the pores with appreciable transmembrane partial vapor pressure difference but negligible total pressure difference (which are the conditions typical of DCMD), the only mass transport mechanisms to be considered are Knudsen transport for Kn > 1, molecular diffusion for Kn < 0.01, and a combination of these in the intermediate (and most common) range.

However, only for the sake of completeness, we mention here that, in order to account for the simultaneous presence of Knudsen transport, molecular diffusion and Poiseuille

 https://doi.org/10.3390/membranes16020064

flow, different models have been proposed, the most popular being the "Dusty gas" model (Figure 16a) and Schofield's model [90] (Figure 16b). A detailed description of the Dusty gas model is given by Khayet and Matsuura [28], see their Equations (10.28) through (10.32). Expressions for the mass fluxes, as derived by both the above models, are given, among others, by Dong et al. [65]; see their Equations (9) and (10).

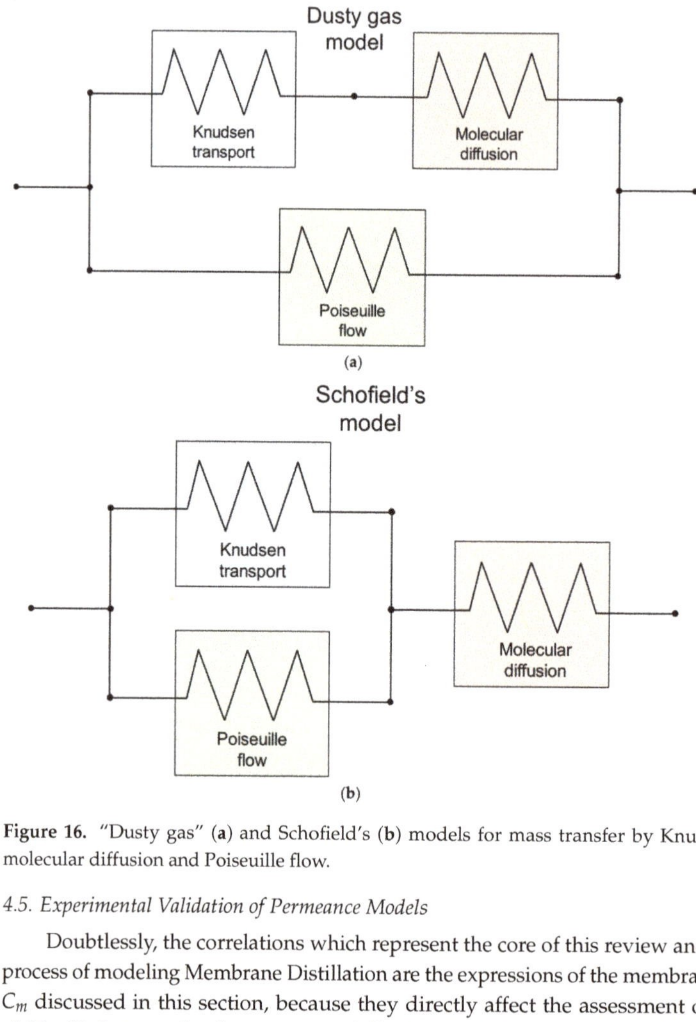

Figure 16. "Dusty gas" (**a**) and Schofield's (**b**) models for mass transfer by Knudsen transport, molecular diffusion and Poiseuille flow.

4.5. Experimental Validation of Permeance Models

Doubtlessly, the correlations which represent the core of this review and of the entire process of modeling Membrane Distillation are the expressions of the membrane permeance C_m discussed in this section, because they directly affect the assessment of the process yield. Unfortunately, they are very difficult to validate experimentally: based on the definition of C_m, validation would require measurements of the local vapor flux J_{vap} and of local temperatures (and, for the feed side, also concentration) at the feed–membrane and permeate–membrane interfaces, from which local values of the vapor saturation pressure p_{sat} on the two interfaces could be reliably calculated. These quantities are hardly accessible, which explains why relevant data are almost non-existent in the literature. In most cases, the available experimental information regards overall water vapor fluxes in a whole module and bulk values of temperatures and concentrations. Therefore, any experimental validation of a model for the membrane permeance can only be indirect and is affected by all the uncertainties associated with the large-scale spatial variation in all quantities within a module and, what is even more relevant, with temperature

 https://doi.org/10.3390/membranes16020064

and concentration polarization (i.e., boundary layer resistance to the transport of heat and mass). Usually, authors presenting experimental values of C_m or using this quantity within predictive models either neglect the boundary layer temperature drops [44] or treat C_m as an adjustable parameter to be tuned on the basis of the overall agreement with experiments [91,92].

5. Transmembrane Heat Transfer

In regard to the transmembrane heat transfer process, most authors, e.g., Khayet and Matsuura [28], Alkhudhiri et al. [8], Olatunji and Camacho [20], Dong et al. [65] and Cancilla et al. [93], represent it as the flow of heat through a combination of resistances, as shown in Figure 17. Three main steps are involved:

(a) Heat transfer from the feed bulk to the feed–membrane interface;
(b) Heat transfer from the feed–membrane interface to the membrane–permeate interface;
(c) Heat transfer from the membrane–permeate interface to the permeate bulk.

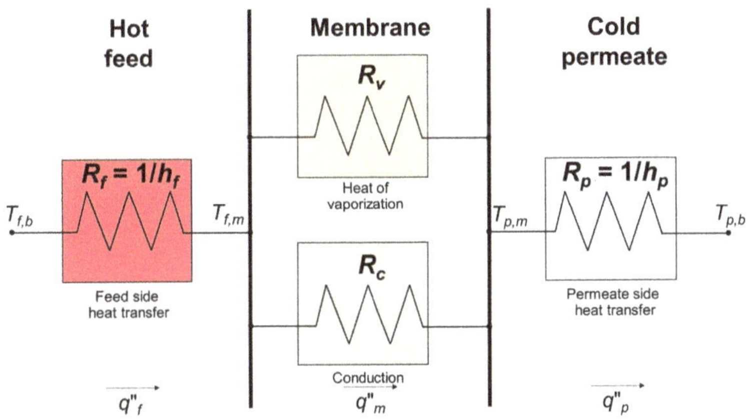

Figure 17. Heat transfer resistance model in DCMD.

The following Equations (65)–(68) express the above heat transfer resistance model in formula form.

(a) The quantity q''_f, expressed in W/m^2, represents the heat flux from the bulk of the feed side channel to the feed–membrane interface:

$$q''_f = h_f\left(T_{f,b} - T_{f,m}\right) \tag{65}$$

in which h_f is the heat transfer coefficient of the feed compartment and $T_{f,b}$ and $T_{f,m}$ are the temperatures at the bulk feed and feed–membrane interface, respectively. The inverse of the heat transfer coefficient h_f represents the thermal resistance R_f associated with the hot feed side compartment in Figure 17.

(b) The quantity q''_m, expressed in W/m^2, represents the heat flux through the membrane. It is given by the parallel of two mechanisms:

- Heat flux due to the conduction across the polymeric membrane material and the gas-filled pores, which is denoted here as q''_c;
- Heat flux due to the latent heat associated with the water vapor, which is denoted here as q''_v.

The resulting total heat flux through the membrane wall can be written as follows:

$$q''_m = q''_c + q''_v = \frac{k_m}{\delta}\left(T_{f,m} - T_{p,m}\right) + J_{vap}H_{fg} \tag{66}$$

https://doi.org/10.3390/membranes16020064

where k_m is the membrane thermal conductivity, δ is the membrane thickness, $T_{p,m}$ is the membrane–permeate interface temperature, J_{vap} is the water vapor flux (Equation (52)) and H_{fg} is the latent heat of vaporization. The ratio k_m/δ in Equation (66) represents the inverse of the thermal resistance R_c associated with the conductive heat transfer through the membrane in Figure 17. The product $J_{vap}H_{fg}$ is inversely proportional to the thermal resistance R_v associated with the latent heat of vaporization in Figure 17. Note that all uncertainties on the membrane thermal conductivity directly affect both the heat flux and the ratio of sensible to latent heat fluxes.

According to Khayet and Matsuura [28], between 50% and 80% of energy is consumed as latent heat for water vapor production (q''_v), while the remainder is lost by thermal conduction (q''_c). In fact, the heat used effectively in DCMD is the latent heat of vaporization associated with the mass flux, whilst the heat transferred by conduction across the membrane is considered as heat loss. This heat loss becomes less significant at higher operating feed temperatures, and the thermal efficiency, η, of a DCMD process defined by the following Equation (67) becomes higher.

$$\eta(\%) = \frac{q''_v}{q''_v + q''_c} \times 100 \tag{67}$$

All temperature-dependent quantities explicitly or implicitly involved in Equation (66) can be evaluated at the membrane arithmetic mean temperature $\langle T_m \rangle = (T_{f,m} + T_{p,m})/2$. As an alternative, following Phattaranawik and Jiraratananon [63], the geometric mean membrane temperature $(T_{f,m} \cdot T_{p,m})^{\frac{1}{2}}$ can be used.

(c) The quantity q''_p, expressed in W/m^2, represents the heat flux from the membrane–permeate interface to the bulk of the permeate compartment:

$$q''_p = h_p \left(T_{p,m} - T_{p,b} \right) \tag{68}$$

in which h_p is the heat transfer coefficient of the permeate compartment and $T_{p,b}$ is the temperature at the bulk permeate. The inverse of the heat transfer coefficient h_p represents the thermal resistance R_p associated with the cold permeate side compartment in Figure 17. Under steady state conditions,

$$q''_f = q''_m = q''_p = q'' \tag{69}$$

Note that, in modules adopting the hollow fiber membrane geometry, the above equations need to be adjusted to reflect the differences in membrane diameter and surface area between the inner (lumen side) and outer (shell side) surfaces.

The mass flux J_{vap} can be expressed (and, if necessary, converted to kg m^{-2} s^{-1}) by Equation (52), using the expressions for the vapor saturation pressure and for C_m discussed in the previous Sections 2 and 4, respectively. Due to the nonlinear dependence of the vapor saturation pressure on the temperature, the resulting set of equations is nonlinear and can only be solved by some iterative procedure. As an alternative, for a low transmembrane bulk temperature difference, many authors [8,28,51,94] linearize the dependence of p_{sat} on T by writing the water vapor flux J_{vap} in Equation (52) as follows:

$$J_{vap} \approx C_m \frac{dp_{sat}}{dT} \left(T_{f,m} - T_{p,m} \right) \tag{70}$$

where the derivative of p_{sat} with respect to T can be expressed by the Clausius–Clapeyron equation as follows:

$$\frac{dp_{sat}}{dT} = \frac{H_{fg}}{R_g T^2} p_{sat}(T) \tag{71}$$

More complex expressions for dp_{sat}/dT, suited for concentrated solutions, are proposed by Schofield et al. [54]. Linearizing p_{sat} as in Equation (70) is only a crude approximation, especially if the temperature varies broadly, but leads to great simplifications, allowing

 https://doi.org/10.3390/membranes16020064

the use of a global heat transfer coefficient incorporating both the latent heat associated with vapor transport and sensible heat associated with conduction. In this way, both terms q''_c and q''_v in Equation (66) become proportional to $T_{f,m} - T_{p,m}$. Thus, accounting for the vapor flux is equivalent to assuming for the membrane an *augmented* thermal conductivity k_{aug}, expressed as follows:

$$k_{aug} = k_m + C_m \delta \left. \frac{dp_{sat}}{dT} \right|_{\langle T_m \rangle} H_{fg} \tag{72}$$

5.1. Temperature Polarization Phenomena

Although the present review is focused on transmembrane heat and mass transfer in Membrane Distillation, a brief mention is appropriate of the heat transfer resistances exhibited by the boundary layers, which are recognized as a source of loss of DCMD efficiency. The temperature polarization coefficient (*TPC*) is generally used to quantify their relative importance with respect to the total heat transfer resistance. With reference to the overall temperature profile shown in Figure 13, *TPC* is defined as follows:

$$TPC = \frac{T_{f,m} - T_{p,m}}{T_{f,b} - T_{p,b}} \tag{73}$$

and reflects the reduction in the driving force (i.e., in the temperature difference $T_{f,m} - T_{p,m}$, resulting in the vapor pressure difference $p_{sat,f} - p_{sat,p}$), which has a negative influence on the DCMD process effectiveness. In the ideal case, *TPC* should be equal to unity, but usually it is significantly lower.

If the film heat transfer coefficients h_f and h_p, associated with the feed and the permeate boundary layers, which appear in Equations (65) and (68), are large, the temperatures at the membrane surfaces approach those of the bulk fluids, *TPC* approaches unity (i.e., the temperature polarization effect is negligible) and the DCMD process is controlled by the mass transfer resistance of the membrane. On the contrary, if the heat transfer coefficients are small, the difference between the temperatures at the membrane surfaces and those of the bulk fluids is large, and the transmembrane temperature difference is low. In this case (*TPC* approaching zero), the temperature polarization effect is very significant, and the DCMD process is controlled by the heat transfer resistances of the boundary layers. Generally, for satisfactory DCMD modules, *TPC* values range between 0.4 and 0.7. This means that between 30% and 60% of the applied temperature difference is spent in the thermal boundary layers.

Several methods have been adopted to minimize the heat transfer resistances of the boundary layers, thus increasing the film heat transfer coefficients. These include the use of spacers (for the spiral wound and plate-and-frame module geometries) or other mixing promoters, or simply the adoption of high feed and permeate flow rates. Notably, spacers promote fluid mixing, thereby enhancing the efficiency of convective heat transfer. Since the pivotal work of Schock and Miquel [95] in 1987, substantial research efforts have been devoted to elucidating the influence of spacer geometry on fluid dynamics and heat or mass transport within membrane systems affected by polarization phenomena. During the years, conventional woven and overlapped spacer designs have been systematically investigated, both experimentally and numerically, to assess their effects on friction as well as on heat and/or mass transfer coefficients [96–99].

More recently, several researchers have also studied unconventional sphere-type spacers using both CFD simulations [100–102] and experiments [103]. CFD has also been used to examine innovative designs, including 3D-printed geometries with hole–pillar arrangements [104], arch-type [105], airfoil-shaped filaments [106] and twisted [107] geometries, and, very recently, pillar–filament configurations [108]. These investigations consistently demonstrate that such advanced designs can achieve superior performance with respect to

 https://doi.org/10.3390/membranes16020064

conventional and commercially available spacers. The body of literature on spacer-filled channels and their effects on polarization phenomena is vast. A comprehensive review of these contributions lies beyond the scope of the present work; readers are referred to dedicated review articles [23,109,110] that focus specifically on this topic.

5.2. Concentration Polarization Phenomena

By analogy to the treatment of temperature polarization and with reference to the overall concentration profile shown in Figure 13, a concentration polarization coefficient (*CPC*) can be defined as follows:

$$CPC = \frac{C_{f,m}}{C_{f,b}} \tag{74}$$

where $C_{f,m}$ and $C_{f,b}$ are the salt concentrations at the feed–membrane interface and at the bulk feed, respectively. Since the membrane retention is ideally unity, the concentration at the permeate–membrane interface ($C_{p,m}$) is equal to that in the permeate bulk ($C_{p,b}$), and both are zero. For this reason, these two concentrations are omitted in Equation (74).

As anticipated in the Introduction, in DCMD, the concentration boundary layers make a smaller contribution to the overall mass transfer resistance, compared with temperature polarization [28]. The reason is that the vapor saturation pressure is strongly dependent on temperature but only weakly dependent on concentration, as shown in Section 2.5. In the MD process, the use of moderate flow rates and high heat transfer coefficients can make the influence of concentration polarization negligible. Khayet et al. [31], using a microporous hydrophobic membrane and a semi-empirical method, reported that concentration polarization was insignificant in DCMD across a range of flow rates and temperatures on both sides of the membrane. The negligible influence of *CPC* relative to *TPC* has also been demonstrated for other membranes in numerous studies [26,28–30].

6. Conclusions

Direct Contact Membrane Distillation (DCMD) is a promising separation technology with a growing relevance in desalination, water treatment and resource recovery. However, despite decades of research, the modeling of heat and mass transfer phenomena across DCMD systems remains fragmented, with significant variability in the models, assumptions, equations and correlations used.

This review has systematically analyzed recent and classical models, highlighting inconsistencies, overlaps and gaps. As anticipated in the Introduction, this comprehensive appraisal reveals that, while numerous models exist, many can be reconciled by recognizing equivalent formulations, excluding non-justified or even non-physically based equations, and acknowledging a few prevalent errors in past studies.

In particular, assumptions which are shared by most of the literature on DCMD are the following:

- The membrane pores are filled with a binary mixture of water vapor (moving from feed to permeate) and still, trapped air;
- The mean free path of water (vapor) molecules in the pores can be evaluated by Equation (54);
- The diffusivity of water (vapor) molecules in the pores can be evaluated by Equation (60) (Fuller equation) in the molecular diffusion regime and by Equation (56) in the Knudsen regime. Also, the resulting expressions for the membrane permeance, Equations (57) and (58), are shared by almost all authors (apart from the choice of units and other details);
- The Liquid Entry Pressure (*LEP*) can be estimated as a function of the surface tension, the pore radius and the solution–membrane contact angle using Equation (50).

 https://doi.org/10.3390/membranes16020064

For other aspects, no general consensus exists in the literature:

- Accounting or not accounting for Poiseuille flow in transmembrane mass transfer; see, e.g., Khayet and Matsuura [28] and Ansari et al. [42] vs. Dong et al. [65] and Field et al. [71];
- Linearizing or not linearizing the dependence of the vapor saturation pressure on temperature in Equation (70) for the water vapor flux; see, e.g., Alkhudhiri et al. [8], Khayet and Matsuura [28] and Field et al. [71] vs. Momeni et al. [41], Olatunji and Camacho [20] and Hitsov et al. [32];
- Choosing the temperature at which to evaluate the membrane properties, i.e., the arithmetic mean of $T_{f,m}$ and $T_{p,m}$ (most authors) or their geometric mean (Phattaranawik and Jiraratananon [63]);
- Predicting the membrane thermal conductivity as a function of porosity and of the conductivities of the polymeric matrix and the filling gas (Equations (31)–(34));
- Predicting the pore tortuosity as a function of porosity (Equation (48) versus Equation (49)).

By providing a comprehensive and rigorous comparative assessment, we aim to establish a common basis for DCMD modeling: a set of recommended equations, parameter values and best practices that researchers and engineers can reliably use for predicting performance. Hopefully, this will not only help researchers to resolve discrepancies and reduce errors in future simulations but also make MD modeling more accessible to scientists and engineers outside the immediate MD community.

A sensitivity analysis of the influence of thermophysical and kinetic correlations on the predicted performance of DCMD modules would require a model of the whole Membrane Distillation process, including the effects of temperature and concentration polarization on overall heat and mass transfer coefficients and the spatial distribution of residence times. This kind of analysis is beyond the scope of the present paper, which focuses on local transmembrane heat and mass transport models, and will be the subject of future studies. In our opinion, small differences between alternative correlations, e.g., in the thermophysical properties of salt water, would have a minor influence on the prediction of water vapor and heat fluxes, whereas other options, e.g., in the estimate of quantities like the pore tortuosity or membrane thermal conductivity, may severely affect the overall process performance.

Possible extensions of the present work, aimed at supporting the scale-up and implementation of DCMD in industrial applications, should include the following:

- A comparison of the proposed models and correlations for transmembrane transport with both experimental results and ab initio molecular dynamics predictions and other advanced approaches;
- An integration of the above results within more complete and fully predictive computational models accounting for the fluid dynamic aspects of the process. This will also make it possible to conduct sensitivity analyses on the influence of different correlations and model options on practical quantities such as freshwater yield and thermal consumption;
- A combination of heat and mass transfer models with economic and environmental assessment tools to support the scale-up of DCMD at a semi-industrial scale level;
- An exploration of new module designs using computational fluid dynamics, which should be evaluated alongside economic considerations such as manufacturing, operating and maintenance costs;
- The use of 3D-printed or nano-engineered modules could offer unprecedented control over membrane properties and module design, allowing for the creation of complex architectures customized for specific applications.

 https://doi.org/10.3390/membranes16020064

Author Contributions: Conceptualization, N.C., L.G., A.C. and M.C.; methodology, N.C., L.G., A.C. and M.C.; software, N.C. and L.G.; validation, A.C. and M.C.; formal analysis, N.C. and L.G.; investigation, N.C., L.G. and M.C.; resources, A.C. and M.C.; data curation, N.C. and L.G.; writing—original draft preparation, N.C., L.G. and M.C.; writing—review and editing, L.G. and A.C.; visualization, N.C., L.G. and M.C.; supervision, A.C. and M.C.; project administration, A.C.; funding acquisition, A.C. All authors have read and agreed to the published version of the manuscript.

Funding: This research received no external funding.

Data Availability Statement: The raw data supporting the conclusions of this article will be made available by the authors on request.

Acknowledgments: The authors are grateful to Alessandra Criscuoli and Giorgio De Luca (CNR—Istituto per la Tecnologia delle Membrane, Rende, Italy) for their constructive discussion of a preliminary draft of this paper.

Conflicts of Interest: The authors declare no conflicts of interest.

Nomenclature

A	Membrane area (m^2)
A_1, B_1	Coefficients in the dynamic viscosity correlation, Equation (11)
A_2, B_2, C_2, D_2	Coefficients in the specific heat capacity correlation, Equation (15)
a_w	Water activity (-)
B	Geometric pore coefficient in Equation (50) (-)
C	Molar concentration, Equation (4) (mol/m^3)
C_m	Membrane permeance ($kg\ m^{-2}\ h^{-1}\ Pa^{-1}$ or $mol\ m^{-2}\ h^{-1}\ Pa^{-1}$)
c_p	Specific heat capacity ($J\ kg^{-1}\ K^{-1}$)
CPC	Concentration polarization coefficient, Equation (74) (-)
D	Diffusivity (m^2/s)
d_p	Pore diameter (m)
h	Heat transfer coefficient ($W\ m^{-2}\ K^{-1}$)
H_{fg}	Latent heat of vaporization ($J\ kg^{-1}$)
ID	Inner fiber diameter of the fiber (m)
J	Mass / molar flux ($kg\ m^{-2}\ h^{-1}$ / $mol\ m^{-2}\ h^{-1}$)
k	Thermal conductivity ($W\ m^{-1}\ K^{-1}$)
K_B	Boltzmann's constant ($J\ K^{-1}$)
Kn	Knudsen number (-)
L	Length of the hollow fiber membrane (m)
l	Pore length of the membrane (m)
LEP	Liquid Entry Pressure, Equation (50) (Pa)
M	Molarity, Equation (3) (mol/L)
m	Molality, Equation (2) (mol/kg)
MW	Molecular weight (g/mol)
OD	Outer fiber diameter of the fiber (m)
p	Pressure (Pa)
p_{sat}	Vapor saturation pressure (Pa or atm)
q''	Heat flux (W/m^2)
R	Heat transfer resistance ($m^2\ K\ W^{-1}$)
R_g	Universal gas constant ($J\ mol^{-1}\ K^{-1}$)
r_{max}	Largest pore radius (m)
S	Salinity (or mass fraction) (kg/kg)
SD_{log}	Standard deviation of lognormal function (-)
T	Absolute temperature (K)
t	Temperature (°C)
TPC	Temperature polarization coefficient, Equation (73) (-)

 https://doi.org/10.3390/membranes16020064

V	Diffusion volume (cm^3/mol)
w	Membrane mass (kg)
x	Molar fraction, Equation (1) (mol/mol)
Greek symbols	
α, β	Coefficients in Equation (42)
Γ	Surface tension (mN/m)
γ_ω	Activity coefficient (-)
δ	Membrane thickness (m)
ε	Porosity of the membrane (-)
ζ	Coefficient in Equations (33) and (34) (-)
η	Thermal efficiency (-)
θ	Contact angle (degree)
λ	Mean free path (m)
μ	Dynamic viscosity (Pa s)
ρ	Density (kg/m^3)
σ	Collision diameter of the molecule (m)
τ	Pore tortuosity of the membrane (-)
Subscripts	
air	Air
aug	Augmented
b	Bulk
c	Conduction
f	Feed
g	Gas
k	Kerosene
m	Membrane
max	Maximum
p	Permeate
pol	Polymeric material
s	Salt
sat	Saturation
sm	Solid matrix
v	Vaporization
vap	Water vapor
w	Water
0	Referred to pure water
1	Referred to before the immersion in kerosene
2	Referred to before the immersion in kerosene
Superscripts	
C	Referred to the transitional (combined between diffusion and Knudsen) region
D	Referred to diffusion region
Kn	Referred to Knudsen region
Averages	
$\langle\, \rangle$	Mean value

Abbreviations

The following abbreviations are used in this manuscript:

CFD	Computational Fluid Dynamics
CPC	Concentration Polarization Coefficient
DCMD	Direct Contact Membrane Distillation
LEP	Liquid Entry Pressure
MD	Membrane Distillation
PES	Polyethersulfone

https://doi.org/10.3390/membranes16020064

PP	Polypropylene
PTFE	Polytetrafluoro-ethylene
PVDF	Polyvinylidene fluoride
TPC	Temperature Polarization Coefficient

References

1. Politano, A.; Al-Juboori, R.A.; Alnajdi, S.; Alsaati, A.; Athanassiou, A.; Bar-Sadan, M.; Beni, A.N.; Campi, D.; Cupolillo, A. D'Olimpio, G.; et al. 2024 Roadmap on Membrane Desalination Technology at the Water-Energy Nexus. *J. Phys. Energy* **2024**, *6* 021502. [CrossRef]
2. Eke, J.; Yusuf, A.; Giwa, A.; Sodiq, A. The Global Status of Desalination: An Assessment of Current Desalination Technologies Plants and Capacity. *Desalination* **2020**, *495*, 114633. [CrossRef]
3. López-Porfiri, P.; Ramos-Paredes, S.; Núñez, P.; Gorgojo, P. Towards the Technological Maturity of Membrane Distillation: The MD Module Performance Curve. *Npj Clean Water* **2023**, *6*, 18. [CrossRef]
4. Ahmed, F.E.; Khalil, A.; Hilal, N. Emerging Desalination Technologies: Current Status, Challenges and Future Trends. *Desalination* **2021**, *517*, 115183. [CrossRef]
5. Nassrullah, H.; Anis, S.F.; Hashaikeh, R.; Hilal, N. Energy for Desalination: A State-of-the-Art Review. *Desalination* **2020**, *491*, 114569. [CrossRef]
6. Zhang, H.; Xian, H. Review of Hybrid Membrane Distillation Systems. *Membranes* **2024**, *14*, 25. [CrossRef] [PubMed]
7. Shahzad, M.W.; Burhan, M.; Ang, L.; Ng, K.C. Energy-Water-Environment Nexus Underpinning Future Desalination Sustainability *Desalination* **2017**, *413*, 52–64. [CrossRef]
8. Alkhudhiri, A.; Darwish, N.; Hilal, N. Membrane Distillation: A Comprehensive Review. *Desalination* **2012**, *287*, 2–18. [CrossRef]
9. Camacho, L.M.; Dumée, L.; Zhang, J.; de Li, J.; Duke, M.; Gomez, J.; Gray, S. Advances in Membrane Distillation for Water Desalination and Purification Applications. *Water* **2013**, *5*, 94–196. [CrossRef]
10. Curcio, E.; Drioli, E. Membrane Distillation and Related Operations-A Review. *Sep. Purif. Rev.* **2005**, *34*, 35–86. [CrossRef]
11. Drioli, E.; Ali, A.; Macedonio, F. Membrane Distillation: Recent Developments and Perspectives. *Desalination* **2015**, *356*, 56–84 [CrossRef]
12. González, D.; Amigo, J.; Suárez, F. Membrane Distillation: Perspectives for Sustainable and Improved Desalination. *Renew. Sustain. Energy Rev.* **2017**, *80*, 238–259. [CrossRef]
13. Alessandro, F.; Macedonio, F. A Critical Review of Membrane Distillation Using Ceramic Membranes: Advances, Opportunities and Challenges. *Materials* **2025**, *18*, 3296. [CrossRef] [PubMed]
14. Patel, R.V.; Yadav, A.; Shahi, V.K. Advances in Membrane Distillation for Wastewater Treatment: Innovations, Challenges, and Sustainable Opportunities. *Sci. Total Environ.* **2025**, *969*, 178749. [CrossRef]
15. Deshmukh, A.; Boo, C.; Karanikola, V.; Lin, S.; Straub, A.P.; Tong, T.; Warsinger, D.M.; Elimelech, M. Membrane Distillation at the Water-Energy Nexus: Limits, Opportunities, and Challenges. *Energy Environ. Sci.* **2018**, *11*, 1177–1196. [CrossRef]
16. Schwantes, R.; Cipollina, A.; Gross, F.; Koschikowski, J.; Pfeifle, D.; Rolletschek, M.; Subiela, V. Membrane Distillation: Solar and Waste Heat Driven Demonstration Plants for Desalination. *Desalination* **2013**, *323*, 93–106. [CrossRef]
17. Gurreri, L.; La Cerva, M.; Ciofalo, M.; Cipollina, A.; Tamburini, A.; Micale, G. Application of Computational Fluid Dynamics Technique in Membrane Distillation Processes. In *Current Trends and Future Developments on (Bio-) Membranes*; Basile, A. Ghasemzadeh, K., Eds.; Elsevier: Amsterdam, The Netherlands, 2022; pp. 161–208.
18. Skuse, C.; Gallego-Schmid, A.; Azapagic, A.; Gorgojo, P. Can Emerging Membrane-Based Desalination Technologies Replace Reverse Osmosis? *Desalination* **2021**, *500*, 114844. [CrossRef]
19. Kiss, A.A.; Kattan Readi, O.M. An Industrial Perspective on Membrane Distillation Processes. *J. Chem. Technol. Biotechnol.* **2018** *93*, 2047–2055. [CrossRef]
20. Olatunji, S.O.; Camacho, L.M. Heat and Mass Transport in Modeling Membrane Distillation Configurations: A Review. *Front. Energy Res.* **2018**, *6*, 130. [CrossRef]
21. Ashoor, B.B.; Mansour, S.; Giwa, A.; Dufour, V.; Hasan, S.W. Principles and Applications of Direct Contact Membrane Distillation (DCMD): A Comprehensive Review. *Desalination* **2016**, *398*, 222–246. [CrossRef]
22. Pangarkar, B.L.; Deshmukh, S.K.; Sapkal, V.S.; Sapkal, R.S. Review of Membrane Distillation Process for Water Purification *Desalin. Water Treat.* **2016**, *57*, 2959–2981. [CrossRef]
23. Ali, A.M.; Shirazi, M.A.; Nthunya, L.; Castro-Munoz, R.; Ismail, N.; Tavajohi, N.; Zaragoza, G.; Quist-Jensen, C.A. Progress in module design for membrane distillation. *Desalination* **2024**, *581*, 117584. [CrossRef]
24. Anvari, A.; Azimi Yancheshme, A.; Kekre, K.M.; Ronen, A. State-of-the-Art Methods for Overcoming Temperature Polarization in Membrane Distillation Process: A Review. *J. Memb. Sci.* **2020**, *616*, 118413. [CrossRef]

 https://doi.org/10.3390/membranes16020064

25. Alsaadi, A.S.; Francis, L.; Amy, G.L.; Ghaffour, N. Experimental and Theoretical Analyses of Temperature Polarization Effect in Vacuum Membrane Distillation. *J. Memb. Sci.* **2014**, *471*, 138–148. [CrossRef]
26. Shang, W.; Chen, H.; Lu, G.; Xu, X.; Cheng, M.; Huang, C.; Guan, X.; An, A.K. In-Situ Temperature Measurement and Non-Linear Interaction Analysis of Temperature Polarization in Direct Contact Membrane Distillation. *J. Memb. Sci.* **2025**, *735*, 124508. [CrossRef]
27. Martínez-Díez, L.; Vázquez-González, M. Temperature and Concentration Polarization in Membrane Distillation of Aqueous Salt Solutions. *J. Memb. Sci.* **1999**, *156*, 265–273. [CrossRef]
28. Khayet, M.; Matsuura, T. *Membrane Distillation-Principles and Applications*; Elsevier: Amsterdam, The Netherlands, 2011. [CrossRef]
29. Schofield, R.W.; Fane, A.G.; Fell, C.J.D.; Macoun, R. Factors Affecting Flux in Membrane Distillation. *Desalination* **1990**, *77*, 279–294. [CrossRef]
30. Martínez-Díez, L.; Vázquez-González, M. Effects of Polarization on Mass Transport through Hydrophobic Porous Membranes. *Ind. Eng. Chem. Res.* **1998**, *37*, 4128–4135. [CrossRef]
31. Khayet, M.; Godino, M.P.; Mengual, J.I. Study of Asymmetric Polarization in Direct Contact Membrane Distillation. *Sep. Sci. Technol.* **2004**, *39*, 125–147. [CrossRef]
32. Hitsov, I.; Maere, T.; de Sitter, K.; Dotremont, C.; Nopens, I. Modelling Approaches in Membrane Distillation: A Critical Review. *Sep. Purif. Technol.* **2015**, *142*, 48–64. [CrossRef]
33. Sharqawy, M.H.; Lienhard, J.H.V.; Zubairb, S.M. Thermophysical properties of seawater: A review of existing correlations and data. *Desal. Water Treat.* **2010**, *16*, 354–380. [CrossRef]
34. El-Dessouky, H.T.; Ettouney, H.M. *Fundamentals of Salt Water Desalination*; Elsevier: Amsterdam, The Netherlands, 2002. [CrossRef]
35. Isdale, J.D.; Spence, C.M.; Tudhope, J.S. Physical properties of sea water solutions: Viscosity. *Desalination* **1972**, *10*, 319–328. [CrossRef]
36. Korosi, A.; Fabuss, B.M. Viscosity of liquid water from 25 °C to 150 °C. *J. Anal. Chem.* **1968**, *40*, 157–162. [CrossRef]
37. Jamieson, D.T.; Tudhope, J.S.; Morris, R.; Cartwright, G. Physical properties of sea water solutions: Heat capacity. *Desalination* **1969**, *7*, 23–30. [CrossRef]
38. Jamieson, D.T.; Tudhope, J.S. Physical properties of sea water solutions–Thermal Conductivity. *Desalination* **1970**, *8*, 393–401. [CrossRef]
39. Yadav, A.; Singh, C.P.; Patel, R.V.; Kumar, A.; Labhasetwar, P.K. Investigations on the effect of spacer in direct contact and air gap membrane distillation using computational fluid dynamics. *Colloids Surf. A Physicochem. Eng. Asp.* **2022**, *654*, 130111. [CrossRef]
40. Khalifa, A.; Ahmad, H.; Antar, M.; Laoui, T.; Khayet, M. Experimental and theoretical investigations on water desalination using direct contact membrane distillation. *Desalination* **2017**, *404*, 22–34. [CrossRef]
41. Momeni, M.; Kargari, A.; Dadvar, M.; Jafari, A. 3D-CFD simulation of hollow fiber direct contact membrane distillation module: Effect of module and fibers geometries on hydrodynamics, mass, and heat transfer. *Desalination* **2024**, *576*, 117321. [CrossRef]
42. Ansari, A.; Galogahi, F.M.; Millar, G.; Helfer, F.; Thiel, D.V.; Soukane, S.; Ghaffour, N. Computational fluid dynamics simulations of solar-assisted, spacer-filled direct contact membrane distillation: Seeking performance improvement. *Desalination* **2023**, *545*, 116181. [CrossRef]
43. Song, L.; Li, B.; Sirkar, K.K.; Gilron, J.L. Direct Contact Membrane Distillation-Based Desalination: Novel Membranes, Devices, Larger-Scale Studies, and a Model. *Ind. Eng. Chem. Res.* **2007**, *46*, 2307–2323. [CrossRef]
44. Bouguecha, S.T.; Aly, S.E.; Al-Beirutty, M.H.; Hamdi, M.M.; Boubakri, A. Solar driven DCMD: Performance evaluation and thermal energy efficiency. *Chem. Eng. Res. Des.* **2015**, *100*, 331–340. [CrossRef]
45. Eykens, L.; Hitsov, I.; De Sitter, K.; Dotremont, C.; Pinoy, L.; Nopens, I.; Van der Bruggen, B. Influence of membrane thickness and process conditions on direct contact membrane distillation at different salinities. *J. Membr. Sci.* **2016**, *498*, 353–364. [CrossRef]
46. Kuang, Z.; Long, R.; Liu, Z.; Liu, W. Analysis of temperature and concentration polarizations for performance improvement in direct contact membrane distillation. *Int. J. Heat Mass Transf.* **2019**, *145*, 118724. [CrossRef]
47. Lai, C.; Zhang, H. An efficient and economic evacuated U-tube solar collector powered air gap membrane distillation hybrid system for seawater desalination. *J. Clean. Prod.* **2023**, *394*, 136382. [CrossRef]
48. Kariman, H.; Mohammed, H.A.; Zargar, M.; Khiadani, M. Performance comparison of flat sheet and hollow fibre air gap membrane distillation: A mathematical and simulation modelling approach. *J. Membr. Sci.* **2025**, *721*, 123836. [CrossRef]
49. ASHRAE. Thermophysical Properties of Refrigerants. In *2017 ASHRAE Handbook: Fundamentals*; ASHRAE: Atlanta, GA, USA, 2017.
50. Wang, Z.; Jia, Z.; Li, R.; Gu, Z. Analysis and Experimental Study on Water Vapor Partial Pressure in the Membrane Distillation Process. *Membranes* **2022**, *12*, 802. [CrossRef] [PubMed]
51. Chang, H.; Ho, C.-D.; Chen, Y.-H.; Chen, L.; Hsu, T.-H.; Lim, J.-W.; Chiou, C.-P.; Lin, P.-H. Enhancing the Permeate Flux of Direct Contact Membrane Distillation Modules with Inserting 3D Printing Turbulence Promoters. *Membranes* **2021**, *11*, 266. [CrossRef]

https://doi.org/10.3390/membranes16020064

52. Bromley, L.A.; Singh, D.; Ray, P.; Sridhar, S.; Read, S.M. Thermodynamic Properties of Sea Salt Solutions. *AIChE J.* **1974**, *20*, 326–335. [CrossRef]
53. Zhang, J.; Dow, N.; Duke, M.; Ostarcevic, E.; Jun-De, L.; Gray, S. Identification of material and physical features of membrane distillation membranes for high performance desalination. *J. Membr. Sci.* **2010**, *349*, 295–303. [CrossRef]
54. Schofield, R.W.; Fane, A.G.; Fell, C.J.D. Heat and mass transfer in membrane distillation. *J. Membr. Sci.* **1987**, *33*, 299–313. [CrossRef]
55. Phattaranawik, J.; Jiraratananon, R.; Fane, A.G. Effect of pore size distribution and air flux on mass transport in direct contact membrane distillation. *J. Membr. Sci.* **2003**, *215*, 75–85. [CrossRef]
56. Garcıa-Payo, M.C.; Izquierdo-Gil, M.A. Thermal resistance technique for measuring the thermal conductivity of thin microporous membranes. *J.Phys. D* **2004**, *37*, 3008–3016. [CrossRef]
57. Stephan, K.; Laesecke, A. The Thermal Conductivity of Fluid Air. *J. Phys. Chem. Ref. Data* **1985**, *14*, 227–234. [CrossRef]
58. National Institute of Standards and Technology. NIST Chemistry WebBook, SRD69. Available online: https://webbook.nist.gov/chemistry/fluid/ (accessed on 23 September 2025).
59. Udoetok, E.S. Thermal Conductivity of binary mixtures of gases. *Front. Heat Mass Transf.* **2013**, *4*, 023008. [CrossRef]
60. El-Bourawi, M.S.; Ding, Z.; Ma, R.; Khayet, M. A framework for better understanding membrane distillation separation process. *J. Membr. Sci.* **2006**, *285*, 4–29. [CrossRef]
61. Lawson, K.W.; Lloyd, D.R. Membrane distillation. *J. Membr. Sci.* **1997**, *124*, 1–25. [CrossRef]
62. Al Nuaimi, R.; Thankamony, R.L.; Boross de Levay, J.-P.B.; Yuan, B.; Lai, Z. Uniformly porous PVDF-co-HFP membranes prepared by mixed solvent phase separation for direct contact membrane distillation. *J. Membr. Sci.* **2024**, *711*, 123175. [CrossRef]
63. Phattaranawik, J.; Jiraratananon, R. Direct contact membrane distillation: Effect of mass transfer on heat transfer. *J. Membr. Sci.* **2001**, *188*, 137–143. [CrossRef]
64. Fan, S.; Nguyen, D.T.; Martinez, J.; Chau, J.; Fung, K.; Sirkar, K.; Straub, A.P.; Ding, Y. The effect of Sharklet patterns on thermal efficiency and salt-scaling resistance of poly (vinylidene fluoride) membranes during direct contact membrane distillation. *J. Membr. Sci.* **2025**, *715*, 123476. [CrossRef]
65. Dong, Y.; Dai, X.; Zhao, L.; Gao, L.; Xie, Z.; Zhang, J. Review of Transport Phenomena and Popular Modelling Approaches in Membrane Distillation. *Membranes* **2021**, *11*, 122. [CrossRef]
66. Khayet, M.; Velázquez, A.; Mengual, J.I. Modelling mass transport through a porous partition: Effect of pore size distribution. *J. Non-Equilib. Thermodyn.* **2004**, *29*, 279–299. [CrossRef]
67. Woods, J.; Pellegrino, J.; Burch, J. Generalized guidance for considering pore-size distribution in membrane distillation. *J. Membr. Sci.* **2011**, *368*, 124–133. [CrossRef]
68. Smolder, K.; Franken, A.D.M. Terminology for Membrane Distillation. *Desalination* **1989**, *72*, 249. [CrossRef]
69. Khayet, M.; Matsuura, T. Preparation and characterization of polyvinylidene fluoride membranes for membrane distillation. *Ind. Eng. Chem. Res.* **2001**, *40*, 5710–5718. [CrossRef]
70. Zou, L.; Zhang, X.; Gusnawan, P.; Zhang, G.; Yu, J. Crosslinked PVDF based hydrophilic-hydrophobic dual-layer hollow fiber membranes for direct contact membrane distillation desalination: From the seawater to oilfield produced water. *J. Membr. Sci.* **2021**, *619*, 118802. [CrossRef]
71. Field, R.W.; Wu, H.Y.; Wu, J.J. Multi-scale modelling of membrane distillation: Some theoretical considerations. *Ind. Eng. Chem. Res.* **2013**, *52*, 8822–8828. [CrossRef]
72. Yasuda, H.; Tsai, J.T. Pore size of microporous polymer membranes. *J. Appl. Polym. Sci.* **1974**, *18*, 805–819. [CrossRef]
73. Srisurichan, S.; Jiraratananon, R.; Fane, A.G. Mass transfer mechanisms and transport resistances in direct contact membrane distillation process. *J. Membr. Sci.* **2006**, *277*, 186–194. [CrossRef]
74. Mackie, J.S.; Meares, E. The Diffusion of Electrolytes in a Cation-Exchange Resin Membrane. I. Theoretical. *Proc. R. Soc. London. Ser. A Math. Phys. Sci.* **1955**, *232*, 498–509. [CrossRef]
75. Ciofalo, M.; Lombardo, S.; Pettinato, D.; Quattrocchi, D. Experimental investigation of two-side heat transfer in spacer-filled channels representative of Membrane Distillation. *Exp. Therm. Fluid Sci.* **2023**, *140*, 110770. [CrossRef]
76. Franken, A.C.M.; Nolten, J.A.M.; Mulder, M.H.V.; Bargeman, D.; Smolders, C.A. Wetting criteria for the applicability of membrane distillation. *J. Membr. Sci.* **1987**, *33*, 315–328. [CrossRef]
77. Lopez, K.P.; Wang, R.; Hjelvik, E.A.; Lin, S.; Straub, A.P. Toward a universal framework for evaluating transport resistances and driving forces in membrane-based desalination processes. *Sci. Adv.* **2023**, *9*, eade0413. [CrossRef]
78. Tomaszewska, M. Preparation and properties of flat-sheet membranes from poly(vinylidene fluoride) for membrane distillation. *Desalination* **1996**, *104*, 1–11. [CrossRef]
79. Sghaier, N.; Prat, M.; Ben Nasrallah, S. On the influence of sodium chloride concentration on equilibrium contact angle. *Chem. Eng. J.* **2006**, *122*, 47–53. [CrossRef]

80. Rácz, G.; Kerker, S.; Kovács, Z.; Vatai, G.; Ebrahimi, M.; Czermak, P. Theoretical and Experimental Approaches of Liquid Entry Pressure Determination in Membrane Distillation Processes. *Period. Polytech. Chem. Eng.* **2014**, *58*, 81–91. [CrossRef]
81. Cussler, E. *Diffusion Mass Transfer in Fluid Systems*, 3rd ed.; Cambridge University Press: Cambridge, UK, 2009. [CrossRef]
82. Mills, A.F. *Mass Transfer*, 2nd ed.; Prentice Hall: Hoboken, NJ, USA, 2001.
83. Ghaleni, M.M.; Bavarian, M.; Nejati, S. Model-guided design of high-performance membrane distillation modules for water desalination. *J. Membr. Sci.* **2018**, *563*, 794–803. [CrossRef]
84. Geankoplis, C.J. *Transport Processes and Separation Process Principles (Includes Unit Operations)*, 4th ed.; Prentice Hall Press: Hoboken, NJ, USA, 2003.
85. Poling, B.E.; Prausnitz, J.M.; O'Connell, J.P. *The Properties of Gases and Liquids*, 5th ed.; McGraw-Hill: New York, NY, USA, 2001.
86. Bosanquet, C.H. *British TA Report BR-507*; London, UK, 1944.
87. Pollard, W.G.; Present, R.D. On gaseous self-diffusion in long capillary tubes. *Phys. Rev.* **1948**, *73*, 762–774. [CrossRef]
88. Krishna, E.; van Baten, J.M. Investigating the validity of the Bosanquet formula for estimation of diffusivities in mesopores. *Chem. Eng. Sci.* **2012**, *69*, 684–688. [CrossRef]
89. Tsai, J.-H.; Quist-Jensen, C.; Ali, A. Multipass hollow fiber membrane modules for membrane distillation. *Desalination* **2023**, *548*, 116239. [CrossRef]
90. Schofield, R.W.; Fane, A.G.; Fell, C.J.D. Gas and vapour transport through microporous membranes. II. Membrane distillation. *J. Membr. Sci.* **1990**, *53*, 173–185. [CrossRef]
91. Song, L.; Ma, Z.; Liao, X.; Kosaraju, P.B.; Irish, J.R.; Sirkar, K.K. Pilot plant studies of novel membranes and devices for direct contact membrane distillation-based desalination. *J. Membr. Sci.* **2008**, *323*, 257–270. [CrossRef]
92. Singh, D.; Li, L.; Obusckovic, G.; Chau, J.; Sirkar, K.K. Novel cylindrical cross-flow hollow fiber membrane module for direct contact membrane distillation-based desalination. *J. Membr. Sci.* **2018**, *545*, 312–322. [CrossRef]
93. Cancilla, N.; Bounou, A.; Ciofalo, M.; Tamburini, A.; Micale, G. CFD prediction of flow and heat transfer in hollow fiber bundles for Membrane Distillation: A comparison of regular and random arrangements. *Int. J. Heat Mass Transf.* **2026**, *254*, 127672. [CrossRef]
94. Ho, C.-D.; Chen, L.; Lim, J.-W.; Lin, P.-H.; Lu, P.-T. Distillate Flux Enhancement of Direct Contact Membrane Distillation Modules with Inserting Cross-Diagonal Carbon-Fiber Spacers. *Membranes* **2021**, *11*, 973. [CrossRef] [PubMed]
95. Schock, G.; Miquel, A. Mass transfer and pressure loss in spiral wound modules. *Desalination* **1987**, *64*, 339–352. [CrossRef]
96. Picioreanu, C.; Vrouwenvelder, J.S.; Van Loosdrecht, M.C.M. Three-dimensional modeling of biofouling and fluid dynamics in feed spacer channels of membrane devices. *J. Membr. Sci.* **2009**, *345*, 340–354. [CrossRef]
97. Koutsou, C.P.; Yiantsios, S.G.; Karabelas, A.J. A numerical and experimental study of mass transfer in spacer-filled channels: Effects of spacer geometrical characteristics and Schmidt number. *J. Membr. Sci.* **2009**, *326*, 234–251. [CrossRef]
98. Tamburini, A.; Renda, M.; Cipollina, A.; Micale, G.; Ciofalo, M. Investigation of heat transfer in spacer-filled channels by experiments and direct numerical simulations. *Int. J. Heat Mass Transf.* **2016**, *93*, 1190–1205. [CrossRef]
99. Ponzio, F.N.; Tamburini, A.; Cipollina, A.; Micale, G.; Ciofalo, M. Experimental and computational investigation of heat transfer in channels filled by woven spacers. *Int. J. Heat Mass Transf.* **2017**, *104*, 163–177. [CrossRef]
100. Chong, Y.K.; Liang, Y.Y.; Lau, W.J.; Fimbres Weihs, G.A. 3D CFD study of hydrodynamics and mass transfer phenomena for spiral wound membrane submerged-type feed spacer with different node geometries and sizes. *Int. J. Heat Mass Transf.* **2022**, *191*, 122819. [CrossRef]
101. Saliakellis, P.; Koutsou, C.; Karabelas, A. The effect of gap reduction on fluid dynamics and mass transfer in membrane narrow channels filled with novel spacers-a detailed computational study. *Membranes* **2023**, *13*, 20. [CrossRef] [PubMed]
102. Ciofalo, M.; Cancilla, N.; Cipollina, A.; Tamburini, A.; Micale, G. Influence of thermal buoyancy on heat transfer in spacer-filled channels for membrane distillation. *J. Phys.–Conf. Ser.* **2025**, *2940*, 012007. [CrossRef]
103. Cancilla, N.; Tamburini, A.; Tarantino, A.; Visconti, S.; Ciofalo, M. Friction and heat transfer in membrane distillation channels: An experimental study on conventional and novel spacers. *Membranes* **2022**, *12*, 1029. [CrossRef] [PubMed]
104. Qamar, A.; Kerdi, S.; Ali, S.M.; Shon, H.K.; Vrouwenvelder, J.S.; Ghaffour, N. Novel hole-pillar spacer design for improved hydrodynamics and biofouling mitigation in membrane filtration. *Sci. Rep.* **2021**, *11*, 6979. [CrossRef]
105. Yu, J.; Chen, D.; Wu, J.J.; Wang, B.; Field, R.W. Arch-type feed spacer with wide passage node design for spiral-wound membrane filtration with reduced energy cost. *Desalination* **2022**, *540*, 115980. [CrossRef]
106. Qamar, A.; Kerdi, S.; Vrouwenvelder, J.S.; Ghaffour, N. Airfoil-shaped filament feed spacer for improved filtration performance in water treatment. *Sci. Rep.* **2023**, *13*, 10798. [CrossRef]
107. Chong, Y.K.; Liang, Y.Y.; Fimbres Weihs, G.A. Validation and characterisation of mass transfer of 3D-CFD model for twisted feed spacer. *Desalination* **2023**, *554*, 116516. [CrossRef]
108. Park, J.E.; Jung, S.Y.; Kang, T.G. Robust design of a stacked-filament feed spacer inducing chaotic advection for enhanced reverse osmosis filtration. *J. Membr. Sci.* **2025**, *734*, 124431. [CrossRef]

https://doi.org/10.3390/membranes16020064

109. Schilling, S.; Glade, H. Review and Analysis of Heat Transfer in Spacer-Filled Channels of Membrane Distillation Systems. *Membranes* **2023**, *13*, 842. [CrossRef]
110. Tabrizi, A.B.; Wu, B. Review of computational fluid dynamics simulation techniques for direct contact membrane distillation systems containing filament spacers. *Desalin. Water Treat.* **2019**, *162*, 79–96. [CrossRef]

 https://doi.org/10.3390/membranes16020064

membranes

Article

Comprehensive Investigation of Iron Salt Effects on Membrane Bioreactor from Perspective of Controlling Iron Leakage

Qiaoying Wang, Bingbing Zhang, Jicheng Sun, Wenjia Zheng, Jie Zhang * and Zhichao Wu

State Key Laboratory of Water Pollution Control and Green Resource Recycling, Shanghai Institute of Pollution Control and Ecological Security, College of Environmental Science and Engineering, Tongji University, Shanghai 200092, China; qywang@tongji.edu.cn (Q.W.); 2331416@tongji.edu.cn (B.Z.); 1631342@tongji.edu.cn (J.S.); 2130558@tongji.edu.cn (W.Z.); wuzhichao@tongji.edu.cn (Z.W.)

* Correspondence: zhangjie130@sina.com; Tel.: +86-21-65980400

Abstract: Although adding iron salts can improve phosphorus removal in membrane bioreactor (MBR) processes, overdosing iron salts may result in excessive iron concentrations in the effluent and pose risks of surface water contamination. In this study, an optimized iron salt dosing method was proposed to comprehensively investigate its effects on the performance of MBRs and the control of iron leakage. The results showed that batch dosing of solid iron salts ($Fe_2(SO_4)_3$) into the influent or activated sludge maintained an effluent Fe^{3+} concentration below 1.0 mg/L and a total phosphorus (TP) concentration below 0.30 mg/L. Long-term operation of the MBR (under conditions of HRT = 4.3 h, SRT = 20 d, and MLSS = 12 g/L) showed that batch dosing of solid iron salts led to an increase in the effluent ammonia–nitrogen (NH_3-N) concentration, and the nitrification effect was restored after supplementing the alkalinity. Iron salts increased the TP removal rate by approximately 40% while inhibiting the biological phosphorus removal capacity. The average Fe^{3+} concentration in the membrane effluent (0.23 ± 0.11 mg/L) met China's Environmental Quality Standard for Surface Water (GB3838-2002). This study demonstrates that batch dosing of solid iron salts effectively controls iron concentration in the MBR effluent while preventing secondary pollution. The mechanisms of the impact of iron salts on MBR performance provide crucial theoretical and technical support for MBR process optimization.

Keywords: membrane bioreactor; iron salts; phosphorus removal; leakage

1. Introduction

With increasingly stringent water quality standards, the membrane bioreactor (MBR) technology has become a research hotspot in the field of water treatment due to its high effluent quality and small footprint [1]. Currently, MBR technology is widely used in municipal and industrial wastewater treatment, as well as in water resource recycling [2–4]. However, the MBR process still faces several technical challenges in practical operation. Chief among these is the need for extensive aeration to control membrane fouling, which not only leads to high energy consumption [5] but also inhibits biological phosphorus removal. Thus, chemical phosphorus removal (CPR) using iron/aluminum salts is usually adopted as a supplementary method in MBR processes [6,7].

In the MBR process, coagulants are added to wastewater to form phosphorus-containing solids, and iron salts are among the most commonly used additives in practical projects [8]. Many researchers have also investigated the effect of iron salts on phosphorus removal in MBRs. Li et al. [9] found that when 10, 15, and 20 mg/L ferric chloride ($FeCl_3$) were added, the TP removal rates of the MBR were 70.6%, 87.7%, and 96.4%, respectively.

Iron salt addition can also effectively mitigate membrane fouling. Yang et al. [10] reported that with a dosage of polymerized ferric chloride (PFC) between 10 and 15 mg/L, the average effluent TP concentration of the MBR reduced to 0.26 mg/L. Previous reports normally focused on the effects of iron salts on TP removal and membrane fouling control, while the leakage of iron salts into the effluent was ignored.

Iron pollution has emerged as a global water contamination issue posing significant potential risks [11]. To meet stricter discharge standards, excessive iron salt dosing for phosphorus removal is widely practiced. However, this approach causes economic losses and poses secondary pollution risks due to residual iron salts in the effluent, which is largely ignored in previous studies. Severe iron pollution in surface and groundwater has been documented, with concentrations reaching 6 mg/L in the Ganges River [12] and widespread exceedances reported elsewhere [13,14]. The influx of Fe^{2+} into aquatic systems facilitates both ferrous sulfide (FeS) formation [15,16] and P release from sediments [17], which intensifies hypoxia and eventually leads to black odor [18,19]. Conventional drinking water treatment processes exhibit limited efficacy in dissolved iron removal. Chronic exposure to iron-contaminated water may induce severe health consequences, including hepatotoxicity, neurodegenerative disorders, and carcinogenic effects [20,21]. Zhuang et al. [22] demonstrated that iron salts readily combine with perfluorinated compounds, subsequently enhancing disinfection by-product formation and compromising water quality safety. Therefore, optimizing iron salt dosing to maintain effluent iron below the 0.3 mg/L threshold specified in China's Environmental Quality Standard for Surface Water (GB 3838-2002) and elucidating the iron retention mechanisms of microfiltration membranes represent research priorities for MBR enhancement.

In a MBR system, iron salt dosing alters mixture characteristics and increases phosphorus removal complexity [23]. Optimizing iron salt dosing requires balancing effluent iron levels, membrane fouling mitigation, and microbial activity maintenance by optimizing key parameters, including dosing method, concentration, and reaction conditions. Gkotsis et al. [24] reported that continuous coagulant dosing achieved a 40% greater TMP reduction compared to intermittent dosing. Li et al. [25] determined that 1.8 mmol/L $FeCl_3$ provided optimal performance with 96.7% TP removal, 21.9% lower TMP, and stable TN removal. However, higher doses impaired PAOs and AOB activity, reducing AOB activity by 60.7%. Existing studies have focused more on the impact of dosing methods on membrane fouling and phosphorus/nitrogen removal, while neglecting the critical indicator of effluent iron concentration.

In contrast to previous studies, this research considers effluent iron concentration as a core evaluation indicator. A systematic comparison of three dosing strategies was conducted: one-time dosing of solid iron salts, continuous dosing of iron salt solution, and batch dosing of solid iron salts. The aim was to identify an optimal approach that ensures MBR performance while minimizing secondary pollution risk. The long-term impacts of the optimal dosing method on effluent quality, membrane fouling, and microbial communities were further evaluated, with emphasis on monitoring effluent iron concentration to assess environmental risks. The results demonstrated that iron salt dosing enhanced TP removal while inhibiting the nitrification process (reversible with alkalinity addition), and appropriate dosing strategies effectively controlled residual iron levels. This study provides a new perspective for the green and sustainable development of MBR technology by achieving synergistic optimization of effluent iron control and water quality improvement. It also offers important insights for addressing iron pollution as a global water environmental issue.

2. Materials and Methods

2.1. Materials

Ferric sulfate ($Fe_2(SO_4)_3$, 21–23% Fe content) and 1,10–phenanthroline were analytical reagent grade and purchased from Aladdin Biotechnology (Shanghai, China). The polyvinylidene fluoride (PVDF) microfiltration membrane used in this study, with an average pore size of 0.2 μm, was prepared in the laboratory. The properties of the PVDF membrane are provided in our previous report [26–28]. The domestic wastewater used in the continuous flow experiment was collected from the effluent of the sand sedimentation tank of the Quyang Wastewater Treatment Plant (WWTP) in Shanghai, China. The key characteristics of the domestic wastewater were as follows: COD: 350 $\pm$ 94 mg/L, NH_3-N: 32.8 $\pm$ 4.6 mg/L, TN: 41.6 $\pm$ 3.7 mg/L, TP: 4.4 $\pm$ 0.6 mg/L. The detailed water quality of the influent during the test period is shown in Table S1. The activated sludge was collected from a laboratory MBR unit that had not been exposed to iron salts. For the long-term operation experiment, the MBR was inoculated with activated sludge from the aerobic tank of the WWTP. The influent quality was identical to the domestic wastewater described above.

2.2. Experimental Methods

2.2.1. Continuous Flow Experiments

A schematic diagram of the continuous flow experimental setup is shown in Figure S1. The size of the reactor was L $\times$ B $\times$ H = 6 $\times$ 6 $\times$ 15 cm with a volume of 0.5 L. A flat-sheet microfiltration membrane module with an effective filtration area of 48 cm^2 was installed in the reactor. Baffles were installed on both sides of the membrane module to form the up-flow zone, with a spacing of 1.0 cm between the baffles and the module. The membrane flux was set to 20 $L/m^2{\cdot}h$ (LMH), and the hydraulic retention time (HRT) was 4.0 h.

Three $Fe_2(SO_4)_3$ dosing methods were set up to investigate their effects on iron leakage and TP removal, and the running time varied according to different test conditions (mainly 4 h and 24 h). The three dosing methods are shown below: (1) One-time dosing of solid iron salts: The total amount of solid iron salts required for chemical phosphorus removal over the entire test duration was added at once. (2) Continuous dosing of iron salt solution: First, a certain concentration of iron salt solution was prepared. To prevent hydrolysis and ensure stability [29,30], H_2SO_4 was added to maintain the pH below 2.0, a standard laboratory procedure that would cause a transient decrease in the pH within the reactor. Finally, the solution was dosed at a controlled rate using peristaltic pumps. (3) Batch dosing solid iron salts: A certain time interval (4 h in the test, equal to 1 HRT) dosing iron salt solids required for chemical phosphorus removal in the corresponding time.

2.2.2. MBR Long-Term Operation Test

The associated designs of two sets of MBR set-ups used in this study are shown in Figure S2, and the design parameters are shown in Table S2. The effective volume was 10.5 L, and the aeration intensity was 1.0 $m^3/m^2{\cdot}min$. To ensure effective aeration, the bottom aeration pipe was positioned 500 mm from the bottom of the membrane modules. The designed operational parameters were as follows: sludge retention time (SRT), 20 d; mixed liquor suspended solids (MLSS) concentration, 12 g/L; membrane flux, 20 LMH; hydraulic retention time (HRT), 4.3 h. The reactor operation mode was "pump 10 stop 2".

The reactor with iron salt dosing was numbered as reactor #1, while the undosed reactor served as the control (reactor #2). Iron salts were dosed daily at approximately 9:00 at a molar Fe/P ratio of 1.5, corresponding to approximately 64 mg/L of $Fe_2(SO_4)_3$. The calculation procedure is provided in the Supporting Information (Text S1). The experiment comprised three stages: Stage 1: Both MBR systems were operated to reach a stable state. Stage 2: Iron salts were added to reactor #1, while reactor #2 still maintained original operating parameters as the control. Stage 3: Dosing was discontinued in reactor #1, while #2 remained unchanged.

2.3. *Analytical Methods*

2.3.1. Routine Chemical Test Items and Methods

The details of routine chemical test items and methods are provided in the Supporting Information (Table S3). Effluent quality was assessed according to the Grade A criteria specified in the Chinese Discharge Standard of Pollutants for Municipal Wastewater Treatment Plant (GB 18918-2002). The specific limit values are listed in Table S4. Rigorous quality control was implemented throughout the study, including the use of standard calibration curves ($R^2 > 0.999$), analysis of parallel samples, and routine blank tests.

2.3.2. Method for the Determination of Fe^{3+}

Fe^{3+} concentrations were determined using the o-phenanthroline colorimetric method. The detection limit of this method was 0.005 mg/L. The procedure was as follows: (1) Total iron, 1 mL of the solution was pipetted into a 25 mL colorimetric tube. Then, 1 mL of reducing agent (10% hydroxylamine hydrochloride), 1 mL of pH buffer (NH_4Ac-HAc), and 1 mL of color reagent (0.1% o-phenanthroline) were added sequentially. The volume was adjusted to 25 mL, shaken, and absorbance was measured at 510 nm after 5 min. (2) Fe^{2+}: The procedure was the same as (1) except that the hydroxylamine hydrochloride was omitted. (3) Samples with flocculation or precipitation: 50 mL of the water sample was added to a 250 mL conical flask. A few drops of HCl (1:1) were added to adjust the pH to <2, followed by heating and boiling for 10 min. After cooling, it was transferred to a 50 mL volumetric flask, and deionized water was added to the mark. The solution was shaken well, and 1 mL was pipetted into a 25 mL colorimetric tube following the procedure outlined in step (1). (4) Fe^{3+}: The Fe^{3+} concentration was calculated by subtracting the Fe^{2+} concentration from the total iron concentration.

2.3.3. Classification of Different Forms of Fe^{3+}

Following the method of Wang et al. [31], iron ions in the solution were divided into three categories using filters with different pore sizes: (1) Total iron (M_1): The total iron content was directly determined from the test sample using the method described in Section 2.3.2. (2) Particulate Fe^{3+} (M_3): The test sample was passed through a 0.80 μm filter, and the Fe^{3+} content of the filtrate, M_2, was measured. The difference between M_1 and M_2 was the particulate Fe^{3+} content M_3. (3) Dissolved Fe^{3+} (M_4): The sample was passed through a 0.22 μm filter, and the Fe^{3+} content of the filtrate, M_4, was measured. (4) Colloidal Fe^{3+} (M_5): The difference between M_2 and M_4 was the content of colloidal Fe^{3+} M_5.

2.3.4. Determination of Phosphorus Release and Phosphorus Absorption Capacity

The phosphorus release rate from activated sludge was tested in a sealed 1 L reactor (Figure S3) under anaerobic conditions, with MLSS controlled to 3–4 g/L. During the test, sludge mixture samples were taken every 10 min and immediately filtered with a 0.45 μm filter. After 1 h, the sampling interval was increased to 20 min, and the total sampling time was 3 h. The soluble orthophosphate concentration in the filtrate was measured immediately, and the phosphorus release rate (PRR) was determined from the slope of its linear time-dependent plot. The specific phosphorus release rate (SPRR) was calculated as the ratio of PRR to MLVSS.

The activated sludge phosphorus absorption rate was measured immediately after the anaerobic phosphorus release process using the device shown in Figure S4. The sludge mixture was aerated through the sand core aeration head to maintain a suspended state, with DO controlled at 6–8 mg/L. The sludge mixture was taken every 10 min (adjusted to

20 min after 1 h), and the sampling time was 3 h in total, with immediate filtration through 0.45 μm membranes. The phosphorus uptake rate (PUR) was determined from the slope of the straight line between filtrate orthophosphate concentration and time. The specific phosphorus uptake rate (SPUR) was calculated as the ratio of PUR to MLVSS.

2.3.5. Determination of Sludge-Specific Nitrification Rate

The specific nitrification rate was measured using the same setup as for the SPUR. The specific test method was as follows: A volume of activated sludge was centrifuged (3000 r/min, 5 min). After discarding the supernatant, the remaining sludge solids were resuspended in PBS (pH 7.0). The sludge concentration was adjusted to 2–3 g/L, supplemented with 100 mg/L $NaHCO_3$ and 76 mg/L NH_4Cl to control the initial NH_3-N concentration at about 20 mg/L, and aerated to maintain 6–8 mg/L DO. Samples of the sludge mixture were collected every 20 min over a 3 h period. The sludge mixture was filtered, and the NH_3-N concentration of the filtrate was measured. The NH_3-N values and time were plotted to generate a straight line. The absolute value of the slope was determined as the nitrification rate (AUR) of the sludge, and the ratio of AUR to MLVSS was calculated as the specific nitrification rate (SAUR).

2.3.6. Determination of Sludge-Specific Oxygen Uptake Rate

The method for determining the specific oxygen uptake rate (SOUR) and the experimental setup are detailed in the Supplementary Information (Text S2, Figure S5).

2.3.7. Testing of Molecular Weight Distribution of Organic Matter

Gel filtration chromatography (Le-10ADVP, Shimadzu) was used to determine the molecular weight distribution of organics in water. Prior to analysis, samples were filtered through a 0.45 μm syringe filter and sonicated to remove air bubbles.

2.3.8. Statistical Analysis

Statistical analyses were conducted using IBM SPSS Statistics 27. Data are expressed as mean ± standard deviation (mean ± s.d.). Differences between the two groups were assessed using independent samples t-tests, paired sample t-tests, or Mann–Whitney U tests. Correlation analysis was used to evaluate associations between variables. $p < 0.05$ was considered significant.

3. Results and Discussions

3.1. *Optimization of Iron Salt Dosing in MBR*

3.1.1. Optimization of Iron Salt Addition in Influent

As shown in Figure 1a, after one-time dosing of solid iron salts to the influent, the TP concentration in the effluent decreased to 0.50 mg/L after 3 h and reached 0.45 mg/L (with a removal rate of 91.3%) after 24 h. Continuous dosing of iron salt solution resulted in an effluent TP concentration of 0.42 mg/L at 0.3 h, which increased to 1.21 mg/L after 24 h, showing a removal rate of only 72.9%. Batch dosing of solid iron salts achieved a 97.48% TP removal rate within 10 min (0.10 mg/L effluent) after the first dosing. Although TP concentration slightly increased to 0.44 mg/L after 4 h, the average TP concentration was 0.09 ± 0.06 mg/L during 5–12 h, with an average removal rate of 97.8%.

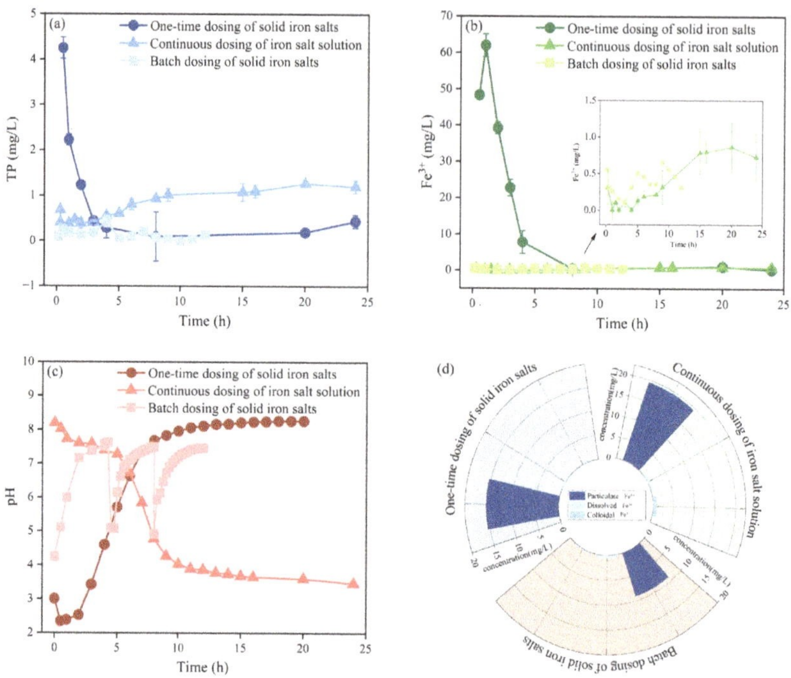

Figure 1. Effects of iron salt dosing in influent on (**a**) TP, (**b**) Fe^{3+}, and (**c**) pH; (**d**) concentrations of particulate Fe^{3+} (>0.8 µm), dissolved Fe^{3+} (<0.22 µm), and colloidal Fe^{3+} (0.22–0.80 µm) in wastewater under three dosing methods; data in (**a**,**b**) are presented as mean ± standard deviation (n = 3).

As shown in Figure 1b, one-time dosing resulted in effluent Fe^{3+} concentration peaking at 62.01 mg/L after 1 h, decreasing to 7.82 mg/L after 4 h, and stabilizing at approximately 0.50 mg/L after 8 h. As shown in Figure 1d, Fe^{3+} in the reactor primarily existed in particulate form (15.45 mg/L, 96.6%) after 24 h, while colloidal and dissolved states had lower concentrations. The flat-sheet microfiltration membrane effectively retained the particulate Fe^{3+}, reducing effluent concentration. The continuous dosing initially yielded low effluent Fe^{3+} (0.78 mg/L at 15 h), then stabilized around 0.90 mg/L. Although maintaining Fe/P = 1.5, Fe^{3+} gradually accumulated and was discharged due to slightly excessive dosage. Fe^{3+} in the reactor was still primarily in particulate form (19.45 mg/L, 93.1%) after 12 h (Figure 1d). When solid iron salts were added in batches, the Fe^{3+} concentration exhibited periodic fluctuations, with a rapid increase at the beginning of each addition, but overall remaining below 1.00 mg/L, followed by a decrease. Fe^{3+} in the reactor primarily existed in particulate form (9.04 mg/L, 92.1%) after 12 h.

As shown in Figure 1c, after one-time dosing, the system pH first decreased to 2.3 (0.5 h) and then gradually increased to 8.1 (11 h). The hydrolysis of the large quantity of added solid iron salts acidified the solution. The low pH inhibited the hydrolysis of iron salts into flocs or precipitates, reducing P removal and the loss of dissolved iron salts. A subsequent increase in pH improved P removal, but as iron salts were consumed, TP concentration rose again, requiring timely replenishment of iron salts. The continuous dosing induced a strongly acidic environment (pH 3.5). This was attributed to both the accumulation and hydrolysis of iron salts, as well as the inherent acidity of the iron salt solution being dosed. Consequently, the effects observed under this dosing mode resulted from the combination of these factors. The batch dosing induced minor pH fluctuations, remaining above 4.0 initially and slowly rising to approximately 8.2 after 4 h. Following each addition, the pH exhibited periodic fluctuations and rapidly recovered to above 8.0, a condition favorable for P removal by iron salts.

3.1.2. Optimization of Iron Salt Dosing in Activated Sludge

In this experiment, the control MBR system without dosing of iron salts was set. The MBR system achieved TP removal, which was inefficient due to the lack of an anaerobic section. As shown in Figure 2a, the control MBR system exhibited poor TP removal performance, with an average effluent TP concentration of 2.34 ± 0.23 mg/L (45.2% removal). The one-time dosing rapidly reduced the effluent TP to below 0.30 mg/L within 2 h but gradually rebounded to 0.89 mg/L after 24 h. In contrast, the continuous dosing resulted in superior P removal. The effluent TP decreased to 0.88 mg/L within 0.5 h, stabilized below 0.30 mg/L, and reached 0.12 mg/L after 24 h, achieving a removal rate of 97.1%. The batch dosing maintained the effluent TP concentration below 0.30 mg/L. Although slight periodic fluctuations were observed, the levels remained stable overall, with an average concentration of 0.14 ± 0.06 mg/L and a removal rate of 97.7%.

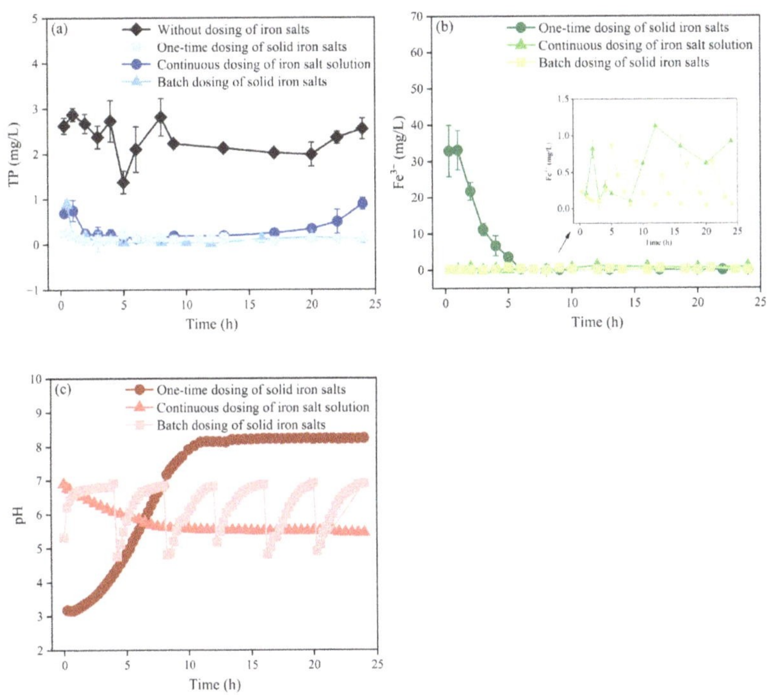

Figure 2. Effects of iron salt dosing in activated sludge on (**a**) TP, (**b**) Fe^{3+}, and (**c**) pH; data in (**a**,**b**) are presented as mean ± standard deviation (n = 3).

As shown in Figure 2b, effluent Fe^{3+} concentration exhibited distinct patterns under different dosing strategies. The one-time dosing caused an initial Fe^{3+} concentration peak (32.91 mg/L) after 1 h, then gradually decreased to 0.31 mg/L at 6 h. The average concentration from 6 to 24 h was 0.14 ± 0.08 mg/L, maintaining a low level. The continuous dosing maintained Fe^{3+} concentration below 1.0 mg/L overall, despite a slight increase in the later stages, with an average concentration of 0.54 ± 0.37 mg/L. Under batch dosing, the Fe^{3+} changed periodically, synchronized with the dosing times. It increased rapidly, immediately after each addition, but remained below 1.0 mg/L overall before declining.

The activated sludge system exhibited greater pH stability than the influent due to abundant microorganisms and buffering substances. As shown in Figure 2c, the one-time dosing caused an initial pH drop to 3.2, then gradually rose to approximately 8.2 due to iron salt efflux and wastewater inflow, and remained stable. Continuous dosing slowly acidified the system to a pH of 5.4, above the normal microbial activity threshold (pH > 5.0) [32], resulting in a minimal impact. In activated sludge systems with sufficient buffering capacity, it is more applicable than dosing directly into the influent. However, the acidic environment it creates may still pose risks to long-term system stability. Therefore, implementing accompanying pH adjustment measures may be necessary in practical applications. The batch dosing showed minimal pH fluctuations, with minimal initial decline and rapid recovery, ultimately stabilizing at 6.8–6.9. This is beneficial for P removal and reduces iron salt loss.

Batch dosing of solid iron salts avoids the high initial Fe^{3+} concentration associated with one-time dosing and mitigates Fe^{3+} accumulation from continuous dosing. Consequently, it achieves improved P removal alongside stable and low effluent Fe^{3+} levels. This mechanism is applicable to various wastewater sources. However, the efficiency of P removal, microbial activity, and iron concentration in the effluent are influenced by wastewater composition and operational conditions. For industrial wastewater with extremely low pH [33], the hydrolysis and precipitation behavior of iron may be altered, necessitating adjustments to the dosing frequency or supplemental pH control. High concentrations of complexing [34] or chelating agents (e.g., EDTA, citrate) may form stable soluble iron complexes, which reduce P removal and increase iron leakage risk. Nevertheless, the batch dosing strategy plays a critical role in controlling iron leakage and provides a reference for a wide range of applications.

3.2. *Effect of Iron Salts on Pollutant Removal in MBR Systems*

3.2.1. Water Quality Analysis

To enable a parallel comparison between the two devices, the effluent quality and operating conditions were kept consistent during the 75 d stabilization period. As shown in Table 1, after an initial 25 d acclimation period, both MBR systems achieved consistent effluent quality (25–75 d), enabling further comparative analysis. As only the aerobic section was set up in the system, the nitrogen and phosphorus removal effects were general, and the TN in the effluent water mainly existed in the form of NO_3-N.

Table 1. Variations in water quality of effluent from reactors #1 and #2 during 25–75 d of operation.

No.	COD (mg/L)	NH_3-N (mg/L)	TN (mg/L)	TP (mg/L)	NO_3-N (mg/L)	NO_2-N (mg/L)
#1	18.31 ± 6.92	0.31 ± 0.39	28.65 ± 6.67	3.60 ± 1.86	28.38 ± 5.77	0.04 ± 0.03
#2	18.38 ± 7.60	0.39 ± 0.37	29.49 ± 7.17	3.33 ± 1.43	29.80 ± 6.26	0.07 ± 0.10

At approximately 75 days, $Fe_2(SO_4)_3$ was added to reactor #1, while reactor #2 maintained its original operating conditions as the control. The effluent quality from both systems is shown in Figure 3. As shown in Figure 3a, effluent COD levels in both reactors remained low and showed no significant difference ($p = 0.474$). This result contrasts sharply with literature reports of an iron salt-enhanced coagulation effect in conventional activated sludge systems. This discrepancy originates from the MBR's inherent microfiltration-level retention effect, which achieves good solid–liquid separation performance without relying on chemical coagulation.

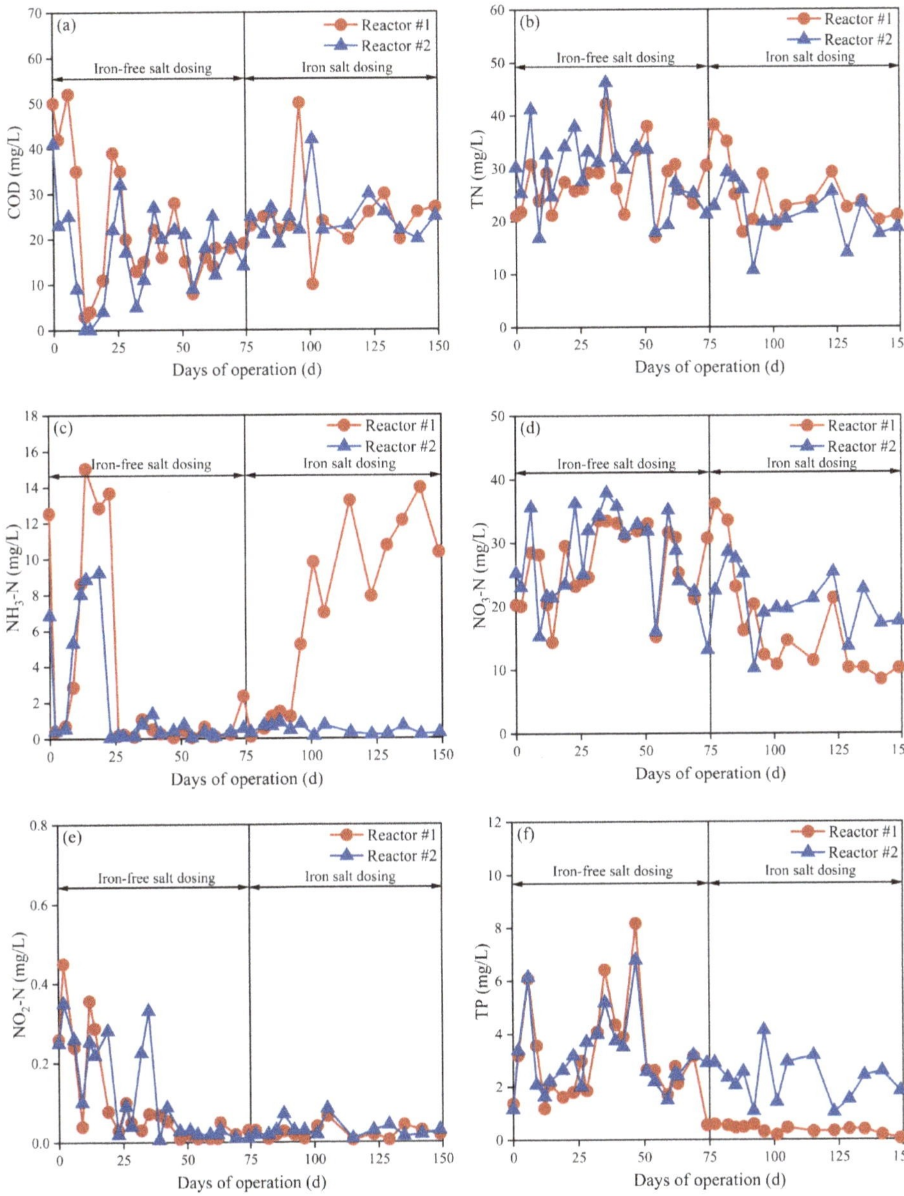

Figure 3. Variations in water quality during operation of reactors #1 and #2: (**a**) COD; (**b**) NH_3-N; (**c**) TN; (**d**) NO_3-N; (**e**) NO_2-N; (**f**) TP.

3.2.2. Nitrogen Removal

Both systems showed limited denitrification capacity, owing to the absence of an anoxic functional zone. As shown in Figure 3b, the effluent TN concentrations showed no significant difference ($p = 0.919$) between the two reactors (#1: 25.27 ± 5.96 mg/L; #2: 21.36 ± 5.03 mg/L). However, as shown in Figure 3c, reaction #1 exhibited a slowly increasing trend in effluent NH_3-N after iron salt dosing, ultimately exceeding the 10 mg/L threshold during later operational stages, while the control group maintained at

0.50 ± 0.26 mg/L. Figure 3d reveals an inverse trend for NO_3-N, with reaction #1 exhibiting 11% lower concentration (18.02 ± 9.20 mg/L) compared to the control. This indicates that iron salts might inhibit the microbial nitrification process.

Previous studies have indicated that iron salts impact nitrification through multiple mechanisms. The structure formed by iron and phosphates enhances the compactness of sludge flocs, hindering the absorption and transfer of nutrients like NH_4^+-N, and limiting oxygen contact with nitrifying bacteria, thereby reducing nitrification efficiency [35,36]. Additionally, when the pH is < 6.0, the NH_3-N concentration in MBR effluent is difficult to stabilize below 1.0 mg/L [23]. Excessive iron salt introduction can lower system pH, inhibiting nitrifying bacteria activity [37]. This further confirms the inhibitory effect of iron salts on nitrification.

This study further revealed the influence of iron salts on nitrification performance by comparing the SAUR of two devices. Figure 4a demonstrated that while the initial SAUR values were comparable between the two reactors, continuous iron salt dosing in reaction #1 caused significant SAUR reduction ($p = 0.002$), declining to 13.26 g NH_3-N/(kg VSS h) by day 142. This value was considerably lower than the 20.27 g NH_3-N/(kg VSS h) of reaction #2. The iron salts dosage showed a significant inverse correlation with SAUR ($p = 0.005$). These results indicate that iron salts directly reduce the nitrification rate, resulting in NH_3-N accumulation, while NO_3-N decreases in reactor #1 effluent. These findings are consistent with previous literature that iron salts inhibit nitrification.

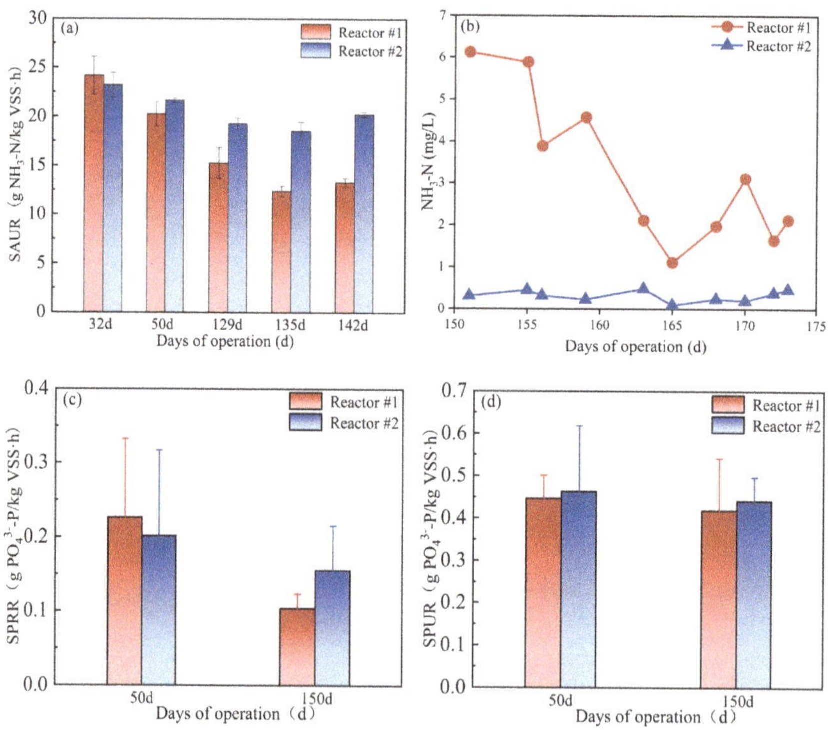

Figure 4. Comparison between reactors #1 and #2: (**a**) SAUR; (**b**) changes in NH_3-N concentration in the effluent after adjusting alkalinity; (**c**) SPRR; (**d**) SPUR. Data in (**a**,**c**,**d**) are presented as mean ± standard deviation ($n = 3$).

According to the stoichiometry of nitrification, oxidizing one unit of NH_3-N to NO_3^- consumes 4.57 units of oxygen and 7.14 units of alkalinity (calculated as $CaCO_3$). The denitrification theoretically can compensate for approximately 50% of nitrification's alkalinity demand. However, the absence of an anoxic zone in this system precluded denitrification, making influent alkalinity the sole source of replenishment. Alkalinity tests revealed lower levels in reaction #1 (92 ± 31 mg/L) than in reaction #2 (178 ± 26 mg/L), indicating that iron salts consumed approximately 48.3% of the system's alkalinity. Theoretical calculations indicate that complete nitrification of the influent NH_3-N (~20 mg/L) to an effluent NH_3-N $\leq$ 1 mg/L would require 136 mg/L alkalinity. However, the measured alkalinity in reaction #1 only supplies 67.6% of this theoretical demand, directly explaining its low nitrification efficiency. As shown in Figure S6, the SOUR values of reactors #1 and #2 showed little difference and both remained within the normal range for microbial metabolic activity. This indicated that iron salt addition did not inhibit overall microbial activity, thereby ruling out DO limitation as the primary cause of nitrification inhibition.

To verify that alkalinity consumption was the dominant mechanism inhibiting nitrification and to improve performance, CaO (35 mg/L) was added to reaction #1 from day 151 to replenish alkalinity. As shown in Figure 4b, this adjustment significantly reduced effluent NH_3-N concentration (p = 0.001) and improved nitrification performance. This result confirms that alkalinity is a key factor limiting nitrification efficiency and clarifies that the inhibition caused by iron salts is reversible, primarily through the pH decrease induced by iron hydrolysis. This finding provides important practical evidence for the synergistic optimization of iron salt phosphorus removal and biological denitrification in MBR systems.

3.2.3. TP Removal

As shown in Figure 3f, the addition of iron salts reduced the effluent TP concentration in reactor #1 to 0.40 ± 0.16 mg/L, meeting the Grade A discharge standard (GB 18918-2002). The average concentration of TP in the effluent of reactor #2 was 2.33 ± 0.85 mg/L, and the contribution of iron salts to the removal of TP in the system reached about 40%.

To investigate the effect of iron salts on the microbial phosphorus removal capacity, activated sludge was sampled from both reactors at the beginning and end of operation. Their phosphorus release and uptake capacities were expressed as SPRR and SPUR, respectively, with the results shown in Figure 4c,d. At 50 d, both reactors were in the late stage of sludge acclimatization prior to iron salt dosing. The activated sludge in both reactors exhibited similar properties, with SPRR values of 0.23 and 0.20 g PO_4^{3-}-P/(kg VSS·h) for reactors #1 and #2, respectively, and corresponding SPUR values of 0.44 and 0.46 g PO_4^{3-}-P/(kg VSS·h). At 150 d, after 76 days of iron salt addition to reactor #1, its SPRR and SPUR values had decreased to 0.10 and 0.41 g PO_4^{3-}-P/(kg VSS·h), respectively. For a contemporaneous comparison to eliminate temperature effects, reactor #2 maintained SPRR and SPUR values of 0.15 and 0.44 g PO_4^{3-}P/(kg VSS·h), respectively, during the same period. The prolonged iron salt dosing in reactor #1 led to a decline in biological phosphorus removal capacity, attributable to a reduction in the relative abundance of known phosphorus-colonizing bacteria [38].

To investigate iron salts' impact on microbial phosphorus removal capacity, dosing was ceased in reactor #1 at 175 d, with subsequent effluent TP concentrations in both reactors shown in Figure 5. The effluent TP concentration remained stable in reactor #2, while reactor #1 initially maintained a low TP concentration following dosing cessation. This may be due to residual Fe^{3+} and their complexes adsorbing phosphate. Although phosphorus concentrations later rose to become consistent with reactor #2, no deterioration beyond the control level occurred. This can be attributed to the effective retention of particulate phosphate by the microfiltration membrane.

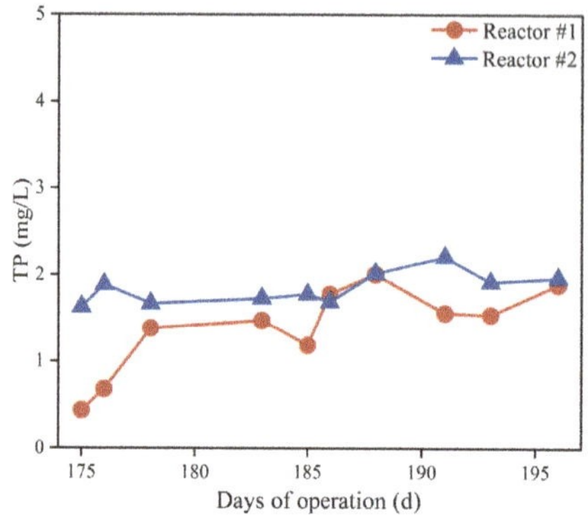

Figure 5. Variations of TP concentration in the effluent from reactors #1 and #2 after ceasing the addition of iron salts.

3.3. Contribution of Membrane Retention to Effluent Water Quality

This study utilized a multi-stage membrane separation technology system to investigate microfiltration membrane's retention characteristics in an iron-enhanced MBR process and its regulatory mechanisms on effluent water quality. Water samples were collected from the membrane effluent, the supernatant of the sludge mixture, and the filtrates obtained by passing the supernatant through 0.80 μm and 0.22 μm membranes. Key water quality indicators were analyzed. The results are shown in Figure 6.

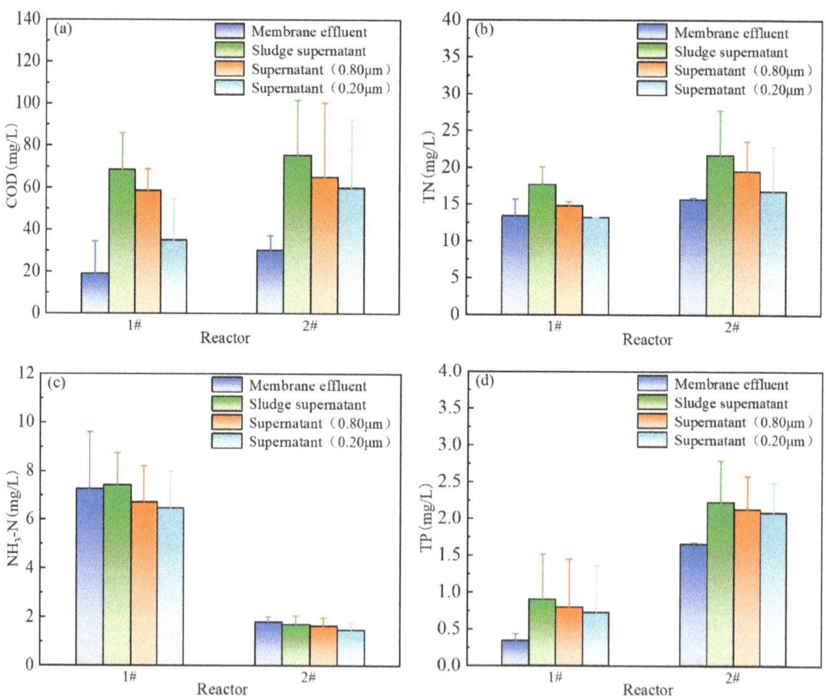

Figure 6. Comparison of membrane effluent, supernatant, and filtrate water quality between reactions #1 and #2: (**a**) COD; (**b**) NH_3-N; (**c**) TN; (**d**) TP. Data in (**a**,**b**) are presented as mean ± standard deviation (n = 3).

A comparative analysis of these water samples (Figure 6a) revealed that the COD concentration in the membrane effluent was lower than that in the supernatant and even the 0.22 μm filtrate. These results demonstrate that the smaller pore size of the membrane effectively retained particulate pollutants, while the sludge layer provided additional organic matter filtration and adsorption, collectively enhancing COD removal. The slightly lower COD values in both the supernatant and effluent of reaction #1 compared to reaction #2 indicate that iron salts reduced the dissolved organic matter content.

Regarding TN, Figure 6b shows a lower concentration in the membrane effluent than in the sludge supernatant. Additionally, the TN concentration in the effluent from different pore sizes also showed a decrease, confirming the microfiltration membrane's retention of particulate nitrogen as a key contributor to enhanced TN removal. Figure 6c revealed comparable NH_3-N concentrations in the membrane effluent and the supernatant, indicating that NH_3-N removal primarily relied on sludge nitrification rather than membrane retention or adsorption.

As shown in Figure 6d, the higher TP concentrations in the filtrates compared to the membrane effluent indicated the microfiltration membrane's retention and adsorption of colloidal and particulate phosphorus in the mixed liquor. Comparison of the TP concentrations in the effluent and supernatant of reactions #1 and #2 further confirmed that iron salt dosing enhanced P removal. This enhancement occurred through two synergistic mechanisms: (1) physical–chemical precipitation of phosphate by iron salts into larger particles; and (2) effective retention of these particles by the microfiltration membrane. This study reveals that the microfiltration membrane selectively retains colloids (0.22–0.80 μm) and elucidates the synergistic purification mechanism between iron salt-modified sludge flocs and the microfiltration membrane.

3.4. The Effect of Iron Salt Addition on Membrane Fouling

TMP variations between reactors #1 and #2 were analyzed to assess the effect of iron salt addition on membrane fouling. Figure 7 shows that while both reactors initially exhibited similar TMP trends, reactor #1 exhibited a slower TMP increase rate after iron salt addition. At the 30 kPa pressure endpoint, reactor #1 exhibited delayed fouling progression compared to reactor #2, extending its average operating cycle from 19.0 d to 21.6 d. These observations preliminarily suggest that iron salts may contribute to the mitigation of membrane fouling.

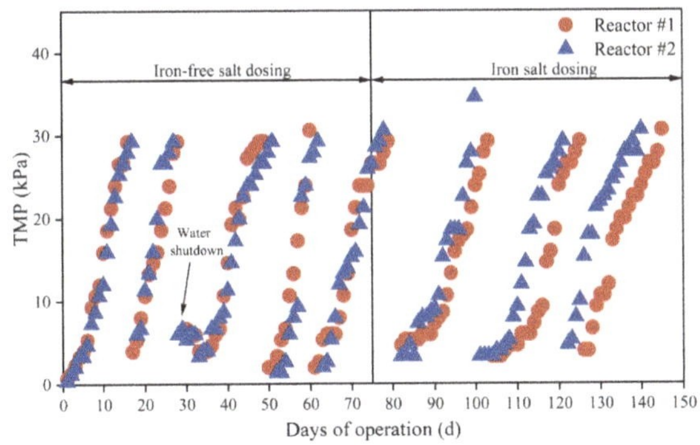

Figure 7. Comparison of TMP between reactors #1 and #2 during operation.

The reduced membrane fouling rate may be attributed to the coagulation and removal of macromolecules by iron salts. The supernatants from both reactors, filtered through 0.45 μm membranes, were analyzed by gel filtration chromatography (GFC) to test the molecular weight distribution (Table 2). The GFC analysis revealed three characteristic peaks (P1–P3) in the supernatant of both reactors, with reactor #1 exhibiting lower Mn values (70.1, 336.3, and 2722.9 kDa) compared to reactor #2 (79.2, 361.0, and 4905.8 kDa). The organic compounds in both reactors showed comparable molecular weights between 104 and 105 Da, but reactor #1 exhibited significantly lower molecular weight than reactor #2 when the molecular weight exceeded 106 Da.

Table 2. Molecular weight distribution characteristics of organic matter in the supernatant of mixed sludge from reactors #1 and #2.

No.	Organic Matter	Mn (kDa)	Mw (kDa)	Mz (kDa)	Mw/Mn
	P1	70.1	72.8	75.8	1.0
#1	P2	336.3	378.8	443.2	1.1
	P3	2722.9	4521.1	8496.9	1.7
	P1	79.2	85.5	92.6	1.1
#2	P2	361.0	405.0	474.0	1.1
	P3	4905.8	6649.8	9582.8	1.4

The MBR process contains higher levels of high-molecular-weight (>104 Da) microbial metabolites (such as SMP and EPS) compared to traditional wastewater treatment systems [39]. It is speculated that the iron salts added to reactor #1 removed macromolecules through coagulation and adsorption, thereby reducing their concentration in the sludge mixture. The lower supernatant COD concentration in reactor #1 than in reactor #2 (Figure 6a) indicated a reduced macromolecular content. These macromolecular and recalcitrant organic compounds are well-documented as significant membrane foulants [40,41].

3.5. Changes in Fe^{3+} Concentration in Membrane Effluent

After adding iron salts to reactor #1, membrane effluent samples were regularly collected at approximately 1 h post-dosing for Fe^{3+} concentration analysis, with results presented in Figure 8. The figure shows that reactor #1 maintained a relatively low Fe^{3+} concentration (0.23 $\pm$ 0.11 mg/L) over approximately 100 d of iron salt addition. The effluent Fe^{3+} concentration complied with China's Environmental Quality Standard for Surface Water (GB 3838-2002) ($\leq$0.30 mg/L), indicating a low environmental risk. Wang et al. [31] reported that effluent Fe^{3+} concentration accounted for only about 10% of that in the 0.2 µm filtrate supernatant, suggesting that fine-scale Fe^{3+} may be retained in the system. Furthermore, the added Fe^{3+} can form monomeric complexes or amorphous ferric oxide (AFOs) [42] with macromolecules like SMP, which are retained in the sludge mixed liquor or on the membrane surface. Concurrently, larger iron salt particles were effectively filtered by the microfiltration membrane, maintaining a low effluent Fe^{3+} concentration.

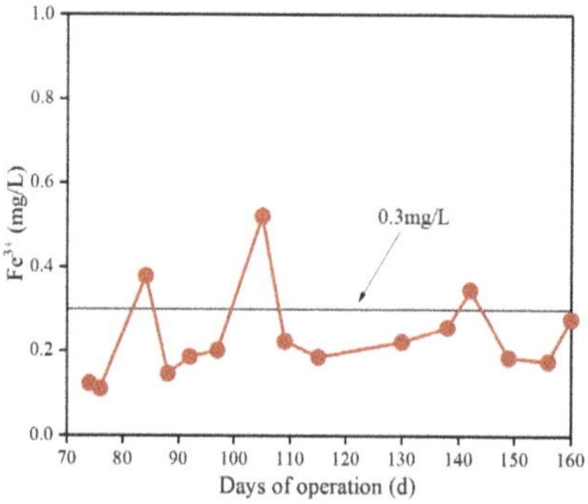

Figure 8. Variations of Fe^{3+} concentration in effluent during iron salt dosing.

Figure 8 also shows periodic fluctuations in effluent Fe^{3+} concentration, with occasional spikes slightly exceeding the 0.30 mg/L limit. Based on cleaning data, this phenomenon primarily occurred immediately after membrane cleaning. Li et al. [43] reported that approximately 98% of system Fe^{3+} was retained in the sludge mixed liquor with minimal distribution in the cake and gel layer. This suggests that the established fouling layer adsorbs iron salts, whereas newly cleaned membranes had not yet formed a significant cake and gel layer, leading to the release of some dissolved Fe^{3+} into the effluent. In Figure 2b, the higher initial effluent Fe^{3+} concentration in the short-term test (24 h) compared to reactor #1 resulted from the membrane surface not having yet formed a fouling layer, further validating the adsorption effect of the fouling layer on Fe^{3+}.

4. Conclusions

This study proposes an optimized iron dosing strategy for MBR systems that ensures efficient phosphorus removal while minimizing iron consumption. Batch dosing of solid iron salts maintained the Fe^{3+} concentration below 1.0 mg/L while achieving 97% TP removal. Long-term operation revealed that iron salt dosing caused effluent NH_3-N exceeding 10 mg/L, but this was effectively mitigated by CaO supplementation. While iron salts contributed approximately 40% to TP removal, SPRR and SPUR data indicated partial inhibition of biological phosphorus removal capacity. The results indicated that iron salts preliminarily mitigated membrane fouling through coagulation of macromolecules in the sludge mixture. The average Fe^{3+} concentration in the MBR effluent was 0.23 ± 0.11 mg/L under the synergistic effect of the optimal dosing strategy and adsorption by the contaminated layer of the microfiltration membrane. The effluent Fe^{3+} concentration consistently complied with China's Environmental Quality Standard for Surface Water (GB 3838-2002), which effectively avoids the risk of secondary pollution caused by over-injection of iron salts. The batch dosing strategy is simple to operate and requires minimal additional equipment, demonstrating potential for application in large-scale MBR projects. The mechanism of controlling iron ion leakage through dosing strategies, as revealed in this study, provides a theoretical basis and methodological framework for MBR applications across various scenarios. For wastewaters of different compositions, parameter optimization is recommended prior to full-scale implementation. Future research should explore the long-term stability and applicability of this strategy in more complex wastewater systems.

Supplementary Materials: The following supporting information can be downloaded at: https://www.mdpi.com/article/10.3390/membranes15100297/s1, Figure S1: Schematic diagram of continuous flow device; Figure S2: Schematic diagram of MBR set-up; Figure S3: Phosphorus-specific release rate measurement device diagram; Figure S4: Phosphorus-specific absorption rate measurement device diagram; Figure S5: Specific oxygen uptake rate measurement device diagram; Figure S6: Comparison of SOUR between reactors #1 and #2; Table S1: Main water quality indicators for the influent of MBR set-up; Table S2: Related design parameters of O-MBR; Table S3: Routine chemical test items and methods; Table S4: Discharge standard of pollutants for municipal wastewater treatment plant; Text S1: Calculation of Fe dosage based on influent TP; Text S2: Determination of sludge-specific oxygen uptake rate.

Author Contributions: Q.W.: conceptualization, funding acquisition, resources, writing—review and editing. B.Z.: data curation, investigation, writing—original draft. J.S.: data curation, formal analysis investigation. W.Z.: formal analysis, software. J.Z.: conceptualization, supervision. Z.W.: resources. All authors have read and agreed to the published version of the manuscript.

Funding: The authors are grateful for the financial support from the National Natural Science Foundation of China (52370045) and the Natural Science Foundation of Shanghai Municipality (23ZR1467500).

Institutional Review Board Statement: Not applicable.

Informed Consent Statement: Not applicable.

Data Availability Statement: The original contributions presented in this study are included in the article/Supplementary Material. Further inquiries can be directed to the corresponding author.

Conflicts of Interest: The authors declare that they have no known competing financial interests or personal relationships that could have appeared to influence the work reported in this paper.

References

1. Judd, S. *The MBR book: Principles and Applications of Membrane Bioreactors for Water and Wastewater Treatment*; Elsevier: Amsterdam, The Netherlands, 2010.
2. Wang, C.; Ng, T.C.A.; Ng, H.Y. Comparison between novel vibrating ceramic MBR and conventional air-sparging MBR for domestic wastewater treatment: Performance, fouling control and energy consumption. *Water Res.* **2021**, *203*, 117521. [CrossRef] [PubMed]
3. Wang, R.; You, H.; Li, Z.P.; Xie, B.H.; Qi, S.J.; Zhu, J.; Qin, Q.Q.; Wang, H.; Sun, J.X.; Ding, Y.; et al. A novel reduced graphene oxide/polypyrrole conductive ceramic membrane enhanced electric field membrane bioreactor: Mariculture wastewater treatment performance and membrane fouling mitigation. *Bioresour. Technol.* **2023**, *376*, 128917. [CrossRef]
4. Lopes, T.S.D.; Hessler, R.; Bohner, C.; Athayde, G.B.; de Sena, R.F. Pesticides removal from industrial wastewater by a membrane bioreactor and post-treatment with either activated carbon, reverse osmosis or ozonation. *J. Environ. Chem. Eng.* **2020**, *8*, 104538. [CrossRef]
5. Sun, J.S.; Liang, P.; Yan, X.X.; Zuo, K.C.; Xiao, K.; Xia, J.L.; Qiu, Y.; Wu, Q.; Wu, S.J.; Huang, X.; et al. Reducing aeration energy consumption in a large-scale membrane bioreactor: Process simulation and engineering application. *Water Res.* **2016**, *93*, 205–213. [CrossRef] [PubMed]
6. Li, R.H.; Wang, W.J.; Li, B.; Zhang, J.Y.; Liu, J.; Zhang, G.J.; Guo, X.C.; Zhang, X.H.; Li, X.Y. Acidogenic phosphorus recovery from the wastewater sludge of the membrane bioreactor systems with different iron-dosing modes. *Bioresour. Technol.* **2019**, *280*, 360–370. [CrossRef] [PubMed]
7. Li, R.H.; Cui, J.L.; Li, X.D.; Li, X.Y. Phosphorus removal and recovery from wastewater using Fe-dosing bioreactor and cofermentation: Investigation by X-ray absorption near-edge structure spectroscopy. *Environ. Sci. Technol.* **2018**, *52*, 14119–14128. [CrossRef]
8. De-Bashan, L.E.; Bashan, Y. Recent advances in removing phosphorus from wastewater and its future use as fertilizer (1997–2003). *Water Res.* **2004**, *38*, 4222–4246. [CrossRef]
9. Li, R.H.; Wang, X.M.; Li, X.Y. A membrane bioreactor with iron dosing and acidogenic co-fermentation for enhanced phosphorus removal and recovery in wastewater treatment. *Water Res.* **2018**, *129*, 402–412. [CrossRef]
10. Yang, X.L.; Song, H.L.; Chen, M.; Cheng, B. Characterizing membrane foulants in MBR with addition of polyferric chloride to enhance phosphorus removal. *Bioresour. Technol.* **2011**, *102*, 9490–9496. [CrossRef]
11. Haldar, D.; Duarah, P.; Purkait, M.K. MOFs for the treatment of arsenic, fluoride and iron contaminated drinking water: A review. *Chemosphere* **2020**, *251*, 126388. [CrossRef]
12. Kumar, V.; Bharti, P.K.; Talwar, M. Studies on high iron content in water resources of Moradabad district (UP), India. *Water Sci.* **2017**, *31*, 44–51. [CrossRef]
13. Abed, E.A.M.; Alaboudi, K.A.N.; Abbas, M.H.H.; Attia, T.M.S.; Abdelhafez, A.A. Iron Contamination in groundwater: Risk assessment and remediation techniques in egypt's new valley. *Water* **2024**, *16*, 1834. [CrossRef]
14. Rahman, M.A.; Hashem, M.A. Arsenic, iron and chloride in drinking water at primary school, Satkhira, Bangladesh. *Phys. Chem. Earth* **2019**, *109*, 49–58.
15. Chen, M.; Li, X.H.; He, Y.H.; Song, N.; Cai, H.Y.; Wang, C.H.; Li, Y.T.; Chu, H.Y.; Krumholz, L.R.; Jiang, H.L. Increasing sulfate concentrations result in higher sulfide production and phosphorous mobilization in a shallow eutrophic freshwater lake. *Water Res.* **2016**, *96*, 94–104. [CrossRef]
16. Xu, W.H.; Wang, W.W.; Deng, B.B.; Liu, Q.X. A review of the formation conditions and assessment methods of black and odorous water. *Environ. Monit. Assess.* **2024**, *196*, 42. [CrossRef] [PubMed]
17. Chen, M.S.; Ding, S.M.; Liu, L.; Xu, D.; Han, C.; Zhang, C.S. Iron-coupled inactivation of phosphorus in sediments by macrozoobenthos (*chironomid larvae*) bioturbation: Evidences from high-resolution dynamic measurements. *Environ. Pollut.* **2015**, *204*, 241–247.
18. Rong, N.; Lu, W.Z.; Zhang, C.Y.; Wang, Y.S.; Zhu, J.L.; Zhang, W.Q.; Lei, P. In situ high-resolution measurement of phosphorus, iron and sulfur by diffusive gradients in thin films in sediments of black-odorous rivers in the Pearl River Delta region, South China. *Environ. Res.* **2020**, *189*, 109918. [CrossRef]
19. Liu, J.J.; Lu, J.; Zhu, G.W. Occurrence characteristics of black patch events and their influencing factors in Lake Taihu during 2009 and 2017. *J. Lake Sci.* **2018**, *30*, 1196–1205. [CrossRef]
20. Zhuang, Y.; Chen, R.; Shi, B.Y. Iron particle formation under chlorine disinfection considering effects of deoxidizers in drinking water. *J. Hazard. Mater.* **2021**, *420*, 126581. [CrossRef]
21. Yang, L.; Cai, X.; Li, R. Ferroptosis induced by pollutants: An emerging mechanism in environmental toxicology. *Environ. Sci. Technol.* **2024**, *58*, 2166–2184. [CrossRef]

22. Zhuang, Y.; Li, D.H.; Shi, B.Y. Perfluorooctanoic acid (PFOA) incorporated into iron particles promoted the formation of disinfection byproducts under drinking water conditions. *Environ. Sci. Technol.* **2023**, *57*, 4863–4869. [CrossRef]
23. Wang, Y.; Tng, K.H.; Wu, H.; Leslie, G.; Waite, T.D. Removal of phosphorus from wastewaters using ferrous salts—A pilot scale membrane bioreactor study. *Water Res.* **2014**, *57*, 140–150.
24. Gkotsis, P.K.; Mitrakas, M.M.; Tolkou, A.K.; Zouboulis, A.I. Batch and continuous dosing of conventional and composite coagulation agents for fouling control in a pilot-scale MBR. *Chem. Eng. J.* **2017**, *311*, 255–264. [CrossRef]
25. Li, J.; Wang, S.L.; Wang, Z.Z.; Zheng, Z.M.; Zhang, J. Assessing biomass activities, sludge characteristics and membrane fouling in the University of Cape Town membrane bioreactor under ferric chloride addition. *Environ. Technol. Innov.* **2021**, *23*, 101796. [CrossRef]
26. Jiang, L.Y.; Liu, Y.; Guo, F.J.; Zhou, Z.; Jiang, J.; You, Z.C.; Wang, Q.Y.; Wang, Z.W.; Wu, Z.C. Evaluation of nutrient removal performance and resource recovery potential of anaerobic/anoxic/aerobic membrane bioreactor with limited aeration. *Bioresour. Technol.* **2021**, *340*, 125728. [CrossRef] [PubMed]
27. Wang, Q.Y.; Li, Y.L.; Wang, Z.W.; Liu, Y.Y.; Zhou, Z.; Hu, W.J.; Lin, L.F.; Wu, Z.C. Effects of microplastics accumulation on performance of membrane bioreactor for wastewater treatment. *Chemosphere* **2022**, *287*, 131968. [CrossRef] [PubMed]
28. Zhang, T.; Zhang, J.; Wang, Q.Y.; Zhang, H.; Wang, Z.W.; Wu, Z.C. Evaluating of the performance of natural mineral vermiculite modified PVDF membrane for oil/water separation by membrane fouling model and XDLVO theory. *J. Membr. Sci.* **2022**, *641*, 119886. [CrossRef]
29. Torapava, N.; Radkevich, A.; Persson, I.; Davydov, D.; Eriksson, L. Formation of a heteronuclear hydrolysis complex in the Th IV–Fe III system. *Dalton Trans.* **2012**, *41*, 4451–4459.
30. Cornell, R.M.; Giovanoli, R.; Schneider, W. Review of the hydrolysis of iron (III) and the crystallization of amorphous iron (III) hydroxide hydrate. *J. Chem. Technol. Biotechnol.* **1989**, *46*, 115–134. [CrossRef]
31. Wang, Y.; Leslie, G.L.; Waite, T.D. Impact of iron dosing of membrane bioreactors on membrane fouling. *Chem. Eng. J.* **2014**, *252*, 239–248. [CrossRef]
32. Zhang, Y.; Bu, D.; Liu, C.G.; Luo, X.; Gu, P. Study on retarding membrane fouling by ferric salts dosing in membrane bioreactors. In Proceedings of the IWA Water Environment-Membrane Technology, Seoul, Republic of Korea, 7–10 June 2004; Volume 753, pp. 7–10.
33. Sun, J.C.; Zhu, D.F.; Fang, R.; Zhang, G.F. Arsenic-containing waste acid treatment—Experimental study on the removal of arsenic from copper smelting wwaste acid by lime-iron salt-ion exchange. *JOM* **2025**, *77*, 5632–5642. [CrossRef]
34. Shi, W.; Li, J.; Gao, F.; Meng, L.J.; Su, X.; Wang, Z.W. Strongly coordinating mediator enables single-step resource recovery from heavy metal-organic complexes in wastewater. *Nat. Commun.* **2024**, *15*, 10828. [CrossRef]
35. Sarah, P.; Korneel, R.; Willy, V. Impact of iron salts on activated sludge and interaction with nitrite or nitrate. *Bioresour. Technol.* **2003**, *88*, 229–239. [CrossRef]
36. Guo, W.; Ngo, H.H.; Vigneswaran, S.; Dharmawan, F.; Nguyen, T.T.; Aryal, R. Effect of different flocculants on short-term performance of submerged membrane bioreactor. *Sep. Purif. Technol.* **2010**, *70*, 274–279. [CrossRef]
37. Clark, T.; Stephenson, T.; Pearce, P.A. Phosphorus removal by chemical precipitation in a biological aerated filter. *Water Res.* **1997**, *31*, 2557–2563. [CrossRef]
38. Huang, X.; Yao, K.; Yu, J.H.; Dong, W.Y.; Zhao, Z.L. Nitrogen removal performance and microbial characteristics during simultaneous chemical phosphorus removal process using Fe^{3+}. *Bioresour. Technol.* **2022**, *363*, 127972. [CrossRef]
39. Wang, Z.W.; Wu, Z.C. Distribution and transformation of molecular weight of organic matters in membrane bioreactor and conventional activated sludge process. *Chem. Eng. J.* **2004**, *150*, 396–402. [CrossRef]
40. Kimura, K.; Yamato, N.; Yamamura, H.; Watanabe, Y. Membrane fouling in pilot-scale membrane bioreactors (MBRs) treating municipal wastewater. *Environ. Sci. Technol.* **2005**, *39*, 6293–6299. [CrossRef] [PubMed]
41. Kimura, K.; Kakuda, T.; Iwasaki, H. Membrane fouling caused by lipopolysaccharides: A suggestion for alternative model polysaccharides for MBR fouling research. *Sep. Purif. Technol.* **2019**, *223*, 224–233. [CrossRef]
42. Wang, X.M.; Waite, T.D. Iron speciation and iron species transformation in activated sludge membrane bioreactors. *Water Res.* **2010**, *44*, 3511–3521. [CrossRef]
43. Li, H.S.; Wen, Y.; Cao, A.S.; Huang, J.S.; Zhou, Q.; Somasundaran, P. The influence of additives (Ca^{2+}, Al^{3+}, and Fe^{3+}) on the interaction energy and loosely bound extracellular polymeric substances (EPS) of activated sludge and their flocculation mechanisms. *Bioresour. Technol.* **2012**, *114*, 188–194. [CrossRef]

membranes

MDPI

Article

Zn^{2+}-Mediated Co-Deposition of Dopamine/Tannic Acid/ZIF-8 on PVDF Hollow Fiber Membranes for Enhanced Antifouling Performance and Protein Separation

Lei Ni [1], Qiancheng Cui [2], Zhe Wang [2], Xueting Zhang [2], Jun Ma [3], Wenjuan Zhang [2,*] and Caihong Liu [4,*]

1 School of Material Science and Engineering, Tiangong University, Tianjin 300387, China
2 Tianjin Key Laboratory of Aquatic Science and Technology, School of Environmental and Municipal Engineering, Tianjin Chengjian University, Tianjin 300384, China
3 State Key Laboratory of Urban Water Resource and Environment, Harbin Institute of Technology, Harbin 150090, China
4 Key Laboratory of Eco-Environments in Three Gorges Reservoir Region, Ministry of Education, College of Environment and Ecology, Chongqing University, Chongqing 400044, China
* Correspondence: wenjuanvivian@126.com (W.Z.); caihong.liu@cqu.edu.cn (C.L.); Tel./Fax: +86-22-23085117 (W.Z.)

Abstract: The inherent hydrophobicity of poly(vinylidene fluoride) (PVDF) ultrafiltration membranes leads to severe membrane fouling when processing proteinaceous solutions and organic contaminants, significantly limiting their practical applications. This study presents a novel metal-ion mediated co-deposition strategy for fabricating high-performance antifouling poly(vinylidene fluoride) (PVDF) hollow fiber ultrafiltration membranes. Through Zn^{2+} coordination-driven self-assembly, a uniform and stable composite coating of dopamine (DA), tannic acid (TA), and ZIF-8 nanoparticles was successfully constructed on the membrane surface under mild conditions. The modified membrane exhibited significantly enhanced hydrophilicity, with a water contact angle of 21° and zeta potential of −29.68 mV, facilitating the formation of a dense hydration layer that effectively prevented protein adhesion. The membrane demonstrated exceptional separation performance, achieving a pure water permeability of 771 L/(m^2·h·bar) and bovine serum albumin (BSA) rejection of 97.7%. Furthermore, it showed outstanding antifouling capability with flux recovery rates exceeding 83.6%, 74.7%, and 71.5% after fouling by BSA, lysozyme, and ovalbumin, respectively. xDLVO analysis revealed substantially increased interfacial free energy and stronger repulsive interactions between the modified surface and protein foulants. The antifouling mechanism was attributed to the synergistic effects of hydration layer formation, optimized pore structure, additional water transport pathways from ZIF-8 incorporation, and electrostatic repulsion from negatively charged surface groups. This work provides valuable insights into the rational design of high-performance antifouling membranes for sustainable water treatment and protein separation applications.

Keywords: hollow fiber membrane; dopamine; tannin; ZIF-8; antifouling

1. Introduction

Ultrafiltration (UF) membranes, with pore sizes ranging from 2 to 100 nm, enable the effective removal of contaminants of microorganisms, natural organic matters, colloidal substances, et al., from sewage [1]. Among UF materials, poly(vinylidene fluoride) (PVDF) stands out due to its exceptional mechanical strength and chemical stability and has been extensively investigated and applied in water treatment [2–5]. However, the inherent hydrophobicity of PVDF membranes induces high interfacial energy at membrane–water interfaces, promoting severe fouling during proteinaceous or organic pollutant filtration. This fouling manifests as pore blockage and surface accumulation, leading to reduced porosity, declined flux, increased energy consumption, and shortened membrane lifespan [6], ultimately limiting PVDF's applications in the food, pharmaceutical, and wastewater treatment sectors [7,8]. Traditional fouling mitigation strategies (e.g., pre-oxidation, coagulation, cleaning procedures) often introduce harmful by-products and elevate operational costs [9,10]. Alternatively, constructing a hydrophilic surface hydration layer can minimize foulant–membrane interactions during initial fouling stages [11]. Recent approaches focus on modifying membranes with hydrophilic moieties (e.g., –OH, $-NH_2$), inorganic nanoparticles, metal-organic frameworks (MOFs), or zwitterionic polymers. MOFs are novel nanoporous particles consisting of metal ions and organic ligands, and exhibiting excellent compatibility with the organic matrix [12–14]. Zeolitic imidazolate framework-8 (ZIF-8), a Zn-based MOF with tunable pores and exceptional stability, has shown promise in PVDF membrane modification [15]. Karimi et al. [16] formed a uniform ZIF-8 layer on a PVDF ultrafiltration membrane, which can significantly improve the water flux and flux recovery rates (FRRs) to 134 L/(m^2 h) and 71.4%, respectively. Guo et al. [17] in situ modified PVDF membrane with a 2 μm ZIF-8 layer obtaining a flux of 137 L/(m^2 h) and protein rejection of 97.5%. However, the stability of nanoporous particles on the UF membrane surface needs to be further improved through other schemes.

Dopamine (DA) exhibits unique adhesion to low-surface-energy polymers (e.g., PVDF, polypropylene) via self-polymerization into polydopamine (PDA), which enhances wettability and biocompatibility through catechol/amine groups [18–22]. However, the DA self-polymerization process is liable to generate large aggregates through non-covalent effects impacting the membrane surface morphology [23–25]. Recent advances have demonstrated accelerated deposition strategies using various catalysts or energy inputs. However, these methods may have their own limitations in terms of coating uniformity and stability. In addition, its residual hydrophobic aromatic domains limit hydrophilicity improvements. Notably, PDA's phenolic hydroxyl and amino groups enable secondary reactions (e.g., Michael addition/Schiff base reactions) with $-NH_2$/–OH/–COOH-containing molecules under an aerobic weakly alkaline condition [26–28], offering avenues for further hydrophilicity enhancement [29,30]. For instance, Zin et al. [31] achieved a super-hydrophilic membrane surface with high oil rejection (>98%) by co-depositing DA/polyethyleneimine on a PVDF microfiltration membrane, while Ren et al. [32] combined PDA with 2-*N*-propylsulfonated chitosan (PCS) to reduce the water contact angle to 39° and achieve 90.4% flux recovery. Tannic acid (TA) is a natural reducing agent with five catechol groups in one TA molecule, and can form more viscous networks than PDA. In this case, compared with the PDA deposition alone, involving TA in the modification process could theoretically improve the adhesion of the modified layer [33]. However, TA's poor self-polymerization necessitates amines (e.g., PEI) or metal ions (e.g., Fe^{3+}, Zn^{2+}) as catalysts [34,35]. Moreover, incorporating nanomaterials into DA/TA layers can reduce cross-linking density while creating additional water transport pathways, mitigating the permeability-selectivity trade-off [36,37].

The hollow fiber membrane (HFM) exhibits greater self-supporting capability owing to its unique tubular hollow structure, which distinguishes it from conventional flat membranes. A high packing density and specific surface area can be readily achieved through simple stacking of hollow fiber filaments—advantages not attainable with ordinary flat membranes [38–40]. However, unlike flat membranes, hollow fibers possess cylindrical and curved surfaces. Consequently, established modification methods designed for flat membranes are not easily transferable to HFMs, often resulting in uneven coating during modification processes, and in some cases, failure to achieve adequate adhesion of the modified layer. Moreover, DA/TA/ZIF-8 composite modifications for PVDF HFMs and their antifouling mechanisms remain unexplored. This study demonstrates a facile co-deposition strategy conducted at room temperature for PVDF HFMs. The Zn^{2+}-mediated chelation with DA/TA accelerates metal-polyphenol network formation and the DA/TA interlayer regulates the ZIF-8 crystallization on the membrane surface, synergistically improving the hydrophilicity and antifouling performance. The physico-chemical properties of membranes before and after modification were thoroughly characterized and analyzed through scanning electron microscopy (SEM), X-ray photoelectron spectroscopy (XPS), attenuated total reflection-Fourier transform infrared spectroscopy (ATR-FTIR), water contact angle (WCA) measuring equipment, and zeta potentiometer. ATR-FTIR spectra were normalized before the data analysis. Filtration performance was evaluated using bovine serum albumin (BSA), lysozyme (LYZ), and ovalbumin (OVA), and the obtained DA/TA/ZIF-8-UF membrane demonstrated superior filtration performance, including high water permeability and excellent protein rejection rates. In addition, antifouling tests against BSA, LYZ, and OVA revealed significantly improved fouling resistance and cleaning recovery properties. The extended Derjaguin–Landau–Verwey–Overbeek (xDLVO) theory analysis provided deep insight into the interfacial interaction energies, quantitatively explaining the antifouling mechanism by revealing reduced adhesion energy between the membrane surface and foulants. This work provides an efficient and scalable surface engineering strategy for constructing multifunctional coatings on hollow fiber membranes, overcoming the limitations of conventional modification techniques suited mainly for flat membranes. In addition, it advances the fundamental understanding of hybrid organic–inorganic antifouling layers and offers a practical pathway for developing high-performance HFMs in applications such as wastewater treatment, protein purification, and biomedical separation.

2. Materials and Methods

2.1. Chemical Reagents and Materials

Commercial hollow fiber PVDF microfiltration membranes (0.2 μm, Tianjin MOTECH, Tianjin, China) were used as substrates for modification. The anti-fouling properties of a commercial hollow fiber UF membrane (PVDF, 30 kDa) from Henan Hongxin Ultrafiltration Membrane Technology Co., Ltd. were measured for comparison. Dopamine hydrochloride ($C_3H_{11}NO_2{\cdot}HCl$, Macklin, Shanghai, China), TA ($C_{76}H_{52}O_{46}$, Macklin, Shanghai, China), 2-Methylimidazole (Hmim, Aladdin, Tianjin, China), ethanol (Aladdin, Tianjin, China) and zinc nitrate hexahydrate ($Zn(NO_3)_2{\cdot}6H_2O$, Aladdin, Tianjin, China) were used as received. Tris buffer with a pH of 8.5 ± 0.1 was prepared from tris (hydroxymethyl)methylaminomethane (THAM, Beijing Solebo Technology, Beijing, China) and dilute hydrochloric acid solution (HCl, 30%, Tianjin, China). BSA, LYZ, and OVA purchased from Aladdin (Tianjin, China) were used for characterization of separation performance and anti-fouling properties of membranes. Deionized (DI) water with a conductivity of 18.2 MΩ cm was produced in the lab.

2.2. Membrane Modification

Before modification, the hollow fiber membranes with a length of 10 cm were submerged in the anhydrous ethanol solution for 12 h to remove the residual solvent, and then washed with DI water thoroughly. The moisture on the membrane surface was absorbed by tissue and both ends of each membrane filament were sealed with glue. Then, the membranes were put into the modification solution (DA: 1 g/L, TA: 4 g/L, and $Zn(NO_3)_2$: 0.015 mol/L) and modified for 6 h at room temperature with a shaking speed of 160 rpm. After that, the samples were thoroughly washed with DI water, and activated with 0.7 mmol/L Hmim solution for 3 h following with the ZIF-8 growth process in ZIF-8 mother solution ($Zn(NO_3)_2 \cdot 6H_2O$:Hmim:H_2O =1:75:6000) for 6 h. Finally, the obtained membranes were washed thoroughly using DI water and dried at 60 °C for 12 h. The schematic illustration for the membrane modification process is shown in Figure 1. Table 1 lists the components of modification solutions. The membrane samples were modified in triplicate.

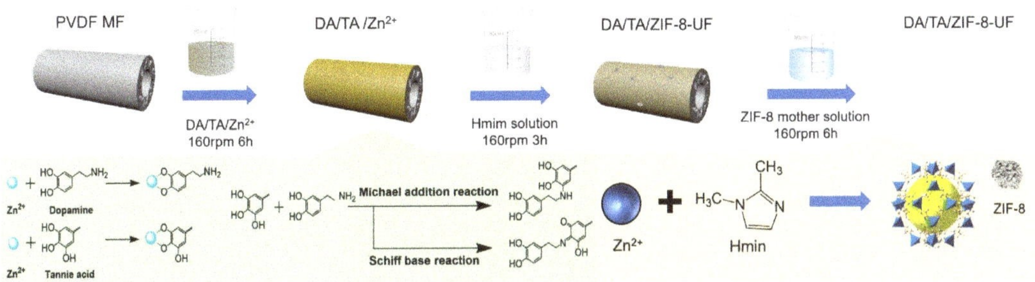

Figure 1. Schematic illustration of membrane modification.

Table 1. Formula of the membrane modification solution.

Membrane	DA (g/L)	TA (g/L)	Zn^{2+} (mol/L)	H_{mim} (mmol/L)	Deposition Time (h)
DA-UF	1	-	-	-	6
TA-UF	-	4	-	-	6
DA/TA-UF	1	4	-	-	6
DA/ZIF-8-UF	1	-	0.015	0.7	6
TA/ZIF-8-UF	-	4	0.015	0.7	6
DA/TA/ZIF-8-UF	1	4	0.015	0.7	6

2.3. Characterization of Membrane Samples

A field emission SEM (Gemini 500, Zeiss, Germany) was applied for observing the morphology of the membrane surface and cross-sections at a working voltage of 10 kV. All film samples were coated with a thin layer (~10 nm) of gold by sputtering at the current of 15 mA for 60 s prior to field emission SEM testing. Membrane surface functional groups before and after modification were analyzed by an ATR-FTIR spectrometer (Nicolet 6700, Thermo Fisher, Waltham, MA, USA) within the spectral range of 650–4000 cm^{-1}. The changes in the membrane surface chemical compositions were determined by an XPS spectrometer (ESCALAB 250Xi, Thermo Scientific, East Grinstead, UK) with a scanning resolution of 0.1 eV. An automatic contact angle meter (DSA100, Kruss, Germany) was used to test the WCA of membrane samples for wettability analysis [41,42]. The hydrophilicity of the whole membrane can be characterized by the water absorption test which was conducted by immersing membranes in DI water for 24 h to obtain wet weight (M_{wet}) after removing surface water by tissues, and to obtain dry weight (M_{dry}) after dried at 50 °C for 3 days in an oven. The water absorption rate (WAR) can be calculated from the formula as follows:

$$WAR = \frac{M_{wet} - M_{dry}}{M_{dry}} \times 100\% \quad (1)$$

The membrane porosity refers to the ratio of the pores' volume to the total volume of the membrane, which was determined by the weight difference method through immersing of membrane samples in *n*-butanol for 24 h. According to the weight difference in the membrane before and after soaking in *n*-butanol, the membrane porosity (ε, %) can be calculated from the following equation:

$$\varepsilon = \frac{\Delta m}{V \cdot \rho} \times 100\% = \frac{m_1 - m_0}{(\pi/4)\left(D^2 - d^2\right) l\rho} \times 100\% \tag{2}$$

where m_0 and m_1 (g) represent the weight of the membrane before and after soaking in *n*-butanol, respectively. D (m) and d (m) are the outer and inner diameters of the membrane, respectively. l (m) is the effective length of the samples, and ρ is the density of the *n*-butanol (0.81 g/cm^3).

2.4. Measurement of Membrane Permeability and Protein Rejection

In this study, a self-made cross-flow filtration device with a length of 30 cm and a radius of 1 cm was used to measure the pure water permeability of the membranes. The PVDF HFMs modules were constructed by sealing 6 hollow fibers (an effective area of 20 cm^2) in a polyurethane (PU) tube and one end of the fibers was sealed with glue. The pressurized DI water permeated from the outside to the inside of the hollow fiber membrane and discharged from the PU tube. The membrane samples were pre-compacted at 4 bar for at least 2 h to stabilize the membrane performance, and then the membrane permeability was measured at 1 bar with the crossflow velocity fixed at 15.0 cm/s. The water permeability (P, L/(m^2·h·bar)) can be determined using the following formula:

$$P = \frac{\mathrm{V}}{\mathrm{AtP}} \tag{3}$$

where V (L) represents the filtration volume, and A (m^2), t (h), and P (bar) are the effective membrane area, filtration time, and the filtration pressure, respectively.

BSA, LYS, and OVA aqueous solutions with a concentration of 1 g/L were prepared for estimation of the membrane separation performance. The protein rejection (R, %) can be calculated via the formula as follows:

$$\mathrm{R} = \left(1 - \frac{C_p}{C_f}\right) \times 100\% \tag{4}$$

where C_p and C_f (g/L) are the protein concentrations of permeate and feed solutions, respectively. The protein concentration was measured through a UV–Vis spectrophotometer at different wavelengths as shown in Table S1.

The membrane stability was tested through water flushing experiments [43] by stirring membrane samples in DI water on a magnetic stirring apparatus at a rotation speed of 1300 rpm for 24 h. And then the water permeability and WCA at different times were tested.

2.5. Organic Fouling Tests

BSA, LYS, and OVA are three different types of proteins applied to investigate the anti-fouling properties of membrane samples, and the molecule properties are listed in Table S1 in supplementary information (SI). Before tests, the membranes were firstly compacted in DI water with the cross-flow filtration device at 4 bar for 2 h, and then the membrane organic fouling tests were carried out through the following procedure two times: (1) filtration with deionized water for 1 h to determine the pure water flux (F_{W1}), (2) filtration of the 1 g·L^{-1} protein solution for 2 h to determine the protein fouling permeate flux (F_{BSA}, F_{LYZ}, and F_{OVA}), (3) backwashing with pure water for 1 h under 1 bar, and (4) filtration with deionized water for 1 h after backwashing to determine the pure water flux (F_{W2}). The parameters of flux recovery rate (FRR), total fouling rate (R_t), reversible fouling rate (R_r), and irreversible fouling rate (R_{ir}) were calculated as follows [22]:

$$FRR = \frac{F_{w2}}{F_{w1}} \times 100\% \tag{5}$$

$$R_t = \frac{F_{w1} - F_x}{F_{w1}} \times 100\% \tag{6}$$

$$R_r = \frac{F_{w2} - F_x}{F_{w1}} \times 100\% \tag{7}$$

$$R_{ir} = \frac{F_{w_1} - F_{w2}}{F_{w1}} \times 100\% \tag{8}$$

where F_x refers to the permeate flux of the contaminants (F_{BSA}, F_{LYZ}, and F_{OVA}).

2.6. Analysis of Membrane Surface Anti-Protein Fouling Mechanism

To analyze the anti-protein fouling mechanism of the modified membrane, the adhesive free energies of the membrane-contaminant and the contaminant-contaminant were calculated using the xDLVO theory. [44] From the xDLVO theory, the total interaction energy (ΔG_{mlf}^{TOT}) between pollutants and the membrane is composed of van der Waals interaction energy (LW, ΔG_{mlf}^{LW}), electrostatic interaction energy (EL, ΔG_{mlf}^{EL}), and polar interaction energy (ΔG_{mlf}^{AB}) [45,46] as expressed in the following formulas:

$$\Delta G_{mlf}^{TOT} = \Delta G_{mlf}^{LW} + \Delta G_{mlf}^{EL} + \Delta G_{mlf}^{AB} \tag{9}$$

$$\Delta G_{mlf}^{LW} = 2\left(\sqrt{\gamma_l^{LW}} - \sqrt{\gamma_m^{LW}}\right)\left(\sqrt{\gamma_f^{LW}} - \sqrt{\gamma_l^{LW}}\right) \tag{10}$$

$$\Delta G_{mlf}^{AB} = 2\sqrt{\gamma_l^+}\left(\sqrt{\gamma_m^-} + \sqrt{\gamma_f^-} - \sqrt{\gamma_l^-}\right) + 2\sqrt{\gamma_l^-}\left(\sqrt{\gamma_m^+} + \sqrt{\gamma_f^+} - \sqrt{\gamma_l^+}\right) - 2\left(\sqrt{\gamma_m^+\gamma_f^-} + \sqrt{\gamma_m^-\gamma_f^+}\right) \tag{11}$$

$$\Delta G_{mlf}^{EL} = \frac{\varepsilon_0\varepsilon_r\kappa}{2}\left(\zeta_m^2 + \zeta_f^2\right)\left[1 - \coth(\kappa d_0) + \frac{2\zeta_m\zeta_f}{\zeta_m^2 + \zeta_f^2}\operatorname{csch}(\kappa d_0)\right] \tag{12}$$

where the subscripts "*f, l, m*" denote foulant, water, and membrane, respectively. ε_0 represents the vacuum permittivity, ε_r represents the relative dielectric constant of aqueous solution, κ represents the reciprocal of the Debye constant, and ζ_m and ζ_f are zeta potentials of membrane surface and pollutant, respectively [47]. The surface tension parameters of γ^{LW}, γ^-, and γ^+ correspond to the tension component of Vander surface, electron donor, and electron acceptor, respectively, and can be obtained from the following equations:

$$(1 + \cos\theta)\gamma_l^{TOT} = 2\left(\sqrt{\gamma_s^{LW}\gamma_l^{LW}} + \sqrt{\gamma_s^+\gamma_l^-} + \sqrt{\gamma_s^-\gamma_l^+}\right) \tag{13}$$

$$\begin{cases} (1 + \cos\theta_W)\gamma_W^{TOT} = 2\left(\sqrt{\gamma_s^{LW}\gamma_W^{LW}} + \sqrt{\gamma_s^+\gamma_W^-} + \sqrt{\gamma_s^-\gamma_W^+}\right) \\ (1 + \cos\theta_G)\gamma_G^{TOT} = 2\left(\sqrt{\gamma_s^{LW}\gamma_G^{LW}} + \sqrt{\gamma_s^+\gamma_G^-} + \sqrt{\gamma_s^-\gamma_G^+}\right) \\ (1 + \cos\theta_D)\gamma_D^{TOT} = 2\left(\sqrt{\gamma_s^{LW}\gamma_D^{LW}} + \sqrt{\gamma_s^+\gamma_D^-} + \sqrt{\gamma_s^-\gamma_D^+}\right) \end{cases} \tag{14}$$

where the subscripts *s* and *l* represent solid and liquid surfaces, respectively, while θ_W, θ_G, and θ_D represent the contact angles of water, glycerol, and diiodomethane, respectively.

3. Results and Discussion

3.1. Membrane Characterization

3.1.1. Morphologies of the Membranes

The visual appearance and SEM images of the pristine and modified membranes are shown in Figure S1 and Figure 2a–g. The unmodified PVDF membrane exhibited a white (Figure S1a), porous structure (Figure 2a), whereas the DA-modified membrane (DA-UF) turned dark brown due to the formation of a PDA layer resembling eumelanin (Figure S1a). In contrast, the TA-modified membrane (TA-UF) displayed a light yellow hue characteristic of TA (Figure S1a). When co-deposited, the DA/TA-UF membrane showed an intermediate light yellow tone, suggesting synergistic interactions between DA and TA [48]. Incorporation of ZIF-8 further altered the membrane coloration: the DA/ZIF-8-UF membrane appeared light gray, while the TA/ZIF-8-UF membrane retained a light yellow tint, both reflecting the inherent white color of ZIF-8 nanoparticles (Figure S1a). Notably, the DA/TA/ZIF-8-UF membrane closely resembled the pristine PVDF in color, indicating uniform coating and minimal light scattering (Figure S1a). Cross-sectional SEM images (Figure S1b–h) revealed that all modified membranes retained the original microstructure of the PVDF substrate. The modifiers (DA, TA, and/or ZIF-8) were predominantly deposited on the outer surface, as the sealed ends of the hollow fibers during modification prevented internal penetration. This surface-localized deposition preserved the structural integrity and porosity of the support layer.

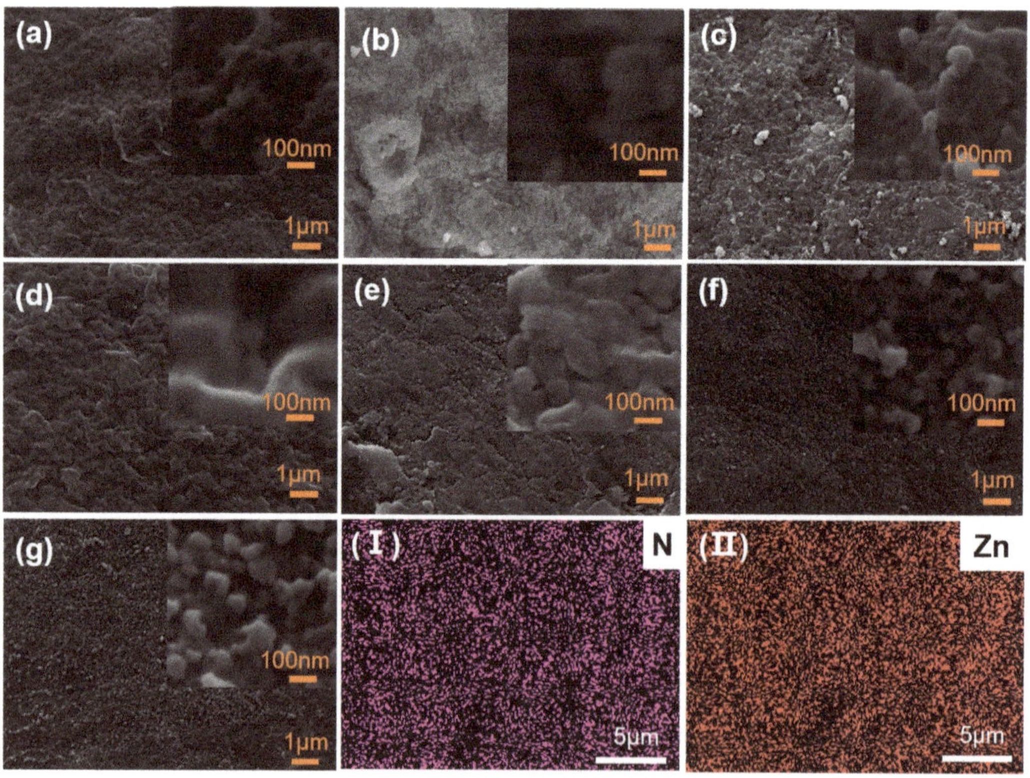

Figure 2. Surface images (**a**–**g**) obtained by SEM for the membranes of (**a**) PVDF substrate, (**b**) TA-UF, (**c**) DA-UF, (**d**) DA/TA-UF, (**e**) TA-ZIF-8-UF, (**f**) DA/ZIF-8-UF, and (**g**) DA/TA/ZIF-8-UF. (**I**) N element and (**II**) Zn element for DA/TA/ZIF-8-UF membrane.

Significant morphological differences were observed on the outer surfaces of the modified membranes (Figure 2a–g). The morphology of the TA-UF membrane (Figure 2b) did not change a lot compared to the pristine PVDF (Figure 2a), owing to TA's limited self-polymerization under alkaline conditions. In contrast, the DA-UF membrane (Figure 2c) displayed unevenly distributed aggregates of varying sizes, characteristic of PDA's spontaneous self-polymerization [23,49], while the DA/TA-UF membrane (Figure 2d) developed a denser, more continuous, and smoother surface layer, resulting from Michael addition and Schiff base reactions between DA and TA. These reactions promoted a highly cross-linked network with improved adhesion [50]. Incorporation of ZIF-8 further altered surface morphology. The TA/ZIF-8-UF membrane (Figure 2e) showed increased deposition compared to TA-UF, due to Zn^{2+}-TA coordination forming a metal-polyphenol network that enhanced TA adhesion. Notably, the DA/TA/ZIF-8-UF membrane (Figure 2g) exhibited a more uniform and continuous ZIF-8 layer with finer crystallites than DA/ZIF-8-UF (Figure 2f), demonstrating that ternary co-deposition yields a homogeneous yet morphologically distinct modification layer without compromising structural integrity. Moreover, the high-resolution SEM image (Figure S2) of the ZIF-8 crystallites on the surface of the DA/TA/ZIF-8-UF membrane reveal the uniformity of deposition. EDX mapping of the DA/TA/ZIF-8-UF membrane (Figure 2I,II) confirmed the uniform distribution of nitrogen (N) and zinc (Zn), characteristic of ZIF-8, further verifying the successful and homogeneous deposition of the composite layer.

3.1.2. Surface Chemical Compositions

The ATR-FTIR spectra of pristine and modified membranes are presented in Figure 3a and Figure S3. The TA-UF membrane exhibited stronger peak intensities than DA-UF, consistent with TA's higher catechol content and superior adhesion [33]. The DA/TA-UF membrane (Figure 3a) showed enhanced absorption at 760 cm^{-1} (phenolic O-H out-of-plane bending), attributed to synergistic reactions between TA-derived quinones and DA's amine groups via Michael addition and Schiff base reactions, alongside DA self-cross-linking, resulting in a highly cross-linked network. With the introduction of ZIF-8, characteristic absorption peaks emerged at 3133 cm^{-1} (C-H stretching vibrations of methyl), 2925 cm^{-1} (C-H stretching vibrations of imidazole rings), 1580 cm^{-1} (the deformation vibration of N-H), 1309 cm^{-1} (asymmetric stretching vibration of C-O-C), 1145 cm^{-1} (C-N stretching vibration in the imidazole ring), and 688 cm^{-1} (out-of-plane bending of the imidazole ring) [51,52]. The DA/TA/ZIF-8-UF membrane exhibited higher intensities for these peaks than TA/ZIF-8-UF or DA/ZIF-8-UF, indicating that the DA/TA network facilitated Zn^{2+} coordination and imidazole ligand binding for formation of metal-phenolic compounds, promoting ZIF-8 nucleation and crystallization. The reduced C=O peak intensity at 1732 cm^{-1} further confirmed ZIF-8 formation on the cross-linked layer [53,54]. Notably, the deposition of ZIF-8 enhances the out-of-plane bending peak of the O-H for the TA phenol hydroxyl group at 760 cm^{-1}, evidence of Zn^{2+}-TA coordination increasing TA's cross-linking density [55,56]. Thus, the ternary DA/TA/ZIF-8 system synergistically formed a composite coating layer on the PVDF substrate.

Figure 3b–e and Figure S4 presented XPS spectra of the wide scan and high resolution of C1s, N1s, and O1s of the PVDF substrate, DA/TA-UF, and DA/TA/ZIF-8-UF, with corresponding atomic percentages summarized in Table 2. Compared to the PVDF substrate, the DA/TA-UF membrane exhibited a sharp decrease in fluorine (F) content (from 33.94% to 0.66%), a substantial increase in oxygen (O) content (from 6.87% to 27.44%) and nitrogen (N) content (from 0.97% to 3.93%), confirming the successful surface coating with polar functional groups (–OH, $–NH_2$) that enhance hydrophilicity. For the DA/TA/ZIF-8-UF membrane, the F content further decreased to 0.55%, and zinc (Zn) emerged with an atomic percentage of 9.47%, directly evidencing ZIF-8 deposition facilitated by Zn^{2+} coordination with gallate groups in TA. The decrease in the O atomic percentage for the DA/TA/ZIF-8-UF membrane, compared to the DA/TA-UF membrane, is attributed to the surface coverage by the ZIF-8 layer, which has a lower oxygen content relative to the oxygen-rich TA. From the O1s peak fitting (Figure S4), the appearance of a Zn-O peak at 531.2 eV in the DA/TA/ZIF-8-UF membrane provides additional evidence for the formation of the ZIF-8 layer on the membrane surface [57,58]. The high N content (17.15%) originated from imidazole ligands in ZIF-8, further confirming crystalline ZIF-8 formation. High-resolution C1s spectra (Figure 3c,d) revealed a reduced C-F_2 peak (289.3 eV) in DA/TA/ZIF-8-UF compared to PVDF, alongside new peaks at 286.0 eV (C-O) and 287 eV (C=N), indicative of Schiff base reactions between DA and TA. The C=O peak at 288.2 eV corresponded to quinone/ester groups from DA oxidation [59–61], aligning with FTIR-proposed cross-linking mechanisms. In the N1s spectrum (Figure 3e), peaks at 398.8 eV (C–N=C) and 406.8 eV (π-π^* bonding) characterized imidazole in ZIF-8. Combined with Zn 2p peaks at 1022 eV and 1046 eV (Figure 3f), these results provide conclusive evidence of ZIF-8 incorporation [62].

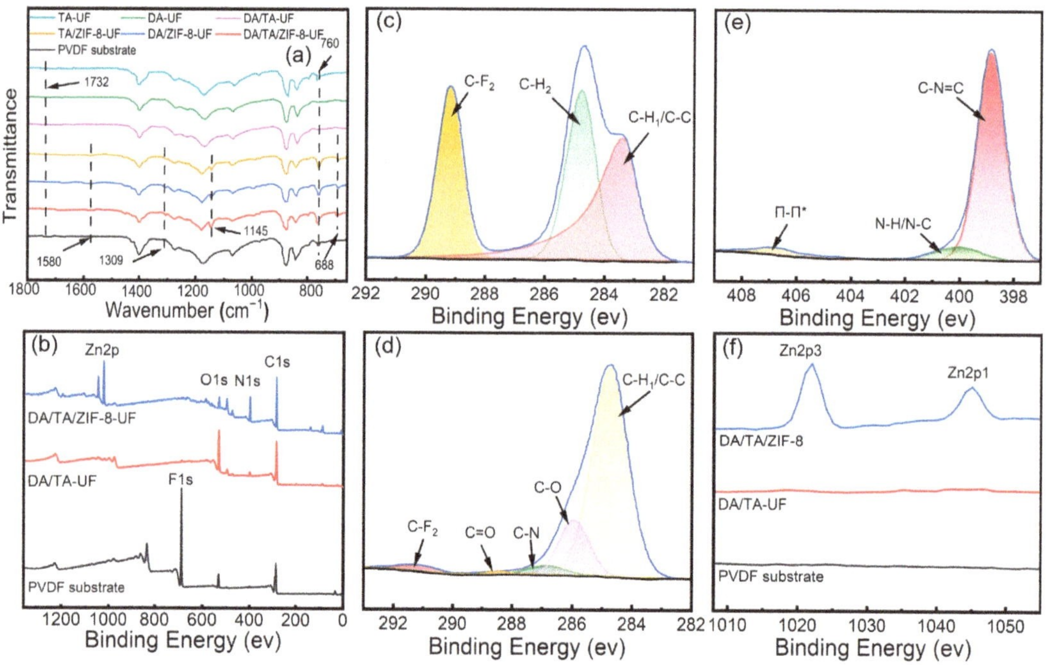

Figure 3. (**a**) ATR-FTIR spectra and (**b**) full-scan XPS spectra of membranes. (**c**,**d**) C1s PVDF substrate and DA/TA/ZIF-8-UF membrane, respectively. (**e**) N1s of DA/TA/ZIF-8-UF membrane. (**f**) Zn2p of membranes before and after modification.

Table 2. Atomic percentages of surface elements on membrane surfaces from XPS analysis.

Membrane	C	N	O	F	Zn
PVDF substrate	58.23	0.97	6.87	33.94	-
DA/TA-UF	67.98	3.93	27.44	0.66	-
DA/TA/ZIF-8-UF	69.44	17.75	6.12	0.55	6.15

3.1.3. Membrane Hydrophilicity and Porosity

The hydrophilicity of the membranes was evaluated by WCA (surface property) and WAR (bulk property), as shown in Figure 4a,b. The pristine PVDF membrane exhibited strong hydrophobicity (WCA~92°). Surface modification with DA or TA significantly improved hydrophilicity, reducing WCAs to 62 ± 1° and 40 ± 1°, respectively, due to their abundant hydrophilic functional groups. The DA/TA-UF membrane showed a further reduction in WCA (18 ± 1°), demonstrating the synergistic effect of cross-linked networks formed between DA and TA. After incorporating ZIF-8, which exhibits inherent hydrophobicity [63], the WCAs of DA/ZIF-8-UF, TA/ZIF-8-UF, and DA/TA/ZIF-8-UF increased to 70 ± 2°, 53 ± 1°, and 21 ± 1°, respectively. Despite this partial hydrophobization, all modified membranes maintained higher hydrophilicity than pristine PVDF. The WAR results correlated well with surface hydrophilicity trends. The DA-UF, TA-UF, and DA/TA-UF membranes showed WAR values of 78%, 71%, and 83%, respectively, significantly higher than PVDF (50%). With ZIF-8 incorporation, WAR values decreased to 58% (DA/ZIF-8-UF), 60% (TA/ZIF-8-UF), and 76% (DA/TA/ZIF-8-UF), yet remained above the PVDF baseline. The enhanced hydrophilicity of modified membranes reduces interfacial energy with aqueous pollutants, thereby improving antifouling performance.

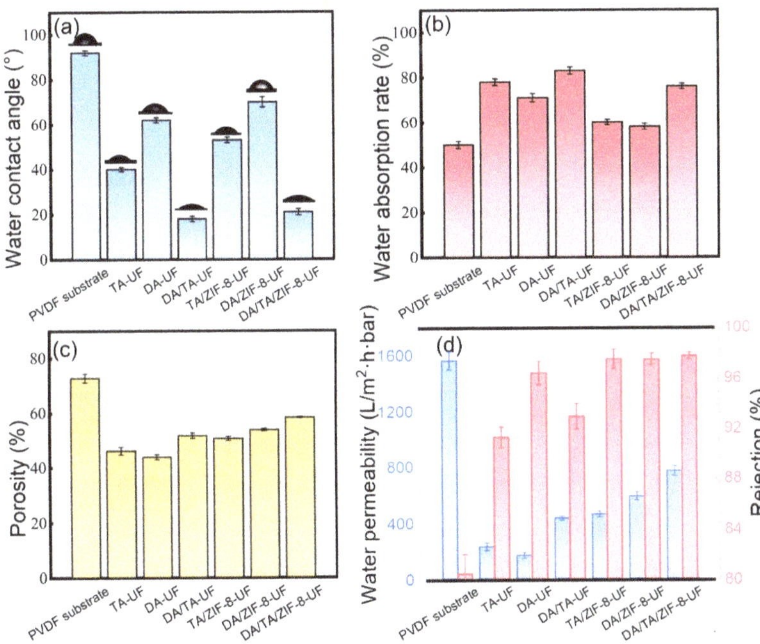

Figure 4. (**a**) Water contact angle, (**b**) water absorption rate, (**c**) porosity, and (**d**) separation performance of membranes before and after modification. Error bars are the standard deviation corresponding to the arithmetic square root of the arithmetic mean of the squared deviations from the mean.

Membrane porosity, a critical factor governing separation and antifouling performance, was quantitatively evaluated (Figure 4c). The pristine PVDF substrate exhibited a porosity of 72.9%. Modification reduced porosity due to polymer deposition on surfaces and within pores. The DA-UF membrane showed the most significant decline (43.9%), attributed to DA's strong self-aggregation tendency, which formed dense, pore-blocking networks. In contrast, the DA/TA-UF membrane maintained higher porosity (51.8%), as the Michael addition/Schiff base reactions between DA and TA yielded a more uniform and open cross-linked structure, mitigating pore blockage [48]. Incorporating ZIF-8 increased porosity relative to polymer-only modifications, leveraging its intrinsic nanoporosity. The DA/TA/ZIF-8-UF membrane achieved the highest porosity (58.4%) among modified membranes, exceeding both DA/ZIF-8-UF (50.8%) and TA/ZIF-8-UF (53.8%). This enhancement stems from dual mechanisms: Zn^{2+}-TA coordination creating metal-phenolic frameworks, and ZIF-8 crystallization introducing additional void spaces. The synergistic DA/TA system thus facilitates a more porous and functionalized surface architecture.

3.2. Separation Performance and Stability of Membranes

The water permeability and BSA rejection of the modified membranes are shown in Figure 5c. The pristine PVDF membrane exhibited high water permeability (1565 ± 63 L/(m^2·h·bar)), attributable to its macroporous surface structure (Figure 2a) and high porosity (72.9%, Figure 4c). Surface modification with DA or TA significantly reduced permeability to 171 ± 21 and 233 ± 27 L/(m^2·h·bar), respectively, due to the formation of dense polymer layers (Figure 2e,f) that partially obstructed surface pores. The TA-UF membrane demonstrated higher permeability but lower BSA rejection (91.3 ± 0.8%) compared to DA-UF (96.4 ± 0.9%), consistent with TA's limited self-polymerization under alkaline conditions, which resulted in incomplete surface coverage and allowed greater water flux through residual macropores. In contrast, the DA/TA-UF membrane showed enhanced permeability (437 ± 16 L/(m^2·h·bar)) relative to DA-UF, owing to improved hydrophilicity (Figure 4a,b) and a more open cross-linked network formed through Michael addition and Schiff base reactions [48,50]. Incorporation of ZIF-8 further improved the separation performance. The DA/TA/ZIF-8-UF membrane achieved a permeability of 771 ± 36 L/(m^2·h·bar) and 97.7 ± 0.2% BSA rejection, representing increases of 76.4% and 5.2% over DA/TA-UF, respectively. This enhancement is attributed to ZIF-8's nanoporosity, which provides additional water transport pathways while maintaining molecular selectivity, thereby mitigating the permeability–rejection trade-off [3,37,64]. The superior performance of DA/TA/ZIF-8-UF over both DA/ZIF-8-UF and TA/ZIF-8-UF membranes aligns with its higher porosity (Figure 4c) and more uniform surface morphology (Figure 2g). The correlation of membrane permeability–porosity is illustrated in Figure S5, which provides strong support for the performance-structure effects induced by the coating of DA, TA, and ZIF-8. Moreover, the higher rejection rate of BSA than OVA (82.5 ± 0.3%) by DA/TA/ZIF-8-UF membrane in Table S1 was ascribed to the size screening effect, while the lowest rejection rate in LYS (71.4 ± 0.5%) was attributed to the electrostatic attraction effect and size screening effect.

Membrane stability was evaluated through a 24 h hydraulic flushing test (Figure 5a). The DA/TA/ZIF-8-UF membrane maintained stable water permeability throughout, while the water contact angle increased only slightly from $20.4 \pm 0.8°$ to $32.5 \pm 1.1°$, demonstrating strong physical stability attributed to the robust adhesion of the modification layer through synergistic metal–polyphenol coordination and covalent cross-linking [65]. These results indicate excellent potential for long-term operational stability in protein separation applications.

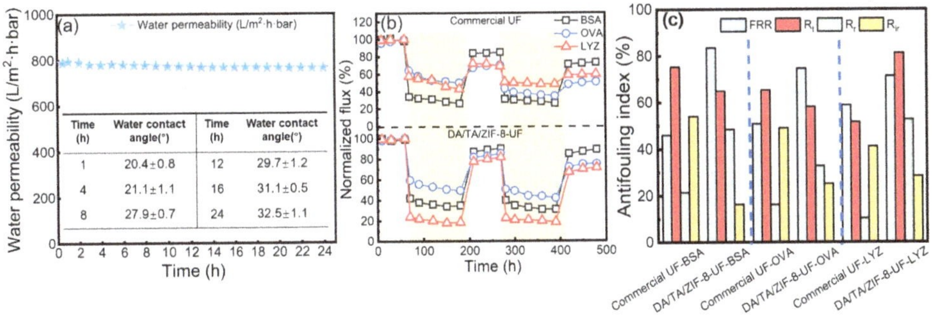

Figure 5. (**a**) Changes of water permeability and water contact angles for DA/TA/ZIF-8-UF membrane in the long-term run. (**b**) The normalized flux variation and (c) fouling parameters for DA/TA/ZIF-8-UF membrane and the commercial ultrafiltration membrane.

3.3. Antifouling Properties of UF Membranes

The antifouling performance of the DA/TA/ZIF-8-UF membrane was evaluated against BSA, OVA, and LYS (1.0 g/L each) and compared to a commercial PVDF UF membrane (Figure 5b,c). All membranes exhibited significant normalized flux decline due to protein adsorption, with the extent of fouling influenced by protein charge and size. The small decline in flux during each cycle of protein filtration (Figure 5b) demonstrated a formation of a fouling layer on the membrane surface hindering the passage of water molecules. The DA/TA/ZIF-8-UF membrane surface, rich in phenolic hydroxyl groups (pKa ~10), carries a strong negative charge [23]. Consequently, the positively charged LYS experienced severe fouling (R_t = 81.2%) due to electrostatic attraction, leading to substantial flux reduction. In contrast, the negatively charged BSA and OVA (Table S1) showed lower total flux decline (R_t = 64.9% and 58.1%, respectively) due to electrostatic repulsion. However, the smaller molecular size of OVA enabled deeper pore penetration and formation of a denser, more irreversible fouling layer (R_{ir} = 25.3%) compared to BSA (R_{ir} = 16.4%), as hydraulic cleaning was less effective for internally deposited OVA [66]. Compared to the commercial PVDF membrane, the DA/TA/ZIF-8-UF membrane demonstrated significantly improved antifouling performance across all tested proteins, exhibiting higher flux recovery ratios (FRRs) and reversible fouling ratios (R_{rs}), along with lower total (R_t) and irreversible (R_{ir}) fouling ratios. This enhancement is attributed to the hydrophilic coating layer, which forms a robust hydration barrier that effectively shields the surface from protein adhesion. The hydration layer minimizes electrostatic interactions and facilitates foulant removal during cleaning, thereby improving fouling reversibility and maintaining long-term filtration performance.

3.4. Anti-Fouling Mechanisms from xDLVO Theory

Membrane fouling is predominantly governed by interfacial interaction energies between foulants and membrane surfaces, which can be quantitatively evaluated using the xDLVO theory [34]. To determine the surface tension parameters, contact angle measurements were performed using three probe liquids: water and glycerol (polar solvents) and dichloromethane (non-polar solvent). The resulting data and zeta potential values are summarized in Table S2 and Table S3. The surface tension components of both the DA/TA/ZIF-8-UF membrane and commercial PVDF UF membrane, along with those of the three protein foulants (BSA, OVA, LYS), are presented in Table 3. The DA/TA/ZIF-8-UF membrane exhibited a significantly higher electron donor parameter (γ^-) compared to the commercial membrane, consistent with its more negative surface charge (−29.68 mV vs. −19.83 mV). This enhanced negativity originates from the abundant phenolic hydroxyl groups of TA and the quinone groups from oxidized DA, which are deprotonated under near-neutral conditions.

Table 3. Surface free energy of membrane and three pollutants.

Materials	γ^{LW}	γ^{+}	γ^{-}	γ^{AB}	γ^{TOT}
Commodity UF	33.65	0.11	2.33	1.01	34.66
DA/TA/ZIF-8-UF	38.18	0.42	64.97	10.45	48.63
BSA	43.22	1.065	47.68	14.25	57.47
OVA	35.57	0.41	50.76	9.12	44.69
LYZ	26.71	0.93	28.31	10.26	36.97

Furthermore, the modified membrane showed higher Lifshitz-van der Waals (γ^{LW}) and acid-base (γ^{AB}) components than the commercial PVDF UF, due to the introduction of hydrophilic functional groups (e.g., amino, catechol, and hydroxyl groups) from DA and TA. These groups facilitate strong hydrogen bonding with water molecules, promoting the formation of a dense hydration layer on the membrane surface [6,67]. This hydration layer acts as a physical and energetic barrier, reducing the adhesion strength of foulants and enhancing fouling reversibility, as schematically illustrated in Figure 5b,c.

The interfacial free energies between the membrane surfaces and protein foulants were calculated based on xDLVO theory, with results summarized in Table 4. The cohesive free energy (ΔG_{SIS}) of LYZ is negative, indicating a strong propensity for self-aggregation and fouling layer formation, while the positive ΔG_{SIS} values for BSA and OVA suggest weaker cohesive tendencies. The total interfacial free energy (ΔG_{mlf}^{TOT}) between the foulants and the commercial PVDF membrane is significantly lower than that for the DA/TA/ZIF-8-UF membrane, indicating stronger foulant-membrane adhesion and more severe fouling on the commercial membrane. In contrast, the DA/TA/ZIF-8-UF membrane exhibits substantially higher ΔG_{mlf}^{TOT} values (less attractive/more repulsive) for all three proteins, demonstrating superior antifouling potential. According to the xDLVO theory, a positive value of the total interfacial free energy (ΔG_{mlf}^{TOT}) indicates a net repulsive interaction, while a negative value signifies a net attractive interaction, which promotes fouling. Notably, the DA/TA/ZIF-8-UF membrane shows higher polar interaction energy (ΔG_{mlf}^{AB}) due to abundant hydrophilic groups, a lower Lifshitz-van der Waals component ΔG_{mlf}^{LW}, and enhanced overall repulsive forces with protein foulants. This energy profile originates from the hydrophilic modification layer rich in polar functional groups (e.g., –OH, $–NH_2$) that form a robust hydration layer through strong hydrogen bonding with water molecules. This hydration barrier requires substantial energy for proteins to disrupt during membrane contact, effectively preventing foulant adhesion and enhancing fouling resistance. The xDLVO analysis provides quantitative thermodynamic evidence for the dramatically improved antifouling performance of the DA/TA/ZIF-8-UF membrane.

Table 4. Interaction energies of membrane and protein pollutants.

	ΔG_{mlf}^{LW}	ΔG_{mlf}^{AB}	ΔG_{mlf}^{EL}	ΔG_{mlf}^{TOT}	ΔG_{SIS}
Commercial UF-BSA	−4.31	24.94	−0.00080	20.63	-
DA/TA/ZIF-8-UF-BSA	−5.75	40.51	−0.00080	34.76	-
BSA-BSA	−2.15	39.77	-	-	37.62
Commercial UF-OVA	−2.93	27.12	−0.00053	24.19	-
DA/TA/ZIF-8-UF-OVA	−3.91	44.82	−0.00053	40.90	-
OVA-OVA	−1.83	52.36	-	-	50.53
Commercial UF-LYZ	−1.13	−36.77	−0.00137	−37.90	-
DA/TA/ZIF-8-UF-LYZ	−1.51	−26.98	−0.00137	−28.49	-
LYZ-LYZ	−3.74	−6.52	-	-	−10.26

Based on comprehensive membrane characterization and xDLVO analysis, the antifouling mechanism of the DA/TA/ZIF-8-UF membrane is illustrated in Figure 6. When protein-containing solutions contact the membrane surface, several synergistic effects occur:

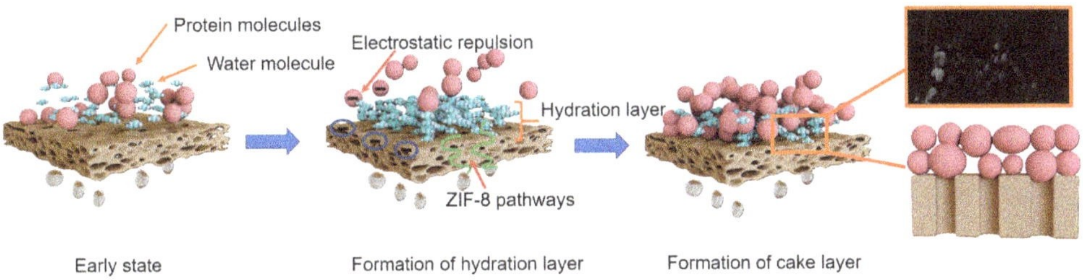

Figure 6. Mechanisms of antifouling properties and fouling model for DA/TA/ZIF-8-UF membrane.

(1) Hydration layer formation: The hydrophilic surface, rich in amino and phenolic hydroxyl groups, rapidly binds water molecules to form a dense hydration layer, preventing direct contact between proteins and the membrane surface and reducing initial adhesion.
(2) Pore size optimization: The metal-polyphenol networks (Zn^{2+}-TA coordination) and DA/TA copolymerization create a tailored surface morphology that reduces effective pore size without significant permeability loss. This structure minimizes internal pore blockage by protein molecules while maintaining rejection performance.
(3) Enhanced water transport: The incorporated ZIF-8 crystals provide additional nanoporous pathways for water molecules, mitigating the permeability-selectivity trade-off and maintaining high flux during operation.
(4) Reversible fouling layer: As protein concentration increases during filtration, foulants primarily form a loose cake layer on the membrane surface rather than penetrating pores. This external fouling is more easily removed through hydraulic cleaning due to the hydration layer's lubricating effect and electrostatic repulsion from the negatively charged surface.

The combination of these mechanisms, hydration barrier effect, optimized pore structure, enhanced water transport, and reversible fouling formation, explains the superior antifouling performance and stability of the DA/TA/ZIF-8-UF membrane compared to conventional PVDF membranes.

4. Conclusions

This study demonstrates a novel and efficient strategy for fabric high-performance antifouling PVDF hollow fiber ultrafiltration membranes through a metal-ion mediated co-deposition of dopamine (DA), tannic acid (TA), and ZIF-8 nanoparticles. The key findings can be summarized as follows:

First, the Zn^{2+}-assisted coordination-driven self-assembly enabled the formation of a uniform, stable, and firmly adherent composite coating on the membrane surface under mild conditions. The metal-phenolic networks between Zn^{2+} and TA not only enhanced the interfacial adhesion but also facilitated the oriented growth of ZIF-8 crystals, creating a robust modification layer. Second, the modified membrane exhibited significantly enhanced surface hydrophilicity, as evidenced by the remarkably reduced water contact angle (21°) and increased surface negativity (zeta potential: −29.68 mV). The abundant polar functional groups (–OH, $-NH_2$) facilitated the formation of a dense hydration layer, which was quantitatively verified by xDLVO analysis showing substantially increased interfacial free energy and stronger repulsive interactions with protein foulants. Third, the DA/TA/ZIF-8-UF membrane achieved exceptional separation performance with a high water permeability of 771 L/(m^2·h·bar) and outstanding BSA rejection of 97.7%. The incorporated ZIF-8 crystals provided additional water transport channels while maintaining molecular selectivity, effectively mitigating the permeability-selectivity trade-off. Fourth, the membrane demonstrated superior antifouling capability against various protein foulants (BSA, LYS, OVA), with flux recovery rates exceeding 71.5% after fouling-cleaning cycles. The antifouling mechanism was attributed to the synergistic effects of hydration layer formation preventing direct foulant–membrane contact, optimized pore structure minimizing internal fouling, enhanced water transport through ZIF-8 pathways, and electrostatic repulsion from negatively charged surface groups.

This work provides valuable insights into the rational design of high-performance antifouling membranes through multifunctional surface engineering, offering promising potential for sustainable water treatment and bioprocessing applications.

Supplementary Materials: The following supporting information can be downloaded at: https://www.mdpi.com/article/10.3390/membranes15090277/s1, Figure S1: (a) Photo images of membranes and SEM cross-sectional images for the membranes of (b) PVDF substrate, (c) TA-UF, (d) DA-UF, (e) DA/TA-UF, (f) TA-ZIF-8-UF, (g) DA/ZIF-8-UF, and (h) DA/TA/ZIF-8-UF; Figure S2: High resolution SEM image of the ZIF-8 crystallites on the surface of (a) DA/TA-UF and (b) DA/TA/ZIF-8-UF membrane; Figure S3: ATR-FTIR spectra of membranes in the wavenumber of 2800-3600 cm^{-1}; Figure S4: O1s peak fitting of (a) DA/TA-UF membrane and (b) DA/TA/ZIF-8-UF membrane; Figure S5: Correlation of permeability-porosity for membranes; Table S1: Properties of protein molecules and rejection rate by DA/TA/ZIF-8-UF membrane; Table S2: Contact angle and Zeta potential of membranes and different protein pollutants; Table S3 The surface tension parameters of the three testing agents.

Author Contributions: L.N.: writing—original draft. Q.C.: writing—original draft. Z.W.: writing—review and editing, data curation. X.Z.: writing—review and editing, data curation. J.M.: project administration. W.Z.: supervision, project administration, investigation. C.L.: visualization, validation, project administration. All authors have read and agreed to the published version of the manuscript.

Funding: This research was funded by the National Natural Science Fund of China (grant number 52100011).

Data Availability Statement: The raw data supporting the conclusions of this article will be made available by the authors on request.

Conflicts of Interest: The authors declare that the research was conducted in the absence of any commercial or financial relationships that could be construed as a potential conflict of interest.

References

1. Deng, J.; Gong, Y.; Jin, Y.; Zhou, J.; Wang, Y. Additive manufacturing of ultrafiltration membranes from block copolymer nanowires. *Sep. Purif. Technol.* **2025**, *368*, 132980. [CrossRef]
2. Tai, Z.S.; Othman, M.H.D.; Tajul Arifin, N.D.; Shazana, N.A.; Iqbal, R.M.; Ab Rahman, M.; Wan Mustapa, W.N.F.; Kadirkhan, F.; Puteh, M.H.; Jaafar, J.; et al. Modification of PVDF hollow fiber membrane with ZnO structure via hydrothermal for enhanced antifouling property in membrane distillation. *J. Membr. Sci.* **2024**, *692*, 122294. [CrossRef]
3. Liu, W.-M.; Chen, X.-R.; Mao, H.; Li, S.-H.; Zhang, Y.-M.; Xu, L.-H.; Zhao, Z.-P. Construction of PDMS@ZIF-8 composite membrane on PVDF hollow fiber inner surface for ultrafast ethanol/water separation. *J. Membr. Sci.* **2024**, *704*, 122889. [CrossRef]
4. Hou, C.; Du, L.; Chen, D.; Zhao, X.; Zhou, J.; Qiao, S. The preparation of electrically-enhanced and hydrophilic CNTs-PVDF composite hollow fiber membranes for fouling resistance. *Sep. Purif. Technol.* **2024**, *351*, 127968. [CrossRef]
5. Alsebaeai, M.K.; Ahmad, A.L.; Ooi, B.S. Construction of wetting resistance surface of highly hydrophobic PVDF-HFP/rGO coated supported PVDF hollow fiber composite membrane for membrane distillation. *Chem. Eng. J.* **2024**, *493*, 152565. [CrossRef]
6. Nie, Z.; Liu, C.; Jiang, X.; Zhou, Y.; Lin, X.; Zhao, X.; He, Q.; Chai, H.; Pang, X.; Ma, J. Dopamine-triggered one-step codeposition of zwitterionic surfactants for anti-fouling polyethersulfone ultrafiltration membrane modification. *Appl. Surf. Sci.* **2022**, *598*, 153871. [CrossRef]

7. Wijers, M.C.; Jin, M.; Wessling, M.; Strathmann, H. Supported liquid membranes modification with sulphonated poly(ether ether ketone)—Permeability, selectivity and stability. *J. Membr. Sci.* **1998**, *147*, 117–130. [CrossRef]
8. Gao, J.; Wang, J.; Xu, Q.; Wu, S.; Chen, Y. Regenerated cellulose strongly adhered by a supramolecular adhesive onto the PVDF membrane for a highly efficient oil/water separation. *Green Chem.* **2021**, *23*, 5633–5646. [CrossRef]
9. Cheng, W.; Lu, X.; Kaneda, M.; Zhang, W.; Bernstein, R.; Ma, J.; Elimelech, M. Graphene Oxide-Functionalized Membranes: The Importance of Nanosheet Surface Exposure for Biofouling Resistance. *Environ. Sci. Technol.* **2020**, *54*, 517–526. [CrossRef]
10. Wang, P.; Li, J.; Zhang, X.; Lu, X.; Liu, Q.; Zhang, T.; Cheng, W.; Ma, J. Utilization of Bidirectional Cation Transport in a Thin Film Composite Membrane: Selective Removal and Reclamation of Ammonium from Synthetic Digested Sludge Centrate via an Osmosis–Distillation Hybrid Membrane Process. *Environ. Sci. Technol.* **2020**, *54*, 10313–10322. [CrossRef]
11. Cheng, W.; Xu, H.; Wang, P.; Wang, L.; Szymczyk, A.; Croué, J.-P.; Zhang, T. Modification Mechanism of Polyamide Reverse Osmosis Membrane by Persulfate: Roles of Hydroxyl and Sulfate Radicals. *Environ. Sci. Technol.* **2022**, *56*, 8864–8874. [CrossRef]
12. Wang, P.; Jin, J.; Hou, R.; Jin, G.; Qin, L.; Chen, X.; Wang, H.; Zhou, Y.; Gao, C. From synergistic integration to advanced substitution: MOF-assisted biocatalytic membranes for emerging contaminants remediation. *J. Environ. Manage* **2025**, *390*, 126261. [CrossRef] [PubMed]
13. Zhang, Y.; Sun, S.; Wang, M.; Shen, Q.; Cong, S.; Zhao, Y.; Peng, J.; Pang, H. Metal-organic framework (MOF) membranes for Mg^{2+}/Li^{+} separation. *Sep. Purif. Technol.* **2025**, *374*, 133737. [CrossRef]
14. Ahmed, E.; Farghali, A.A.; Hmamm, M.F.M. Development of Cu-MOF@PVDF–PS hybrid membranes for high-temperature proton exchange membranes: Electrospinning, characterization, and fuel cell performance. *RSC Adv.* **2025**, *15*, 29777–29798. [CrossRef]
15. Karthikeyan, G.; Devadiga, V.; Mohan, S.; Balakrishna, R.G. Europium MOFs-enabled fluorescent membranes with optimized flux, filtration, and fouling resistance for wastewater treatment: A strategy to probe MOF distribution in membranes. *J. Environ. Chem. Eng.* **2025**, *13*, 118679. [CrossRef]
16. Karimi, A.; Vatanpour, V.; Khataee, A.; Safarpour, M. Contra-diffusion synthesis of ZIF-8 layer on polyvinylidene fluoride ultrafiltration membranes for improved water purification. *J. Ind. Eng. Chem.* **2019**, *73*, 95–105. [CrossRef]
17. Guo, Y.; Wang, X.; Hu, P.; Peng, X. ZIF-8 coated polyvinylidenefluoride (PVDF) hollow fiber for highly efficient separation of small dye molecules. *Appl. Mater. Today* **2016**, *5*, 103–110. [CrossRef]
18. Li, F.; Bhushan, B.; Pan, Y.; Zhao, X. Bioinspired superoleophobic/superhydrophilic functionalized cotton for efficient separation of immiscible oil-water mixtures and oil-water emulsions. *J. Colloid. Interface Sci.* **2019**, *548*, 123–130. [CrossRef]
19. Wang, Z.; Yang, H.-C.; He, F.; Peng, S.; Li, Y.; Shao, L.; Darling, S.B. Mussel-Inspired Surface Engineering for Water-Remediation Materials. *Matter* **2019**, *1*, 115–155. [CrossRef]
20. Huang, S.; McDonald, J.A.; Kuchel, R.P.; Khan, S.J.; Leslie, G.; Tang, C.Y.; Mansouri, J.; Fane, A.G. Surface modification of nanofiltration membranes to improve the removal of organic micropollutants: Linking membrane characteristics to solute transmission. *Water Res.* **2021**, *203*, 117520. [CrossRef] [PubMed]
21. Cheng, X.Q.; Zhang, C.; Wang, Z.X.; Shao, L. Tailoring nanofiltration membrane performance for highly-efficient antibiotics removal by mussel-inspired modification. *J. Membr. Sci.* **2016**, *499*, 326–334. [CrossRef]
22. Li, R.; Liu, J.; Shi, A.; Luo, X.; Lin, J.; Zheng, R.; Fan, H.; Selasie, S.V.; Lin, H. A facile method to modify polypropylene membrane by polydopamine coating via inkjet printing technique for superior performance. *J. Colloid. Interface Sci.* **2019**, *552*, 719–727. [CrossRef]
23. Wang, Z.; Zhang, W.; Wen, S.; Wang, L.; Wang, S.; Wang, Y.; Lu, J.; Ma, J.; Cheng, W. Rapid co-deposition of dopamine and polyethyleneimine triggered by $CuSO_4/H_2O_2$ oxidation to fabricate nanofiltration membranes with high selectivity and antifouling ability. *Sep. Purif. Technol.* **2023**, *305*, 122409. [CrossRef]
24. Zhang, T.; Zhang, J.; Wang, Q.; Zhang, H.; Wang, Z.; Wu, Z. Evaluating of the performance of natural mineral vermiculite modified PVDF membrane for oil/water separation by membrane fouling model and XDLVO theory. *J. Membr. Sci.* **2022**, *641*, 119886. [CrossRef]
25. Li, B.; Hou, D.; Li, C.; Yun, Y. Mussels-inspired design a carbon nanotube based underwater superoleophobic/hydrophobic Janus membrane with robust anti-oil-fouling for direct contact membrane distillation. *Sep. Purif. Technol.* **2022**, *294*, 121163. [CrossRef]
26. Pz, A.; Wl, A.; Sr, A.; Yj, A.; Qin, S.A.; Cf, B.; Nk, A.; Hm, A. Modification of PVDF hollow fiber membrane by co-deposition of PDA/MPC-co-AEMA for membrane distillation application with anti-fouling and anti-scaling properties. *J. Membr. Sci.* **2021**, *636*, 119596. [CrossRef]
27. Ye, W.; Liu, H.; Lin, F.; Lin, J.; Zhao, S.; Yang, S.; Hou, J.; Zhou, S.; Van der Bruggenf, B. High-flux nanofiltration membranes tailored by bio-inspired co-deposition of hydrophilic g-C_3N_4 nanosheets for enhanced selectivity towards organics and salts. *Environ. Sci. Nano* **2019**, *6*, 2958–2967. [CrossRef]
28. Fang, S.; Zhang, Z.; Yang, H.; Wang, G.; Zhu, L. Mussel-inspired hydrophilic modification of polypropylene membrane for oil-in-water emulsion separation. *Surf. Coat. Techol.* **2020**, *403*, 126375. [CrossRef]

29. Varol, H.S.; Herberger, T.; Kirsch, M.; Mikolei, J.; Veith, L.; Kannan-Sampathkumar, V.; Brand, R.D.; Synatschke, C.V.; Weil, T.; Andrieu-Brunsen, A. Electropolymerization of Polydopamine at Electrode-Supported Insulating Mesoporous Films. *Chem. Mater.* **2023**, *35*, 9192–9207. [CrossRef]
30. Alfieri, M.L.; Weil, T.; Ng, D.Y.W.; Ball, V. Polydopamine at biological interfaces. *Adv. Colloid. Interface Sci.* **2022**, *305*, 102689. [CrossRef] [PubMed]
31. Wu, J.; Rezzadori, K.; Petrus, J.; Luccio, M.; Li], Q. Modification of hydrophobic commercial PVDF microfiltration membranes into superhydrophilic membranes by the mussel-inspired method with dopamine and polyethyleneimine. *Sep. Purif. Technol.* **2019**, *212*, 641–649. [CrossRef]
32. Ren, J.; Xia, W.; Feng, X.; Zhao, Y. Surface modification of PVDF membrane by sulfonated chitosan for enhanced anti-fouling property via PDA coating layer. *Mater. Lett.* **2021**, *307*, 130981. [CrossRef]
33. Zhan, K.; Kim, C.; Sung, K.; Ejima, H.; Yoshie, N. Tunicate-inspired gallol polymers for underwater adhesive: A comparative study of catechol and gallol. *Biomacromolecules* **2017**, *18*, 2959–2966. [CrossRef]
34. Xie, H.; Shen, L.; Xu, Y.; Hong, H.; Yang, L.; Li, R.; Lin, H. Tannic Acid (Ta)-Based Coating Modified Membrane Enhanced by Successive Inkjet Printing of Fe^{3+} and Sodium Periodate (Sp) for Efficient Oil-Water Separation. *J. Membr. Sci.* **2022**, *660*, 120873. [CrossRef]
35. Song, Y.Z.; Kong, X.; Yin, X.; Zhang, Y.; Sun, C.C.; Yuan, J.J.; Zhu, B.; Zhu, L.P. Tannin-inspired superhydrophilic and underwater superoleophobic polypropylene membrane for effective oil/water emulsions separation. *Colloids Surf. Physicochem. Eng. Aspects* **2017**, *522*, 585–592. [CrossRef]
36. Xu, Y.; Xiao, Y.; Zhang, W.; Lin, H.; Liao, B.Q. Plant polyphenol intermediated metal-organic framework (MOF) membranes for efficient desalination. *J. Membr. Sci.* **2021**, *618*, 118726. [CrossRef]
37. Zhao, B.; Guo, Z.; Wang, H.; Wang, L.; Qian, Y.; Long, X.; Ma, C.; Zhang, Z.; Li, J.; Zhang, H. Enhanced water permeance of a polyamide thin-film composite nanofiltration membrane with a metal-organic framework interlayer. *J. Membr. Sci.* **2021**, *625*, 119154. [CrossRef]
38. Ye, L.; Liu, X.; Chen, Y.; Xu, L.; Fu, S.; Wu, X. Fe^{3+}/tannic acid assisted preparation of dynamic assembled hollow fiber nanofiltration membrane with salt tolerance. *J. Environ. Chem. Eng.* **2025**, *13*, 118378. [CrossRef]
39. Zhao, Z.; Gao, W.; Zhao, W.; Yang, Y.; Shen, H.; Chang, Y.; Zhao, S. Surface-functionalized composite hollow fiber membrane for extracorporeal membrane oxygenation. *J. Membr. Sci.* **2025**, *733*, 124377. [CrossRef]
40. Mohmed, A.H.; Isawi, H.; Elbehary, S.A.; El-Aassar, A.H.M.; Shawky, H.A.; Mehany, M.A.S.; Khalil, M.M.H. Fabrication and characterization of nanocomposite hollow fiber membranes for wastewater treatment. *Mater. Sci. Eng. B* **2025**, *321*, 118557. [CrossRef]
41. Varol, H.S.; Seeger, S. Fluorescent Staining of Silicone Micro- and Nanopatterns for Their Optical Imaging. *Langmuir* **2022**, *38*, 231–243. [CrossRef]
42. Wu, T.; Zhou, B.; Zhu, T.; Shi, J.; Xu, Z.; Hu, C.; Wang, J. Facile and low-cost approach towards a PVDF ultrafiltration membrane with enhanced hydrophilicity and antifouling performance via graphene oxide/water-bath coagulation. *RSC Adv.* **2015**, *5*, 7880–7889. [CrossRef]
43. Li, M.-P.; Zhang, X.; Zhang, H.; Liu, W.-L.; Huang, Z.-H.; Xie, F.; Ma, X.-H.; Xu, Z.-L. Hydrophilic yolk-shell ZIF-8 modified polyamide thin-film nanocomposite membrane with improved permeability and selectivity. *Sep. Purif. Technol.* **2020**, *247*, 116990. [CrossRef]
44. Zhao, X.; Zhang, R.; Liu, Y.; He, M.; Su, Y.; Gao, C.; Jiang, Z. Antifouling membrane surface construction: Chemistry plays a critical role. *J. Membr. Sci.* **2018**, *551*, 145–171. [CrossRef]
45. Yu, X.; Lin, T.; Xu, H.; Tao, H.; Chen, W. Ultrafiltration of up-flow biological activated carbon effluent: Extracellular polymer biofouling mechanism and mitigation using pre-ozonation with H_2O_2 backwashing. *Water Res.* **2020**, *186*, 116391. [CrossRef]
46. Yuan, X.-T.; Wu, L.; Geng, H.-Z.; Wang, L.; Wang, W.; Yuan, X.-S.; He, B.; Jiang, Y.-X.; Ning, Y.-J.; Zhu, Z.-R.; et al. Polyaniline/polysulfone ultrafiltration membranes with improved permeability and anti-fouling behavior. *J. Water Process Eng.* **2021**, *40*, 101903. [CrossRef]
47. Zhang, B.; Tang, H.; Huang, D.; Gao, X.; Zhang, B.; Shen, Y.; Shi, W. Effect of pH on anionic polyacrylamide adhesion: New insights into membrane fouling based on XDLVO analysis. *J. Mol. Liq.* **2020**, *320*, 114463. [CrossRef]
48. Chen, M.; Li, J.; Liu, Y.; Cao, K.; Zhu, M.; Liu, H.; Tang, Y.; Wang, Q. Self-cleaning PDA/TA/FeOOH-PVDF membrane for efficient separation of volatile oil from water of traditional Chinese medicine. *J. Membr. Sci.* **2025**, *734*, 124467. [CrossRef]
49. Zhang, W.; Wang, Z.; Zhao, Z.; Wang, P.; Wang, S.; Ma, J.; Cheng, W. High-stable bimetallic AgCu nanoalloys with core-shell structures for sustainable antibacterial and biofouling mitigation in nanofiltration. *Water Res.* **2025**, *271*, 122986. [CrossRef]
50. Xiao, F.; Ge, H.; Wang, Y.; Bian, S.; Tong, Y.; Gao, C.; Zhu, G. Novel thin-film composite membrane with polydopamine-modified polyethylene support and tannic acid-Fe^{3+} interlayer for forward osmosis applications. *J. Membr. Sci.* **2022**, *642*, 119976. [CrossRef]
51. Thekkayil, R.; John, H.; Gopinath, P. Grafting of self assembled polyaniline nanorods on reduced graphene oxide for nonlinear optical application. *Synth. Met.* **2013**, *185–186*, 38–44. [CrossRef]

52. Jang, H.; Song, D.H.; Kim, I.C.; Kwon, Y.N. Fouling control through the hydrophilic surface modification of poly(vinylidene fluoride) membranes. *J. Appl. Polym. Sci.* **2014**, *132*, 41712. [CrossRef]
53. Zhou, H.; Wang, X.; Tang, J.; Yang, Y.W. Surface Immobilization of pH-Responsive Polymer Brushes on Mesoporous Silica Nanoparticles by Enzyme Mimetic Catalytic ATRP for Controlled Cargo Release. *Polymers* **2016**, *8*, 277. [CrossRef]
54. He, G.; Li, Z.; Liu, Y.; Liu, M.; Zhu, C.; Zhang, L.; Zhang, H. A novel lithium ion-imprinted membrane with robust adsorption capacity and anti-fouling property based on the uniform multilayered interlayer. *Desalination* **2022**, *539*, 115973. [CrossRef]
55. Zhang, R.; Jiang, J.; Liu, Y.; Li, Y.; Zhao, C.; Yang, Z.; Cai, L.; Yu, J.; Chen, Z.; Hu, Z. Tannic acid/ethylenediamine/succinic acid graft modified PVDF anti-pollution membrane and its application in the field of organic pollutant separation. *J. Mater. Sci.* **2022**, *57*, 20156–20173. [CrossRef]
56. Chen, F.; Dong, S.; Wang, Z.; Xu, J.; Xu, R.; Wang, J. Preparation of mixed matrix composite membrane for hydrogen purification by incorporating ZIF-8 nanoparticles modified with tannic acid. *Int. J. Hydrogen Energy* **2020**, *45*, 7444–7454. [CrossRef]
57. Battistoni, C.; Dormann, J.L.; Fiorani, D.; Paparazzo, E.; Viticoli, S. An XPS and Mössbauer study of the electronic properties of $ZnCrxGa_2{-}xO_4$ spinel solid solutions. *Solid. State Commun.* **1981**, *39*, 581–585. [CrossRef]
58. Tan, H.; Tang, Y.; Hou, Z.; Yang, P.; Liu, C.; Xie, Z.; Li, S. Antimicrobial polymer-based zeolite imidazolate framework composite membranes for uranium extraction from wastewater and seawater. *J. Colloid. Interface Sci.* **2025**, *677*, 435–445. [CrossRef] [PubMed]
59. Li, S.; Zhang, M.; Sun, J.; Sun, J.; Wang, Y. Preparation and characterization of superior hydrophilic PVDF/DA membranes by the self-polymerization approach of dopamine. *Front. Chem.* **2023**, *11*, 1162348. [CrossRef] [PubMed]
60. Kan, J.; Zhang, Z.; Ren, L.; Han, J.; Zhang, H.; Li, J.; Wu, H.; Chen, J. Hydrophilic modification of PVDF membranes for oily water separation with enhanced anti-fouling performance. *J. Appl. Polym. Sci.* **2023**, *140*, e53738. [CrossRef]
61. Jiao, Y.; Yang, J.; Zhang, J.; Li, J.; Qin, S.; Wu, X.; Cui, Z. Removal of heavy metal ions from acidic wastewater by constructing positively charged hollow fiber nanofiltration separating-layer based on Fe (III)/co deposition-quaternization. *J. Water Process Eng.* **2023**, *56*, 104450. [CrossRef]
62. Yang, J.; Sun, W.; Ju, J.; Tan, Y.; Yuan, H. Facile Fabrication of Superwetting PVDF Membrane for Highly Efficient Oil/Water Separation. *Polymers* **2023**, *15*, 327. [CrossRef]
63. Wang, Z.; Zhang, W.; Wang, W.; Wang, P.; Ni, L.; Wang, S.; Ma, J.; Cheng, W. Amine-Modified ZIF Composite Membranes: Regulated Nanochannel Interactions for Enhanced Cation Transport and Precise Separation. *Environ. Sci. Technol.* **2025**, *59*, 4199–4209. [CrossRef] [PubMed]
64. Han, B.; Sun, Z.; Jiang, H.; Sun, X.; Ma, J.; He, M.; Zhang, W. Thin and defect-free ZIF-8 layer assisted enhancement of the monovalent perm-selectivity for cation exchange membrane. *Desalination* **2022**, *529*, 115637. [CrossRef]
65. Bi, Q.; Li, Q.; Tian, Y.; Lin, Y.; Wang, X. Hydrophilic modification of poly(vinylidene fluoride) membrane with poly(vinyl pyrrolidone) via a cross-linking reaction. *J. Appl. Polym. Sci.* **2012**, *127*, 394–401. [CrossRef]
66. Yang, L.; Wang, Z.; Zhang, J. Zeolite imidazolate framework hybrid nanofiltration (NF) membranes with enhanced permselectivity for dye removal. *J. Membr. Sci.* **2017**, *532*, 76–86. [CrossRef]
67. Liu, C.; Song, D.; Zhang, W.; He, Q.; Huangfu, X.; Sun, S.; Sun, Z.; Cheng, W.; Ma, J. Constructing zwitterionic polymer brush layer to enhance gravity-driven membrane performance by governing biofilm formation. *Water Res.* **2020**, *168*, 115181. [CrossRef]

membranes

MDPI

Article

Investigations on the Particle Fouling and Backwash Efficiency During Microplastic Microfiltration–Particle Size Aspects

Saeedeh Saremi [1,2], Leonie Marie Scheer [1], Gerhard Braun [1], Marcus Koch [1], Markus Gallei [2,3] and Matthias Faust [1,*]

1 Institute for Physical Process Technology, Saarland University of Applied Sciences, 66117 Saarbrücken, Germany; saeedeh.saremi@htwsaar.de (S.S.); gerhard.braun@htwsaar.de (G.B.); marcus.koch@htwsaar.de (M.K.)
2 Polymer Chemistry, Saarland University, 66123 Saarbrücken, Germany; markus.gallei@uni-saarland.de
3 Saarene, Saarland Center for Energy Materials and Sustainability, Saarland University, 66123 Saarbrücken, Germany
* Correspondence: matthias.faust@htwsaar.de

Abstract: The characteristics of polystyrene (PS) microplastic (MP) microfiltration by a cellulose acetate (CA) membrane were investigated within this study. Particle sizes and pore sizes were selected in a comparable range in order to challenge the dead-end microfiltration. Backwashing experiments round up the investigations. Microfiltration characteristics and particle size measurements, as well as a particle fouling analysis by different methods, were applied in the study in order to provide a comprehensive picture of particle deposition and particle fouling structuring. The particle removal efficiency was particle-size-dependent, and especially small particles were further reduced during the proceeding filtration, while the larger particles were already removed within the first minutes of filtration. This observation was attributed to the pore blocking (internal and/or complete) and build-up of the filter cake. The difference in the particle-fouling structure at low and elevated filtration pressure significantly influences the backwashing efficiency. The particle fouling resulting from low-pressure filtration was completely removed due to the backwashing procedure applied, while an increased filtration pressure resulted in a different particle-fouling structure, which negatively influenced the backwashing efficiency. This knowledge of the formation and structure of the MP particle fouling and its removal by backwashing is a prerequisite for further process development.

Keywords: microplastic separation; microfiltration; particle-fouling analysis; water treatment process

1. Introduction

The increasing pollution of the environment by microplastics (MPS) is a global challenge that endangers aquatic ecosystems and potentially compromises the safety of drinking-water resources. The tiny plastic particles enter the environment from a wide variety of sources and via multiple pathways and are now ubiquitously detectable in water bodies and other ecosystems [1,2].

Numerous articles regarding the origin and measurement of microplastics (1 μm–5 mm) in water, as well as possible removal strategies for microplastics from water, were published within the last decade. Most of the papers focus on the hazardous properties of microplastics and/or nanoplastics and on engineering processes and cleaning strategies, e.g., sedimentation, coagulation/flocculation, degradation, or filtration [2–8]. Filtration, and especially membrane filtration, seems to be a promising separation technology for the removal of microplastics from drinking water, pre-treated water from industry processes, or pre-cleaned water from sewage plants [2–7,9–16]. Nearly all studies in the literature focus on the environmental technology aspects, such as material, average size, associated biofouling, or efficiency of removal technologies. However, there are only a few studies that analyze filtration and particle-fouling characteristics experimentally from a particle-engineering perspective.

For example, Enfrin et al. investigated the usage of ultrafiltration membranes for the removal of micro/nanoplastics (NP) in detail. They found that hydrophilic membranes have a lower adsorption rate of NP/MP compared to hydrophobic membranes [17,18]. Investigations on the usage and efficiency of gas scouring for the reduction of NP/MP fouling were also performed by Enfrin et al. [18]. Experimental investigations regarding the usage of microfiltration membranes for the separation of microplastics were performed by Pizzichetti et al. [19,20]. The results showed that the microfiltration membranes are, in principle, suitable for the controlled removal of microplastics from water. Pizzichetti et al. used plastic particles in a size range between 20 and 300 μm. Microfiltration membranes with a nominal pore size of 5 μm were used during these experiments and received nearly complete removal of the particles [19,20]. The results confirmed that microfiltration could be a suitable tool for the removal of microplastics in many applications. Pizzichetti et al. also studied the PS and PA (polyamide) particle fouling on a CA membrane. Due to the use of relatively large particles compared to the small CA membrane pores, a complete pore blockage followed by a build-up of a filter cake was observed and evaluated in detail during their investigations [20].

Cellulose acetate (CA) is a sustainable polymer that is used in a number of applications, e.g., separation membranes, packaging films, medical devices, or composite materials. Besides its low costs, one of the main advantages of CA is its relatively easy recycling in comparison to most other polymers [21]. CA membranes are produced for nearly all membrane operations, e.g., reverse osmosis, nanofiltration, ultrafiltration, and microfiltration [22].

The selection and the efficiency of a filtration system usually depend on the feed stream composition and on the required filtrate characteristics.

Due to sustainability reasons, the membrane material, area, pore size, and thickness are usually selected by the operator depending on, e.g., the required flux, the operation time, the fouling mechanism, and possible membrane-cleaning opportunities [23].

In large-scale microfiltration operations, the pore size is usually selected to be smaller than the particle size. However, especially in applications in environmental technology, the particle size can vary. In order to choose an optimized set-up for our investigations, the membrane pores are in the same size range as the filtered particles, which challenges the membrane. This approach could be a powerful tool to gain a better understanding of microplastics deposition and particle fouling during filtration and gives important hints on the filtration mechanism, which is a prerequisite for possible anti-fouling membrane optimizations [24]. Backwashing is a common method in order to remove deposits on and in the membrane. Apart from flushing, backwashing is a physical treatment that requires only elevated process conditions and suitable process design. No additional hazardous chemicals are necessary for this method. Therefore, backwashing is a sustainable way to adjust and increase the microfiltration membranes' operation time. It is a prerequisite for the scale-up of many membrane applications. Backwashing-efficiency tests round off the investigations because the backwash possibility is often a prerequisite for the usage of specific membranes in the field.

2. Materials and Methods

Polystyrene particles were provided by Coating Products OHZ (Osterholz-Scharmbeck, Germany). Very narrowly distributed polystyrene (PS) MS-5 FHC microparticles (1–20 μm) with a median diameter of $d_{50,3}$ 4–6 μm were selected for our investigations in order to challenge the CA microfiltration membrane (Sartorius Stadim Biotech GmbH, Goettingen, Germany), which is characterized by a pore size of 5 μm given by the producer. For validation, the median pore size of the surface pores was calculated to be $d_{50,0}$ 4.0 μm (scanning electron microscopy, see Supplementary Information). This particle–membrane combination enables the study of the pore-blocking mechanism as well as the structuring of the top layer (filter cake) during filtration, which is dependent on the trans-membrane pressure (TMP). The filtration efficiency was calculated via a comparison of the particle size distribution and the turbidity in the feed suspension and the filtrate (after defined time intervals) at different filtration pressures. In order to study the effect of the backwash, a second microfiltration test series was performed under equal conditions. Subsequent to the filtration experiment, the set-up was switched to backwash mode, and a 5 min backwash was carried out. The backwash efficiency was checked via the comparison of the initial pure water flux with the pure water flux after backwashing.

Besides the particle analysis and turbidity measurement in feed and filtrate, the fresh membranes, as well as the used membranes and the membranes after backwashing, were holistically investigated via light microscopy, SEM (scanning electron microscopy), contact angle measurement, and thermogravimetric analysis.

Thermogravimetric analysis (TGA) is a common method used to determine the thermal properties of materials (e.g., decomposition temperature). However, to our knowledge, there are only a few publications that focus on the usage of TGA for the investigation of membrane fouling [25,26]. Within these studies, a sample of the fouled membrane is dynamically heated in a N_2 flow, while the mass of the sample is constantly recorded. Tay et al. [25] described the method for the investigation of membrane fouling in 2003. According to their observation, a difference in the decomposition temperatures of the membrane material (polymer) and the foulant (albumin in their experiments) is a prerequisite for the distinction between weight loss derived from the membrane and weight loss derived from the foulant during the TG analysis. Reactions between the foulant and the membrane must be excluded. Therefore, TGA measurements of the membrane material and of the foulant are required before the analysis of the fouled membrane. The publication by Kamarudin et al. [26] confirmed the suitability of TGA as a tool for the investigation of membrane fouling in 2022, presupposing that the decomposition temperatures of the materials deployed are different. Weight loss derived from the fouling and weight loss derived from the membrane were clearly observed during both studies. Due to the appearance of the thermogram, it was even possible to state which kind of fouling mechanism occurred during the prior filtration experiments. The two studies detected natural organic foulants, which implies the advantage that the fouling-decomposition temperature is very low in comparison to the decomposition temperature of the polymer membranes. During our investigations, we examined the potential usage of TGA for the analysis of microplastic fouling on membranes. We came to the conclusion that TGA could be a promising method to study fouling on inorganic membranes with high decomposition temperatures, which should be far above the decomposition temperature of the plastic particles. However, due to many advantages, it is more probable that polymeric membranes will be used for the separation of MPs in future applications. The elevated decomposition temperatures of the plastic particles, which are usually very close to the decomposition temperature of the polymeric membrane, make it very difficult to distinguish between a weight loss derived by the foulant or from the membrane. We investigated the decomposition characteristics of different MP materials and different membrane materials. In comparison to most membrane materials, CA is characterized by a relatively low decomposition temperature of less than 400 °C [27,28], while the foulant PS decomposes at around 400 °C or even higher temperatures [29–31]. Based on this research, we came to the conclusion that a detection of fouling by TGA could be feasible for the combination of the CA membrane and PS foulant.

2.1. Preparation of MP Particle–Water Suspensions

Previous to the filtration experiments, tap water was filtered through two microfiltration filter cartridges, followed by an ultrafiltration cartridge in order to provide comparable water conditions for the generation of the particle–water suspensions. The particle size distribution within the feed water was measured prior to each experiment. Homogeneous PS particle–feed water suspensions (5 mg PS/kg water) were generated by continuously stirring the PS water mixture for 30 min using a propeller agitator (Heidolph, Schwabach, Germany) at 500 RPM. The particle size distribution was measured by Klotz Syringe particle analysis system (Markus Klotz GmbH, Bad Liebenzell, Germany, LDS 30/30, equipped with a laser sensor, optical counting method, measurement range 0.9–139 µm, syringe size 10 mL) after different durations in order to ensure accurate and comparable feed suspensions for all filtration experiments. At least three measurements were applied for each sample. A magnetic stirrer integrated into the base of the particle measurement device ensures homogeneous particle distribution during the measurement. Data is recorded using the Protrend software (Version 3.3.1). SEM images of the PS particles were analyzed for verification of the particle measurement system results. For the turbidity measurement, samples were collected from the feed suspension and from the filtrates in three independent cuvettes at predefined time intervals (5 min, 30 min, and 60 min) and analyzed using the Hach 2100N device (Hach, Loveland, CO, USA). The measurement mode 0-1 NTU with a maximum resolution of 0.001 NTU was selected for the turbidity analysis.

2.2. Microfiltration Experiments and Backwash Procedure

A self-built experimental set-up was used for the dead-end microfiltration experiments (Figure 1). The flat sheet membranes were placed in a stainless steel filter holder (E-3, Sartorius Lab Instruments GmbH & Co. KG, Goettingen, Germany, model 16254, membrane diameter 47 mm, effective diameter 40 mm), which was equipped with an additional feed side needle valve (V-02) for the removal of air from the system (during the filling procedure). A volume of 2 L of the feed suspension (E-1) was continuously stirred by a magnetic stirrer in order to prevent any sedimentation. During dead-end operation, the needle valve (V-01) on the filtrate side remained closed. The feed stream was pumped through the membrane by a SHURflo diaphragm pump (E-2, Pentair Shurflo, Costa Mesa, USA, model 8015-114-111, which was controlled by a Conrad electronics laboratory power supply, 0–24 V, Conrad Electronics, Hirschau, Germany). Two pressure sensors (Hydac Electronic GmbH, Saarbruecken, Germany), in combination with a data-recording system (Hydac HMG 3000), ensure the online recording of the pressure, which was used to control the constant trans-membrane pressure during the experiment. After passage of the membrane, the filtrate was reintroduced into the feed suspension. The filtrate flux was measured after defined time intervals (5 min, 30 min, and 60 min) gravimetrically (E-4) and compared with the pure water flux of the respective membrane, which was measured prior to each experiment. At least three measurements were performed for each measurement point. Particle size distributions and turbidity in the filtrate were measured using a Klotz Syringe particle analysis system, LDS 30/30, and a Hach 2100N device. The mass removal efficiency (MRE%) as well as the particle removal efficiency (NpRE%) were calculated from the particle size distributions of the particles in the feed suspension and in the filtrate (mathematical equations are included in the Supplementary Information). The analysis of the particle size distributions and the calculation of the mass and particle number removal rates enable a valid evaluation of the membrane performance. The results are decisive for the optimization of filtration processes and offer starting points for further work.

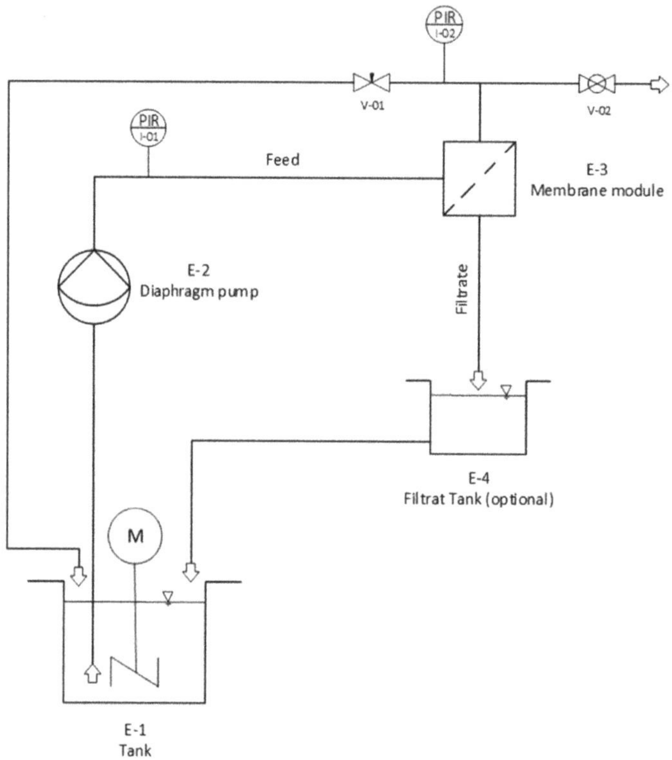

Figure 1. Flow sheet of the experimental filtration set-up.

The used membranes were carefully removed from the filter housing and dried an air atmosphere for one night before further investigations. The set-up was rinsed with clean water after each experiment, and all tubing was renewed before a new microfiltration experiment started.

For backwashing, it was necessary to reverse the water flow through the system at constant pumping conditions (24 V power supply) subsequent to the microfiltration. Therefore, the pump was installed on the filtrate side. The backwash was carried out for 5 min (0.44 bar), while the water was collected in an additional beaker. Subsequent to the completion of the backwash procedure, the configuration was transitioned back to the filtration mode, and the membrane's pure water flow rate was evaluated gravimetrically. This flow was compared to the pure water flow prior to the filtration experiment in order to characterize the backwashing efficiency.

2.3. *Characterization of the Particle Fouling*

For visualization of the particles on top and in cross-sections of the membranes, light microscopy (LM) and scanning electron microscopy (SEM) were carried out. Additionally, water contact angle, Fourier Transform Infrared Spectroscopy (FTIR), and TGA measurements were performed in order to characterize the membrane particle fouling.

2.4. Microscopy

For light microscopy, the membranes were dried under ambient conditions and imaged in transmission mode using an Olympus CX 30 (Evident Corporation, Tokyo, Japan) equipped with a Jenoptik Progres C3 CCD camera (Jenoptik AG, Jena, Germany).

A JEOL JSM-6460LV scanning electron microscope (SEM) (JEOL Ltd., Tokyo, Japan) was used at 15 kV accelerating voltage in high vacuum. Secondary electron images of the top layer and cross-sections were obtained after coating the dried samples with gold by sputter deposition (Emitech K575X, Laughton, UK). Cross-sections of the samples were prepared by placing the membranes in a flat metallic cylinder filled with liquid ethanol, freezing the ethanol by dipping the metallic cylinder in a Dewar vessel filled with liquid nitrogen, and using a scalpel to cut the membranes embedded in frozen ethanol into two pieces. After warming up, the membrane pieces were dried on a tissue and placed on the side of a rectangular Al block using double-sided carbon tape to image the inner structure of the samples.

2.5. Water Contact Angle Measurement

To measure the water contact angle, a self-built contact angle goniometer was used, which consisted of an inclined Zeiss Standard 16 light microscope (Carl Zeiss AG, Oberkochen, Germany) with a horizontally movable sample stage and a Jenoptik Progres C3 CCD camera for lateral visualization of the water droplets. Therefore, a 5 μL water droplet was carefully placed on top of the membrane, the contact angle was measured after 10 s. ImageJ Ver. 1.54 was used to analyze the shape of the droplet using the plugin "drop_analysis", and the water contact angle was determined.

2.6. Fourier Transform Infrared Spectroscopy (FTIR)

FTIR and micro-FTIR are common methods for the characterization of microplastics [32]. FTIR was performed with Bruker Alpha (Bruker Optics, Ettlingen, Germany) at wave numbers between 400 and 4000 cm^{-1} with a resolution better than 2 cm^{-1}. A standard ATR (attenuated total reflection) method was used for the analysis.

2.7. Thermogravimetric Analysis (TGA)

The MP fouling was also determined via thermogravimetric analysis (TGA) using a Netzsch TG 209 (NETZSCH-Gerätebau GmbH, Selb, Germany). Accurate measurement conditions were achieved by an elaborate preparation before each analysis. The Al_2O_3 crucible was heated in air and subsequently in N_2 up to 800 °C prior to each measurement. For calibration, the crucible was then heated in N_2 atmosphere (100 mL/min) using a dynamic heating rate of 10 K/min from 20 to 800 °C, followed by the analysis of the sample under equal conditions. The weight loss was monitored, and DTG (differential thermogravimetric) curves were calculated. A detailed understanding of the thermal decomposition of the pure membrane and fouling materials is a prerequisite for the analysis of fouling by TGA. Especially for CA, structural changes can occur during treatment in an aqueous environment [33]. Therefore, TGA of CA membranes and PS particles, which were exposed to equal aqueous process conditions (without filtration), were also analyzed for comparison (Supplementary Information).

3. Results

3.1. Preparation of the Suspension

Homogeneous microplastic/water feed suspensions are a prerequisite for filtration experiments at accurate test conditions. The number particle size distributions of six exemplary feed suspensions are given in Figure 2. Average particle sizes, standard deviation of the size distribution, and results of the turbidity measurements (Table 1) confirmed the generation of homogenous suspensions as well as the suitability of the preparation method. The median particle size $d_{50,0}$ was calculated to be 3.0 µm for the MP particles.

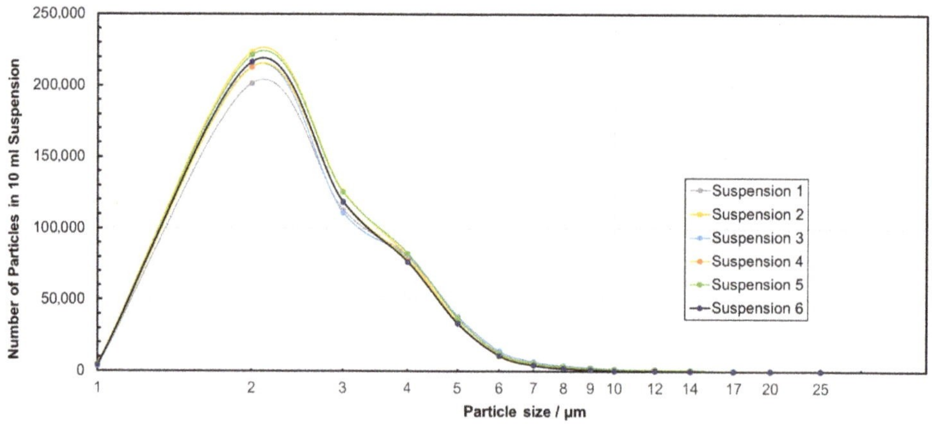

Figure 2. Particle size distributions of PS microplastic/water feed suspensions for the microfiltration experiments (lines are guides for the eyes).

Table 1. Characteristics of the PS microplastic water feed suspensions.

Suspension ID	Average Particle Size/µm	Standard Deviation of Particle Size Distribution/µm	Turbidity/NTU
1	3.20	1.60	2.74 ± 0.04
2	3.02	1.41	2.52 ± 0.04
3	3.20	1.71	2.92 ± 0.03
4	3.11	1.60	2.83 ± 0.04
5	3.12	1.58	3.09 ± 0.09
6	3.00	1.36	2.71 ± 0.02

3.2. Microfiltration Experiments

For all filtration experiments, a significant decrease in the turbidity was measured after five minutes of filtration time. Taking into account the reference value and the standard deviations in the measurements, the turbidity values for all filtrates were 0.0 NTU. This result indicates that the particle concentration was already reduced to a very high extent after only five minutes of filtration time. As the turbidity measurement does not provide any detailed information about the remaining particle sizes in the filtrate, the filtration process is analyzed in more detail below using the particle size distributions.

As described in Section 2, both MRE% and NpRE% were calculated for each filtration experiment after 5, 30, and 60 min. Very high, filtration-time-dependent particle removal rates between 99 and 100% with very narrow standard deviations were achieved for all experiments, independent of the TMP applied (Table 2). The average values of the particle size classes in the filtrate for all tested TMP, but after different filtration times, are shown in Figure 3. In comparison to the feed suspension, a time-dependent decrease in the number of particles was detected. Particles smaller than the nominal pore size of the membrane were also removed in a remarkably high proportion, which indicates that a top layer (filter cake) formation and/or pore blockage significantly influenced the filtration mechanism. Taking into account the frequency of the particle sizes (Figure 4), it becomes obvious that the percentage of smaller particles (e.g., 1 µm) in the filtrate increased in comparison to the percentage in the feed suspension. The retention is therefore not initially the same for all particle sizes, so that smaller particles tend to remain in the filtrate, while larger particles are separated more quickly, which is characteristic of a filter cake. The calculation of the median particle sizes $d_{50\cdot 0}$ for the feed and for the filtrate after different filtration times confirms these observations (Table 2). The slight increase in the median particle size in the filtrate after 60 min could indicate that smaller particles are also retained more effectively in the further course of filtration due to the increasing filter cake. Overall, the two figures enable a comprehensive analysis of the filtration behavior by considering both absolute particle retention and changes in the size distribution.

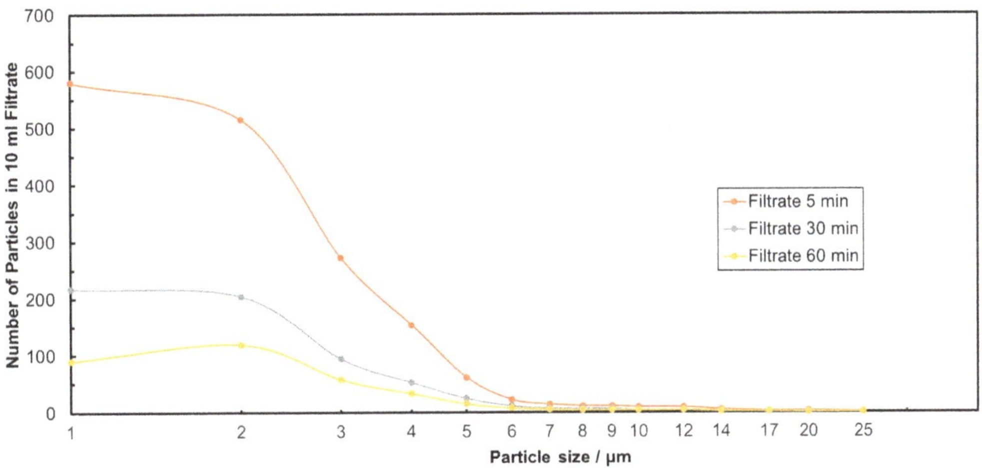

Figure 3. Particle size distributions of PS microplastic in filtrate after different filtration durations.

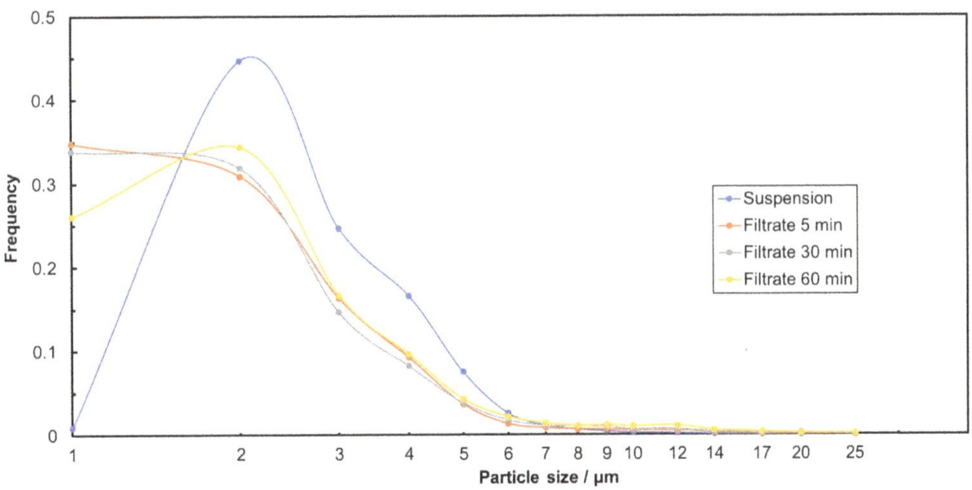

Figure 4. Comparison of the particle size distribution (frequency) of PS microplastic in the feed suspension and in the filtrate after different filtration durations.

Table 2. Median particle size of PS microplastic in suspension and removal efficiency with standard deviation during all MP filtration experiments.

Sample	Median Particle Size/μm	MRE%	NpRE%
Feed suspension	3.00	/	/
Filtrate after 5 min	2.00	99.56 ± 0.27	99.65 ± 0.45
Filtrate after 30 min	2.00	99.76 ± 0.14	99.87 ± 0.09
Filtrate after 60 min	2.33	99.84 ± 0.11	99.93 ± 0.04

In order to further examine the time dependency of the removal rate, an average value of the MRE% and NpRE% and the standard deviation were calculated for all filtration experiments at each of the three time points (Table 2, Figures 5 and 6). A nearly complete removal of the particles was reached during these experiments. The very small variation in the results for all filtration experiments was confirmed by the very low standard deviation (Table 2).

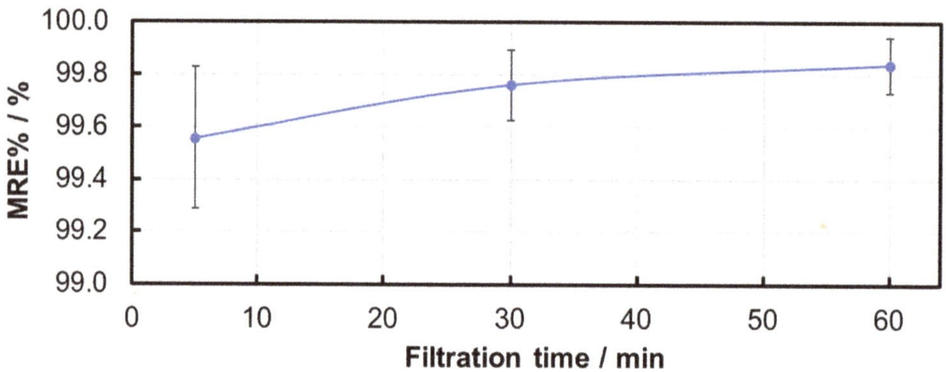

Figure 5. Time dependency of the average MRE% and standard deviation during all PS MP filtration experiments.

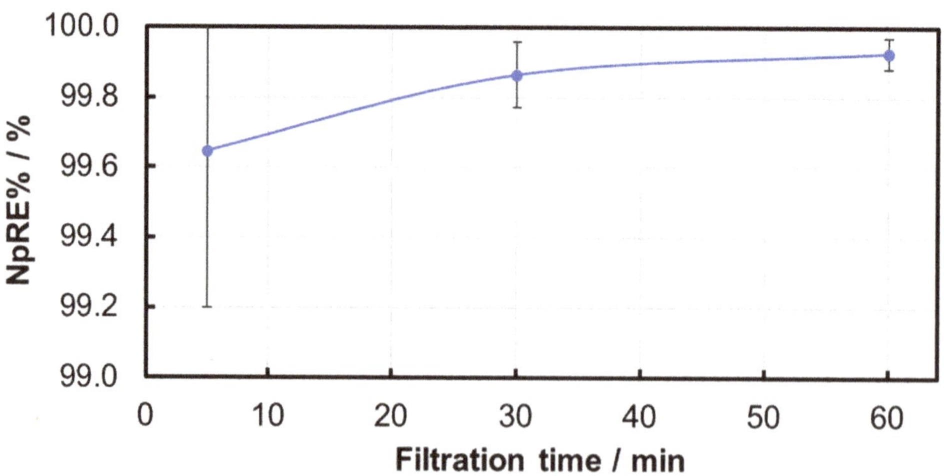

Figure 6. Time dependency of the average NpRE% and standard deviation during all PS MP filtration experiments.

In addition, the NpRE% is determined individually for each particle size class at each of the three measurement times in order to identify any difference in the separation of different particle sizes. The results confirmed that the removal rate for the small particles increased over time, which is characteristic of the buildup of a filter cake on the membrane. Small particles that were able to pass the nominal pore size at the beginning of the filtration were retained (by the filter cake) within the filtration duration. Larger particles were widely separated by the size effect. The number of larger particles (>10 μm) in the suspension was relatively low in comparison to the number of smaller particles in the suspension. Therefore, the passage of the membrane by a small number of larger particles could significantly influence the removal rate calculation. As an example, the particle size-dependent NpRE% for 0.2 bar pressure is shown in Figure 7.

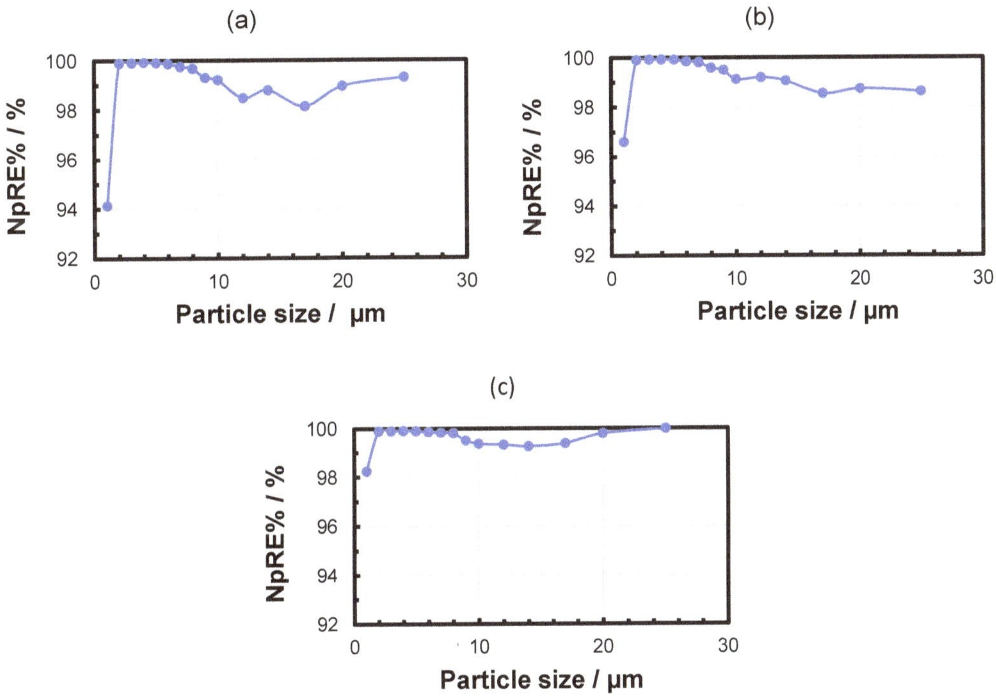

Figure 7. Time dependency of the NpRE% of different particle sizes during the PS MP filtration experiment at 0.2 bar after 5 min (**a**), 30 min (**b**), and 60 min (**c**).

The flux was measured after different filtration durations and at different TMPs. The results of six filtration experiments were compared to the pure water flux of the respective fresh membrane in order to calculate the flux decline (Figures 8–10). Increased TMP resulted in an increased flux. The flux decline over time during the filtration experiment was reduced after a short time (Table 3). Only a very small relative flux decline was observed between 30 and 60 min of filtration time. This observation could be attributed to a fast build-up of the filter cake (see Section 3.4. The difference in relative flux decline between 5 and 60 min decreased with increased pressure. This could be attributed to a faster deposition of particles within the first minutes of filtration due to the increased filtration pressure and elevated fluid velocity.

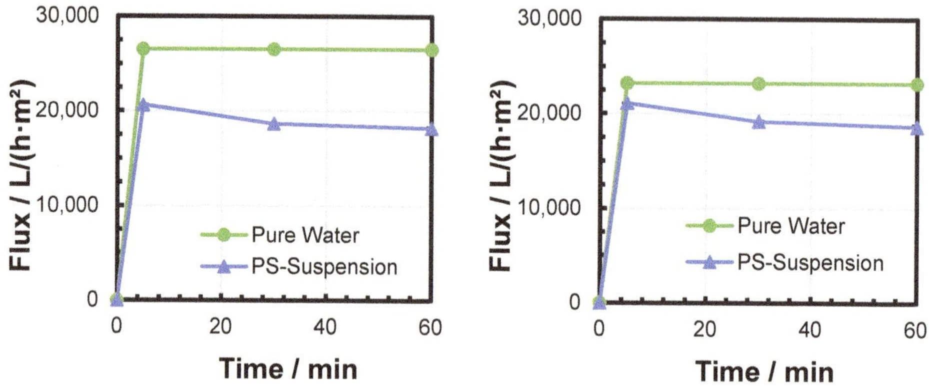

Figure 8. Evaluation of flux decline during two PS MP filtration experiments at 0.1 bar.

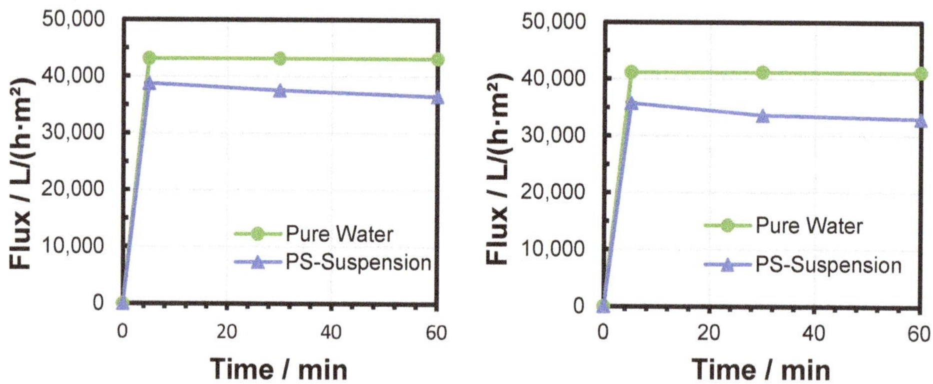

Figure 9. Evaluation of flux decline during two PS MP filtration experiments at 0.2 bar.

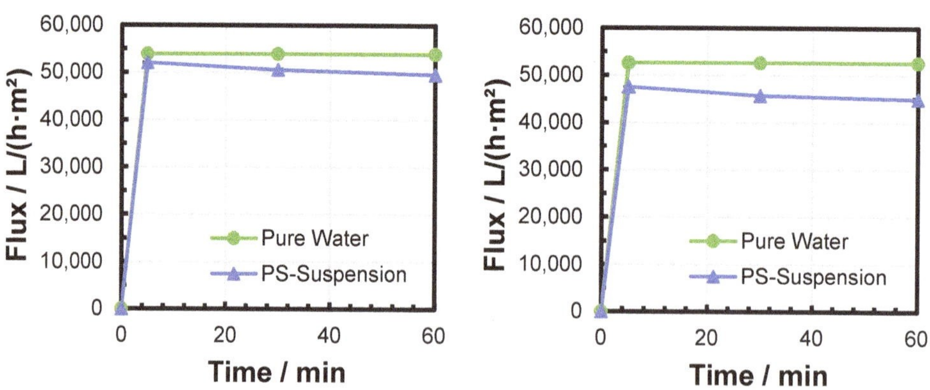

Figure 10. Evaluation of flux decline during two PS MP filtration experiments at 0.3 bar.

Table 3. Relative flux decline for PS suspension microfiltration experiments at different filtration pressures.

Filtration Experiment	Pressure/bar	Flux Decline After 5 min/%	Flux Decline After 30 min/%	Flux Decline After 60 min/%
Filtration 1	0.1	−22.2	−29.7	−31.4
Filtration 2	0.1	−9.1	−17.4	−19.7
Filtration 3	0.2	−10.2	−13.1	−15.4
Filtration 4	0.2	−13.3	−18.4	−19.9
Filtration 5	0.3	−3.4	−6.2	−8.0
Filtration 6	0.3	−9.6	−13.2	−14.5

3.3. Backwashing Experiments

The pure water flux was measured before microfiltration and after backwashing in order to observe the effectiveness of the backwash procedure. The results are presented in Figures 11–13 and clearly confirm that particle fouling from microfiltration at low TMP could be removed by the backwashing procedure applied. The results also show that the flux was not completely recovered after microfiltration at 0.3 bar with the backwash procedure applied. This could be addressed to a difference in the particle-fouling mechanism dependent on the TMP (see Section 3.4).

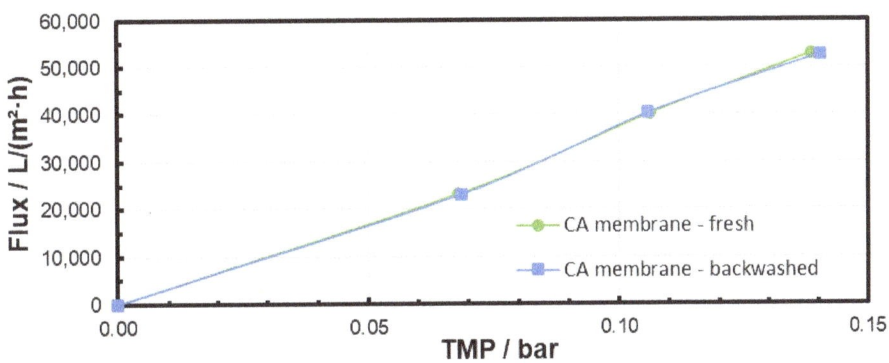

Figure 11. Pure water flux through the CA membrane before microfiltration and after 60 min microfiltration of the PS particle suspension at 0.1 bar and subsequent backwash.

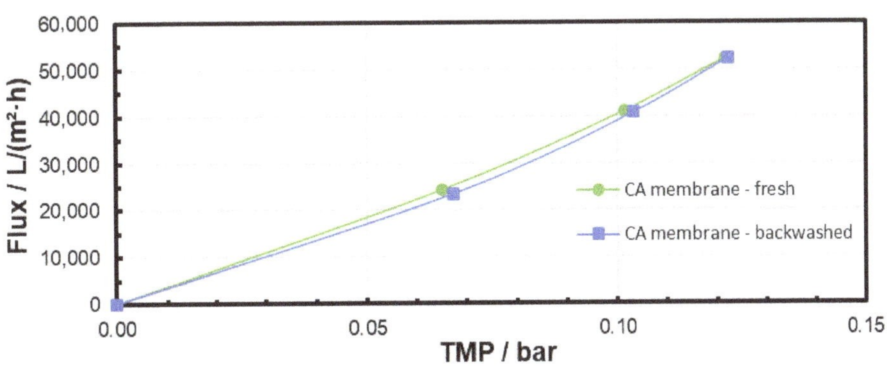

Figure 12. Pure water flux through the CA membrane before microfiltration and after 60 min microfiltration of the PS particle suspension at 0.2 bar and subsequent backwash.

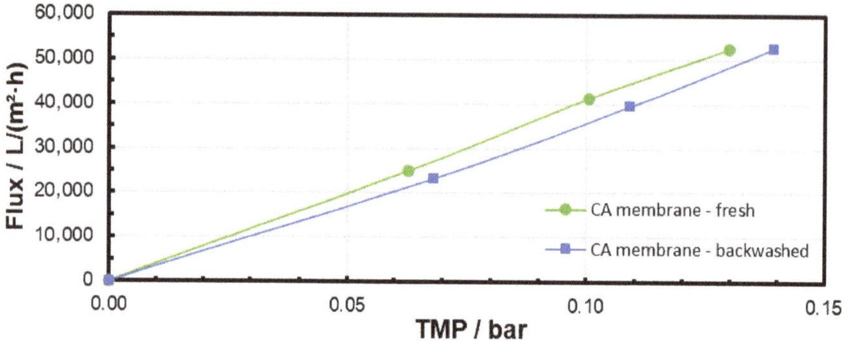

Figure 13. Pure water flux through the CA membrane before microfiltration and after 60 min microfiltration of the PS particle suspension at 0.3 bar and subsequent backwash.

3.4. *Particle-Fouling Analysis*

In order to further explain the particle-fouling mechanism, the used membranes were extensively analyzed using SEM, light microscopy, TGA, and water contact angle measurements.

3.5. *Particle-Fouling Analysis by SEM and Light Microscopy*

The build-up of a filter cake subsequent to the blockage (or partial closure) of pores is expected to describe the particle-fouling mechanism for many filtration processes [34–36]. Depending on the particle size and the membrane pore size, particle transport inside the pores can occur or not. This internal blockage of pores is not expected if the particles are larger than the pores. Our challenging microfiltration conditions (membrane pores and particle sizes on the same scale) enable the transport of particles inside larger pores. Therefore, the observation of the particle deposition in the pores, as well as the observation of the filter cake, gives indications of the fouling mechanics. PS microparticles can be assumed to be incompressible, and particle deformation can be excluded under the given conditions; however, particle movement within the filter cake, variation in the packaging, and changes in the cake height cannot be excluded and need to be identified for the microfiltration process [34–36]. SEM images of the membrane surface clearly show particle deposition during all TMPs (Figures 14 and 15). The cross-sections are characterized by typical dead-end filter cakes. Increasing the TMP resulted in a decreased filter cake height. However, due to our analysis of the membrane surface, we could also identify more heterogeneous and irregular loading at increased TMP. This was observed, in particular, for the highest pressure applied (Figure 15). Increasing the TMP results in a decreased cake height, which is partly explained by filling voids with smaller particles at elevated TMP [35]. On the other hand, particles were also observed inside the membrane pores (Figure 15). Up to 99% of all particles smaller than the median pore size were already removed from the feed suspension after 5 min of filtration time. Due to our experiment design (particle and pore sizes) and in accordance with PMMA model filtration experiments by Dersoir et al. [37], we conclude that the smaller MP particles entered the pores and initiated an internal pore blockage, which led to a first flux decline. It is known that agglomerates were usually built up in the next blocking step [37]. Together with complete pore blockage by larger particles, this sequence initiated the build-up of the filter cake on the membrane surface. The number of particles inside the pores depends on the TMP. Elevated TMP resulted in an increased number of particles inside the pores. Agglomerates were only observed at the highest pressure tested (0.3 bar). However, the number of particles and agglomerates inside the pores was relatively low compared to the number of particles on the membrane surface, which led to the assumption that most of the small particles and agglomerates were already flushed out of the pores by water after a short time. It is very probable that this small amount of small particles was deposited on the filter cake within the duration of the experiment.

Figure 14. SEM image of the PS filter cake after 60 min filtration at 0.1 bar (top view and cross-section).

Figure 15. SEM image of the PS filter cake after 60 min filtration at 0.3 bar (top view and cross-section).

Almost all of the filter cake was removed by backwashing (Figures 16 and 17). Due to the low forces between MP particles and the membrane, the cleaning by backwash is very efficient. Only isolated particles remained inside the membrane, which also confirmed the low interparticle forces within the agglomerates inside the pores.

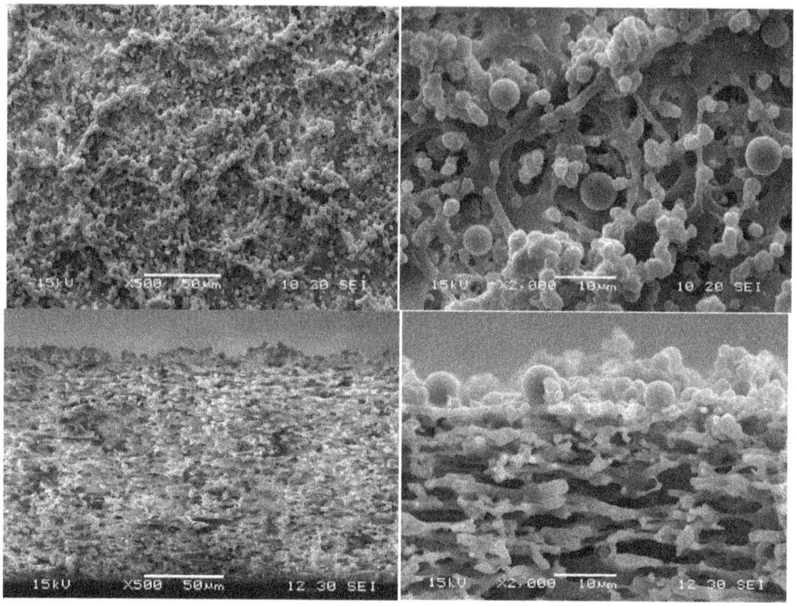

Figure 16. SEM image of the PS filter cake after 60 min filtration at 0.1 bar and subsequent backwashing (top view and cross-section).

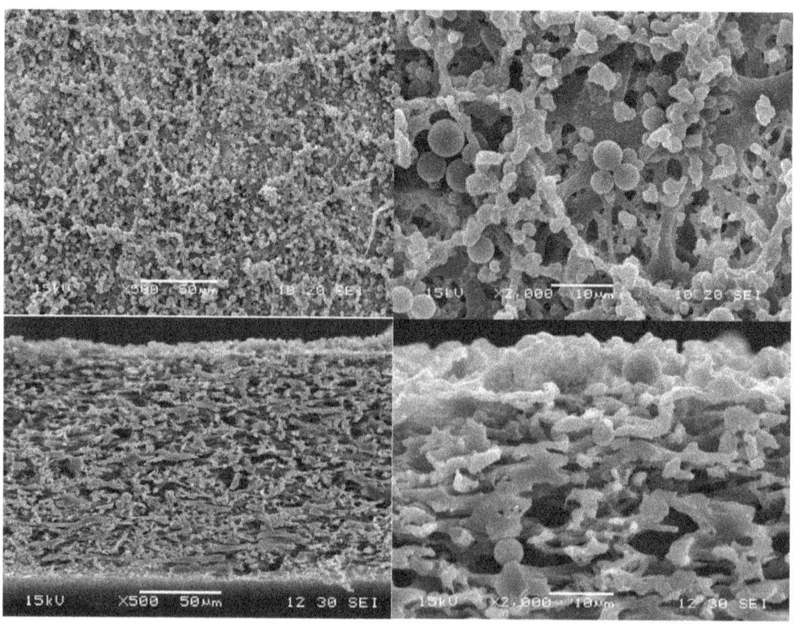

Figure 17. SEM image of the PS filter cake after 60 min filtration at 0.3 bar and subsequent backwashing (top view and cross-section).

The pore-blocking procedure, cake morphology (structure), and cake size (height) are decisive factors for the efficiency of the backwashing procedure. The experiments of backwashing after filtration of identical MP suspensions at different TMPs show that more heterogeneous cake morphologies and a higher number of blocked pores after filtration impede the backwashing. Especially for the highest TMP, flux recovery was not completely possible after the backwash (see Figures 13 and 17). However, for all TMPs, the flux recovery clearly confirmed that backwashing is an efficient MP particle-fouling cleaning technology at our experimental conditions.

3.6. *Particle-Fouling Analysis by TGA*

The TG analysis of a CA membrane after a 60 min PS microfiltration using 0.3 bar filtration pressure is shown in Figure 18. As expected, the CA membrane decomposed in the temperature range 280–380 °C (first weight loss step). Afterwards, a second decomposition step was clearly observed, which is characteristic of the decomposition of the deposited PS between ~390 and 430 °C. The DTG graphs confirm the existence of two decomposition steps, which can be attributed to membrane (CA) and particle foulant/filter cake (PS).

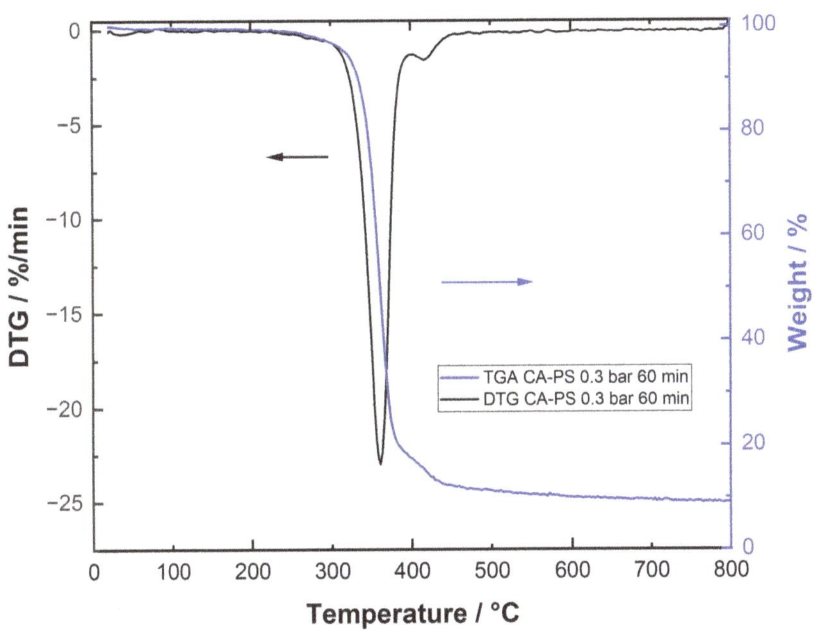

Figure 18. TGA and DTG signals of the CA membrane after 60 min microfiltration of the PS suspension at 0.3 bar.

For comparison, a TG analysis was also performed for a membrane after a 60 min PS filtration at 0.3 bar and subsequent backwashing. No PS signal was monitored during this measurement. Only one peak (CA) was observed during the DTG, which also indicates that most of the PS was removed during the backwashing (Figure 19).

Due to inhomogeneous particle deposition and complex sample preparation, it is hard to provide a statement on whether TGA is suitable for the analysis of MP particle fouling on CA membranes. Our results show that TGA is suitable for the detection of MP on CA membranes. We also came to the conclusion that accurate sample preparation and detailed knowledge about the behavior of the included materials at process conditions are decisive for the usage of TGA for fouling analysis. Partly overlapping residual decomposition of small amounts cannot be excluded (see DTG analysis of pure materials in Supplementary Information). In comparison to our clean lab experiments, combinations of different MPs will be filtered in the field. This will increase the number of degrees of freedom and will complicate the TG analysis. On the other hand, the different decomposition temperatures of CA and most MP foulants could be a key for MP detection and identification on the membrane. Other analytical methods may hold advantages for extensive characterization.

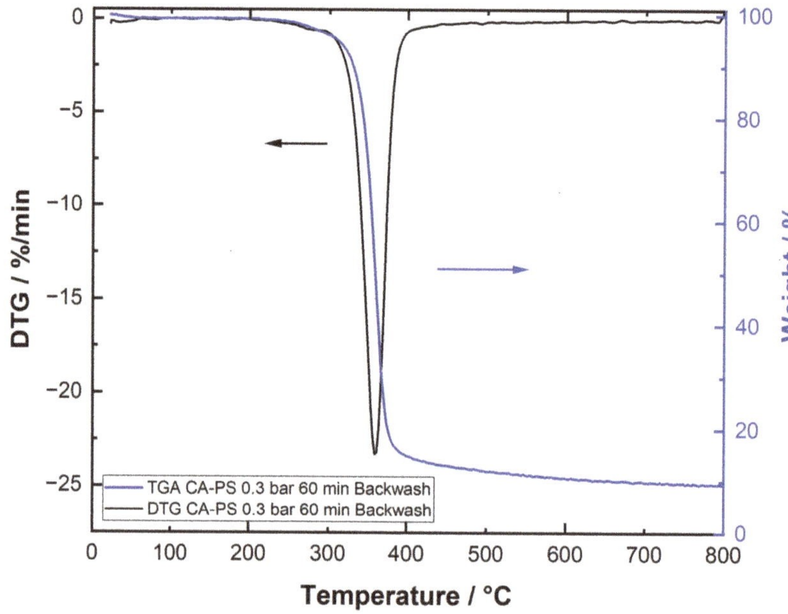

Figure 19. TGA and DTG signal of the CA membrane after 60 min microfiltration of PS suspension at 0.3 bar and subsequent backwashing procedure.

3.7. Particle-Fouling Analysis by Water Contact Angle Measurements

The results of the water contact angle measurement are presented in Figure 20. The fresh CA membrane was characterized by a water contact angle of 0°, indicating a hydrophilic surface, which is consistent with values reported in the literature [20,38]. After wetting and subsequent drying, the contact angle increased to 96 ± 2°, which indicates that the original hydrophilicity was reduced by this treatment. A significantly higher contact angle of 124 ± 3° was measured for the PS powder, which is clearly attributed to its hydrophobic properties. After filtration, the membranes were characterized by a contact angle of 126 ± 5°, which is almost identical to that of the PS powder. It can be concluded that the top layer of PS particles completely covered the membrane and that its hydrophobic properties dominated on the surface. After backwashing, the contact angle dropped to 103 ± 9°, which is between the values of the membranes after wetting and after filtration. This indicates that most of the PS layer is removed by backwashing, but the membrane does not completely return to its original state. The high standard deviation also shows that the cleaning was carried out rather unevenly, so that inhomogeneous residues remain on the membrane. Overall, the measured values show that the filtration resulted in a hydrophobic PS top layer, which was only partially removed by backwashing. The membrane was not completely regenerated at the selected backwashing conditions.

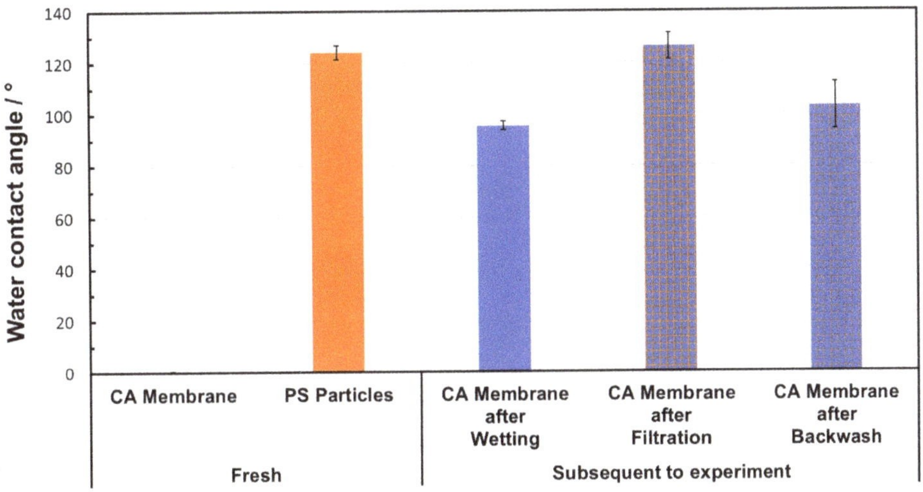

Figure 20. Results of water contact angle measurements for fresh materials and the used CA membrane.

PS particles and fresh, water-treated, and fouled membranes were also analyzed using ATR-FTIR. However, due to the overlapping of the characteristic peaks, no clear detection of PS fouling on CA was possible with ATR-FTIR. The results are provided in the Supplementary Information.

4. Discussion and Conclusions

The focus of this study was to experimentally investigate the microfiltration characteristics of a commercial CA microfiltration membrane for the removal of PS microplastic particles from water during dead-end operation with a special view on particle size aspects. A comparable membrane pore size/PS particle size combination was selected in order to challenge the microfiltration. The particle fouling mechanism was investigated by different methods. Backwashing efficiency experiments round off the investigations.

Within our experiments, most of the microplastic particles were already removed from the feed suspension after a short microfiltration duration. The removal efficiency was particle size-dependent, and especially small particles were further reduced during the ongoing microfiltration, while the larger particles were already removed within the first minutes of microfiltration. This observation was attributed to the pore blocking (internal and/or complete) and the build-up of the filter cake. Increased filtration pressure resulted in decreased flux decline, which was attributed to more heterogeneous particle layer formation at elevated pressure. On the other hand, a thinner filter cake height and an increased number of particles within the membrane pores were detected at elevated pressure. This difference in particle fouling structure at low and elevated filtration pressure significantly influences the backwashing efficiency. The particle fouling resulting from low-pressure filtration was almost completely removed due to the backwashing procedure applied, while an increased filtration pressure resulted in a different particle fouling structure, which negatively influenced the backwashing efficiency at given conditions.

Microfiltration could be a decisive technology for the removal of MP from, e.g., industrial process wastewater or within the purification of drinking water. The results of our lab-scale investigations show that material characteristics, particle size and pore size, and process conditions influence the MP removal and backwashing efficiency. Knowledge on MP–membrane interactions and resulting particle-fouling mechanisms is a decisive key for effective MP removal and is a prerequisite for further process development. For example, a subsequent treatment of the concentrated MP backwash solution could be sedimentation, precipitation, additional filtration, concentrating, or further conditioning steps.

The spherical microplastic particles are well suited for our investigation. However, microplastics also appear in different morphologies, e.g., fibers or films. Therefore, we will also take into account the particle morphology within our future investigations. A lab-scale dead-end filtration mode was addressed during the investigations presented in this paper. For many water purification applications, cross-flow filtration is very common on an industrial scale. Therefore, future examinations will also focus on cross-flow mode in order to compare the microfiltration and particle-fouling characteristics, which is a prerequisite for process development.

Supplementary Materials: The following supporting information can be downloaded at https://www.mdpi.com/article/10.3390/membranes15090272/s1, Figure S1. SEM images of PS particles; Figure S2. Light microscopy (left) and SEM images (middle, right) of fresh CA membrane; Figure S3. Light microscopy: PS filter cake after 60 min filtration at 0.1 bar (left), membrane surface after subsequent backwashing (right); Figure S4. Light microscopy: PS filter cake after 60 min filtration at 0.2 bar (left), membrane surface after subsequent backwashing (right); Figure S5. Light microscopy: PS filter cake after 60 min filtration at 0.3 bar (left), membrane surface after subsequent backwashing (right); Figure S6. SEM image of the PS filter cake after 60 min filtration at 0.2 bar (top view); Figur S7. SEM image of the PS filter cake after 60 min filtration at 0.2 bar and subsequent backwashing (top view); Figure S8. SEM image of the PS filter cake after 60 min filtration at 0.2 bar (cross section); Figure S9. SEM image of the PS filter cake after 60 min filtration at 0.2 bar and subsequent backwashing (cross section); Figure S10. FTIR spectra of CA membrane, CA membrane after PS filtration, CA membrane after PS filtration and subsequent backwashing and FTIR spectrum of PS particles; Figure S11. TGA and DTG signals of the CA membrane after 60 min flushing clean H2O at 0.3 bar; Figure S12. TGA and DTG signals of the PS particles after treatment in H2O.

Author Contributions: S.S.: Conceptualization, methodology, validation, formal analysis, investigation, writing—review and editing, visualization. L.M.S.: Methodology, investigation, writing—review and editing, visualization. G.B.: Conceptualization, methodology, writing—review and editing, supervision. M.K.: Investigation, methodology, writing—review and editing, visualization. M.G.: Conceptualization, methodology, writing—review and editing, project administration, funding acquisition. M.F.: Conceptualization, methodology, investigation, writing—review and editing, visualization, supervision, project administration, funding acquisition. All authors have read and agreed to the published version of the manuscript.

Funding: The project "Switchable Membrane Systems—SWIMEMSYS" is co-financed by the European Regional Development Fund and Saarland funds. S.S., M.G., and M.F. would like to thank the European Union and the Ministry of Economics, Innovation, Digitalisation, and Energy for the financial support. M.K. and M.F. would like to thank the Deutsche Forschungsgemeinschaft (DFG, German Research Foundation) for the founding project number 533266269.

Institutional Review Board Statement: Not applicable.

Data Availability Statement: Data are contained within the article or within the Supplementary Material. Further inquiries can be directed to the corresponding author.

Acknowledgments: We would like to thank Olivia Freitag-Weber (htw Saar) for support with the SEM set-up. We also thank Jörg Meyer (KIT) for help with TGA equipment.

Conflicts of Interest: The authors declare no conflicts of interest.

References

1. Tajwara, M.; Gazi, M.J.; Sahaa, S.K. Characterization and Spatial Abundance of Microplastics in the Coastal Regions of Cox's Bazar, Bangladesh: An Integration of Field, Laboratory, and GIS Techniques. *Soil Sediment Contam. Int. J.* **2021**, *31*, 57–80. [CrossRef]

1. Park, H.; Park, B. Review of Microplastic Distribution, Toxicity, Analysis Methods, and Removal Technologies. *Water* **2021**, *13*, 2736. [CrossRef]
2. Xu, Y.; Ou, Q.; Wang, X.; van der Hoek, J.P.; Liu, G. Mass Concentration and Removal Characteristics of Microplastics and Nanoplastics in a Drinking Water Treatment Plant. *ACS EST Water* **2024**, *4*, 3348–3358. [CrossRef]
3. Ma, B.; Xue, W.; Ding, Y.; Hu, C.; Liu, H.; Qu, J. Removal characteristics of microplastics by Fe-based coagulants during drinking water treatment. *J. Environ. Sci.* **2019**, *78*, 267–275. [CrossRef] [PubMed]
4. Acarer, S. A review of microplastic removal from water and wastewater by membrane technologies. *Water Sci. Technol.* **2023**, *88*, 199. [CrossRef]
5. Di Bella, G.; Corsino, S.F.; De Marines, F.; Lopresti, F.; La Carrubba, V.; Torregrossa, M.; Viviani, G. Occurrence of Microplastics in Waste Sludge of Wastewater Treatment Plants: Comparison between Membrane Bioreactor (MBR) and Conventional Activated Sludge (CAS) Technologies. *Membranes* **2022**, *12*, 371. [CrossRef]
6. Egea-Corbacho, A.; Martín-García, A.P.; Franco, A.A.; Quiroga, J.M.; Andreasen, R.R.; Jørgensen, M.K.; Christensen, M.L. Occurrence, identification and removal of microplastics in a wastewater treatment plant compared to an advanced MBR technology: Full-scale pilot plant. *J. Environ. Chem. Eng.* **2023**, *11*, 109644. [CrossRef]
7. Rius-Ayra, O.; Biserova-Tahchieva, A.; LLorca-Isern, N. Surface-functionalised materials for microplastic removal. *Mar. Pollut. Bull.* **2021**, *167*, 112335. [CrossRef]
8. Murray, A.; Örmeci, B. Removal Effectiveness of Nanoplastics (<400 nm) with Separation Processes Used for Water and Wastewater Treatment. *Water* **2020**, *12*, 635. [CrossRef]
9. Akrsu, C.; Kumbur, H.; Kideys, A.E. Removal of microplastics from wastewater through electrocoagulation-electroflotation and membrane filtration processes. *Water Sci. Technol.* **2021**, *84*, 1648. [CrossRef]
10. Ali, I.; Tan, X.; Mustafa, G.; Gao, J.; Peng, C.; Naz, I.; Duan, Z.; Zhu, R.; Ruan, Y. Removal of micro- and nanoplastics by filtration technology: Performance and obstructions to market penetrations. *J. Clean. Prod.* **2024**, *470*, 143305. [CrossRef]
11. Kundu, A.; Shetti, N.P.; Basu, S.; Reddy, R.R.; Nadagouda, M.N.; Aminabhavi, T.M. Identification and removal of micro- and nano-plastics: Efficient and cost-effective methods. *Chem. Eng. J.* **2021**, *421*, 129816. [CrossRef]
12. Li, J.; Wang, B.; Chen, Z.; Ma, B.; Chen, J.B. Ultrafiltration membrane fouling by microplastics with raw water: Behaviors and alleviation methods. *Chem. Eng. J.* **2021**, *410*, 128174. [CrossRef]
13. Golgoli, M.; Khiadani, M.; Shafieian, A.; Sen, T.K.; Hartanto, Y.; Johns, M.L.; Zargar, M. Microplastics fouling and interaction with polymeric membranes: A review. *Chemosphere* **2021**, *283*, 131185. [CrossRef] [PubMed]
14. Shen, M.; Zhao, Y.; Liu, S.; Hu, T.; Zheng, K.; Wang, Y.; Lian, J.; Meng, G. Recent advances on micro/nanoplastic pollution and membrane fouling during water treatment: A review. *Sci. Total Environ.* **2023**, *881*, 163467. [CrossRef] [PubMed]
15. Ghasemi, S.; Yan, B.; Zargar, M.; Ling, N.N.A.; Fridjonsson, E.O.; Johns, M.L. Impact of microplastics on organic fouling of hollow fiber membranes. *Chem. Eng. J.* **2023**, *467*, 143320. [CrossRef]
16. Enfrin, M.; Wang, J.; Merenda, A.; Dumèe, L.F.; Lee, J. Mitigation of membrane fouling by nano/microplastics via surface chemistry control. *J. Membr. Sci.* **2021**, *633*, 119379. [CrossRef]
17. Enfrin, M.; Lee, J.; Fane, A.G.; Dumée, L.F. Mitigation of membrane particulate fouling by nano/microplastics via physical cleaning strategies. *Sci. Total Environ.* **2021**, *788*, 147689. [CrossRef]
18. Pizzichetti, A.R.P.; Pablos, C.; Alvarez-Fernandez, C.; Reynolds, K.; Stanley, S.; Marugan, J. Evaluation of membranes performance for microplastic removal in a simple and low-cost filtration system. *J. Environ. Chem. Eng.* **2021**, *3*, 100075. [CrossRef]
19. Pizzichetti, A.R.P.; Pablos, C.; Alvarez-Fernandez, C.; Reynolds, K.; Stanley, S.; Marugan, J. Kinetic and mechanistic analysis of membrane fouling in microplastics removal from water by dead-end microfiltration. *J. Environ. Chem. Eng.* **2023**, *11*, 109338. [CrossRef]
20. Slejko, E.A.; Tuan, A.; Scuor, N. From waste to value: Characterization of recycled cellulose acetate for sustainable waste management. *Waste Manag. Bull.* **2024**, *1*, 67–73. [CrossRef]
21. Islam, M.D.; Uddin, F.J.; Rashid, T.U.; Shahruzzaman, M. Cellulose acetate-based membrane for wastewater treatment—A state-of-the-art review. *Mater. Adv.* **2023**, *4*, 4054–4102. [CrossRef]
22. Peonemann, K.-V.; Nunes, S.P. *Membranes for Water Treatment*; Wiley-VCH: Weinheim, Germany, 2010.
23. Zhao, X.; Zhanga, R.; Liua, Y.; Hea, M.; Sua, Y.; Gaob, C.; Jianga, Z. Antifouling membrane surface construction: Chemistry plays a critical role. *J. Membr. Sci.* **2018**, *551*, 145–171. [CrossRef]
24. Tay, J.-H.; Liu, J.; Sun, D.D. Quantification of membrane fouling using thermogravimetric method. *J. Membr. Sci.* **2003**, *217*, 17–28. [CrossRef]
25. Kamarudin, D.; Hashim, N.A.; Ong, B.H.; Faried, M.; Suga, K.; Umakoshi, H.; Mahari, W.A.W. Alternative fouling analysis of PVDF UF membrane for surface water treatment: The credibility of silver nanoparticles. *J. Membr. Sci.* **2022**, *661*, 120865. [CrossRef]
26. da Conceicao, M.; Lucena, C.; de Alencar, A.E.V.; Mazzeto, S.E.; Soares, S.A. The effect of additives on the thermal degradation of cellulose acetate. *Polym. Degrad. Stab.* **2003**, *80*, 149–155. [CrossRef]

28. Shaikh, H.M.; Anis, A.; Poulose, A.M.; Al-Zahrani, S.M.; Madhar, N.A.; Alhamidi, A.; Aldeligan, S.H.; Alsubaie, F.S. Synthesis and Characterization of Cellulose Triacetate Obtained from Date Palm (*Phoenix dactylifera* L.) Trunk Mesh-Derived Cellulose. *Molecules* **2022**, *27*, 1434. [CrossRef]
29. Marcilla, A.; Beltrán, M. Kinetic Study of the thermal decomposition of polystyrene and polyethylene-vinyl acetate graft copolymers by thermogravimetric analysis. *Polym. Degrad. Stab.* **1995**, *50*, 117–124. [CrossRef]
30. Faravelli, T.; Pinciroli, M.; Pisano, F.; Bozzano, G.; Dente, M.; Ranzi, E. Thermal degradation of polystyrene. *J. Anal. Appl. Pyrolysis* **2001**, *60*, 103–121. [CrossRef]
31. Jang, B.N.; Wilkie, C.A. The thermal degradation of polystyrene nanocomposite. *Polymer* **2005**, *46*, 2933–2942. [CrossRef]
32. Corami, F.; Rosso, B.; Bravo, B.; Gambaro, A.; Barbante, C. A novel method for purification quantitative analysis and characterization of microplastic fibers using Micro-FTIR. *Chemosphere* **2020**, *238*, 124564. [CrossRef]
33. Yadav, N.; Hakkarainen, M. Degradation of Cellulose Acetate in Simulated Aqueous Environments: One-Year Study. *Macromol. Mater. Eng.* **2022**, *307*, 2100951. [CrossRef]
34. Mackley, M.R.; Sherman, N.E. Cross-Flow Cake Filtration Mechanism and Kinetics. *Chem. Eng. Sci.* **1992**, *47*, 3067–3084. [CrossRef]
35. Vyasa, H.K.; Mawsona, A.J.; Bennett, R.J.; Marshall, A.D. A new method for estimating cake height and porosity during crossflow filtration of particulate suspensions. *J. Membr. Sci.* **2000**, *176*, 113–119. [CrossRef]
36. Chellam, S.; Xu, W. Blocking laws analysis of dead-end constant flux microfiltration of compressible cakes. *J. Colloid Interface Sci.* **2006**, *301*, 248–257. [CrossRef] [PubMed]
37. Dersoir, B.; Schofield, A.B.; Robert de Saint Vincent, M.; Tabuteau, H. Dynamics of pore fouling by colloidal particles at the particle level. *J. Membr. Sci.* **2019**, *573*, 411–424. [CrossRef]
38. Ahmad, A.L.; Mat Yasin, N.H.; Derek, C.J.C.; Lim, J.K. Microfiltration of Chlorella sp.: Influence of material and membrane pore size. *Membr. Water Treat.* **2013**, *4*, 143–155. [CrossRef]

membranes

MDPI

Article

Exploration of Methods for In Situ Scale Removal During Magnesium Hydroxide Membrane Crystallization

Ester Komačková , Lukáš Sedlák , Ivan Červeňanský * and Jozef Markoš

Institute of Chemical and Environmental Engineering, Faculty of Chemical and Food Technology, Slovak University of Technology, Radlinského 9, 812 37 Bratislava, Slovakia; ester.komackova@stuba.sk (E.K.); lukas.sedlak@stuba.sk (L.S.); jozef.markos@stuba.sk (J.M.)

* Correspondence: ivan.cervenansky@stuba.sk

Abstract: In coastal countries facing a shortage of drinking water, seawater desalination is essential for the production of potable water. During desalination, a large volume of waste stream, known as brine, is generated. This stream contains high concentrations of salts, particularly those of economic importance to the European Union, such as magnesium and calcium. By further processing this stream, these materials can be recovered. One method studied for separating magnesium from wastewater is membrane crystallization (MCr). The MCr process developed in this work utilizes ion-exchange membranes that separate the model brine solution from a precipitating agent, which is a solution of sodium hydroxide. During the process, the membrane allows the transport of anions between the two solutions, enabling the reaction between OH^- anions and Mg^{2+} cations, which leads to the formation of a magnesium hydroxide precipitate. The formed precipitate can then be filtered out of the brine solution, which now has decreased salinity due to crystallization facilitated by the ion-exchange membrane. However, precipitation occurs near the membrane surface, resulting in the deposition of magnesium hydroxide onto the outer surface of the membrane. The aim of this study is to investigate methods for effectively removing magnesium hydroxide from the membrane surface, with a primary focus on maximizing the yield of magnesium hydroxide crystals in suspension. Crystal removal was induced by circulation of hydrochloric acid, followed by circulation of demineralized water through the membrane module after crystallization. In this study, a membrane module made of hollow-fiber anion-exchange membranes was employed. The production cost of these membranes is approximately 50% lower per square meter compared to flat-sheet membranes commonly used in electrodialysis, demonstrating strong potential for commercial application. More than 85% magnesium conversion was achieved during the process, yet the majority of the crystals remained attached to the membrane. Circulation of hydrochloric acid and demineralized water after the crystallization process caused detachment of the crystals into suspension, nearly doubling their yield.

Keywords: membrane crystallization; magnesium hydroxide; hollow fiber anion-exchange membrane

1. Introduction

The global human population continues to grow, and projections indicate that this trend will persist [1]. With the rising number of inhabitants, demands for a higher standard of living are also increasing. In many coastal regions, people have begun constructing desalination plants to process seawater for the production of clean drinking water. These

https://doi.org/10.3390/membranes15090267

facilities primarily utilize membrane technologies, especially reverse osmosis (RO) [2]. However, processing seawater through RO generates a waste stream with a high concentration of dissolved salts, several times higher than in the original seawater [3,4]. Table 1 shows a comparison of cation concentrations between sea water and different types of exhausted brines.

Table 1. Cation composition of exhausted brines.

Source	Concentration of Main Cations [g/L]			
	Na^+	K^+	Ca^{2+}	Mg^{2+}
Sea water [3]	10.8	0.39	0.41	1.29
Mine brine [5]	36.84	0.65	2.52	2.84
Trapani saltworks brine [4]	48.30	8.62	0.46	36.06
Ionic exchange resins brine [6]	26.15	-	24.12	3.08

Approximately 140 million m^3 of brine are produced globally each day by reverse osmosis or mining [7]. Common methods for brine disposal include discharging into the ocean [8], surface water [9], and sewage system [10], which poses significant environmental problems. These methods often involve high costs and environmental risks, underscoring the need for more sustainable and cost-effective solutions [11].

The European Union annually determines a list of critical raw materials that are in high demand on the European market and carry a high risk of supply disruption to the EU. Among these are substances found in brines [12]. One such valuable material is magnesium, [13,14]. Magnesium, in the form of magnesium hydroxide, has a wide range of industrial applications. It can be used not only as a flame-retardant [15–17] but also as an antibacterial agent [18] or for reducing water pollution caused by heavy metals [19–21].

Research into the reactive crystallization of magnesium hydroxide from magnesium chloride solution with sodium hydroxide as a precipitating agent began almost 50 years ago [22,23]. These early experiments showed that direct contact between magnesium chloride solution and precipitating agent produces very fine particles, which tend to agglomerate, forming a gelatinous suspension that is difficult to separate. In subsequent years, studies explored the effects of operational conditions on reactive crystallization to achieve the smallest possible agglomerates [5,24,25].

Cipollina et al. [4] proposed a desalination process concept for exhausted brine from saltworks in Trapani, Italy, which included the precipitation of magnesium in the form of $Mg(OH)_2$. The authors then continued their research using a semi-batch crystallizer and a continuous crystallizer [26]. However, purity of produced $Mg(OH)_2$ crystals is significantly impacted by co-precipitation. The precipitation of $Mg(OH)_2$ may be accompanied by the precipitation of calcium in the form of $Ca(OH)_2$, occurring at a higher pH within the same process. Vassallo et al. [6] tested a pilot plant for the crystallization of magnesium and calcium. But in reactive crystallization, the direct contact between the processed brine and the precipitating agent can cause unintended issues in the precipitation process, such as co-precipitation, of calcium in the form of $Ca(OH)_2$, occurring at a higher pH within the same process which affects the product's quality [27].

Membrane technologies, such as electrodialysis using ion-exchange membranes, are also employed for the valorization of raw materials from waste brines. This led to the idea of combining membrane technology with the precipitation crystallization process for magnesium recovery, which has been patented by La Corte et al. [27] as an ion-exchange membrane crystallizer. This ion-exchange membrane crystallizer consisted of a flat-sheet anion-exchange membrane that separated the brine from the alkaline reactant which flowed through the serpentine channel on opposite sides of the membrane. The basic principle of the process operation is depicted in Figure 1. The anion-exchange membrane selectively allowed only anions to pass through, causing hydroxide anions to migrate to the side of the brine rich in magnesium cations. Supersaturation of the solution with OH^- ions was achieved rapidly, and at a pH of approximately 9.8 [6,28] an immediate reaction occurred between the hydroxide ions according to the following reaction:

$$Mg^{2+}(aq)+2OH^-(aq) \rightarrow Mg(OH)_2(s) \quad (1)$$

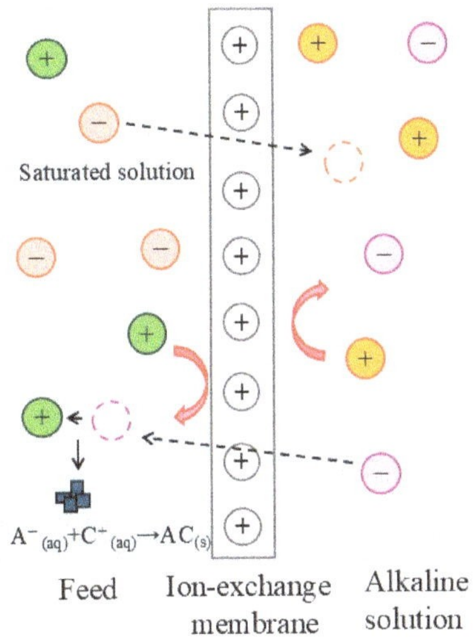

Figure 1. Reactive membrane crystallization using ion-exchange membrane.

At the same time, chloride anions from the brine migrated to the alkaline reactant side to maintain electroneutrality. In general, it is assumed that ions with the same charge as the membrane were repelled based on the Donnan exclusion theory [29,30]. Ion transport across the membrane occurs not only due to the concentration gradient within it, but also due to the electrochemical potential gradient, which arises from the maintenance of Donnan equilibrium at the membrane-solution interface [28].

One of the significant advantages of membrane crystallization is that there is no direct contact between the alkaline solution and the brine, thus preventing the co-precipitation of other substances that could contaminate the product and affect its final quality. The second advantage is decreasing brine conductivity and salinity due to the crystallization of magnesium ions. In contrast, the direct addition of a precipitating agent to the brine also introduces sodium ions. Furthermore, the integration of membrane crystallization with electrodialysis using bipolar membranes could enhance the sustainability of the process by enabling the regeneration of the precipitating agent while simultaneously producing hydrochloric acid, which is also required for the proposed membrane crystallization process. Although studies report achieving high magnesium recovery efficiency, the issue of membrane fouling is only marginally mentioned. Upon reaching supersaturation of the solution with OH- ions, most magnesium cations react with hydroxide anions near the membrane surface, where the supersaturation is highest. Consequently, most of the product may remain trapped on the membrane rather than forming in the solution. La Corte et al. [27] achieved with their membrane crystallizer high magnesium recovery efficiency and high product purity, while also reporting that a thin layer of magnesium hydroxide was attached to the membrane. They concluded that the formed layer did not affect the subsequent experiments, which were performed without previous cleaning of the membrane. Vassallo et al. [28] also mentioned the issue of membrane fouling but mitigated it by positioning the membrane module horizontally and adding cleaning balls, which were circulating in the brine channel, whose purpose was to hinder the formation of scaling during the experimental run.

Membrane fouling has been extensively investigated, primarily in the context of membrane crystallization using membrane distillation, whereas reactive membrane crystallization remains largely unexplored in this regard. Fouling mitigation strategies employed in membrane crystallization with membrane distillation include pre-treatment, membrane bubbling, flushing, backflow application, chemical cleaning, and membrane modification [31]. The application of pre-treatment methods depends on the brine composition and works on the principle of reducing the concentration of species that cause fouling. These methods include processes such as low-pressure membrane filtration, coagulation and flocculation, adsorption, pH adjustment, and the addition of anti-scalants. Gas bubbling can mitigate external fouling by increasing the shear rate and enhancing fluid dynamics on the membrane surface [32]. Membrane flushing commonly uses deionized water to remove solutes from the membrane surface [31]. The application of backflow reverses the driving force of the process, thereby reducing surface interactions between the membrane and scalants [33]. Chemical cleaning is commonly employed to disrupt foulant–membrane interactions [32]. Membrane distillation crystallization relies on strongly hydrophobic membranes that inherently resist fouling; thus, surface modification offers an effective strategy to further minimize scaling. An interesting approach is the development of self-cleaning membranes capable of sustaining stable separation for multiple cycles [34]. Since reactive membrane crystallization with ion-exchange membranes operates differently from membrane distillation, fouling mitigation should primarily rely on removing scale deposits from the external surface, as ions of the same charge are repelled by the membrane and thus the formation of crystals inside the membrane is unlikely. An alternative approach could be to prevent the formation of scales in the first place, through improved control of nucleation, which could be significantly altered by changing the volume-to-surface ratio, as shown by Odua et al. [35] for membrane distillation crystallization.

Previous studies have confirmed that magnesium can be separated in the form of magnesium hydroxide crystals with high purity and high recovery efficiency, or with high conversion rates, through reactive membrane crystallization. They also reported that crystals form on the other surface of the membrane but do not significantly impede the anion-exchange process. However, these studies do not report the percentage of crystals present in the suspension and therefore do not provide insight into the extent of membrane fouling that may occur during the process. In this work, an experimental hollow fiber anion exchange membrane is applied within the reactive membrane crystallization setup. The main objective is not only to recover magnesium hydroxide from model solution but also to investigate in situ cleaning methods for removing crystals from the membrane surface, thereby increasing their yield in suspension. These methods incorporate chemical cleaning with hydrochloric acid solution and demineralized water, which was delivered directly inside the hollow fiber in a backflow configuration, meaning that the protons needed to diffuse from the internal membrane surface through the anion-exchange membrane itself to the external membrane surface, where partial dissolution of deposited $Mg(OH)_2$ occurs.

2. Materials and Methods

2.1. Materials

In this work, the solution of magnesium chloride was used as a model solution of brine. Magnesium chloride hexahydrate $MgCl_2{\cdot}6H_2O$ (technical grade, purity > 98%), NaOH pellets (technical grade) and 6 M solution of HCl were all supplied by CentralChem, Slovakia. The experimental apparatus was built from Eppendorf (Eppendorf

SE, Hamburg, Germany) BioFlo® 320 control unit, multimeter Thermo Scientific (Waltham, MA, USA) PC Eutech 450 for pH measurement, laboratory scale KERN PCB 3000-2, IKA MICROSTAR 15 control equipped with Rushton turbine for solution mixing.

Experimental, commercially unavailable anion-exchange membranes in the form of hollow fibers, developed by MemBrain s.r.o. (Stráž pod Ralskem, Czech Republic), were used in this study. Their fabrication is similar to that of flat-sheet RALEX® membranes: the ion-exchange resin is ground, mixed with polyethylene, homogenized, and melted to prepare a granulate consisting of a uniform mixture of both components. This granulate is then fed to the extrusion head in a melt-spinning process, producing hollow fiber membranes generally regarded as dense. A scanning electron microscope (SEM) image of the cross-section of the experimental anion-exchange hollow fiber membrane is shown in Figure 2.

Figure 2. SEM image showing the cross-section of the ion-exchange hollow fiber manufactured by MemBrain s.r.o.

These membranes present a promising option for large-scale commercial implementation, as their cost per square meter is only half that of flat-sheet membranes produced by MemBrain s.r.o. The membrane module consists of a single hollow fiber membrane with a length of 1.5 m. The properties of the membrane are shown in Table 2.

Table 2. Anion-exchange membrane properties and module characteristics.

Item	Value	Unit
Outer diameter	2	mm
Inner diameter	1.5	mm
Membrane thickness	0.25	mm
Total length	150	cm
Ion-exchange capacity	2	meq/g
Active surface of the dry module	94.25	cm^2
Active surface of the swollen module	100.85	cm^2

All other components of the membrane module were fabricated from polypropylene using a BambuLab 3D printer. The membrane module used in presented experiments is shown in Figure 3 and it consisted of a single hollow fiber wound around custom 3D-printed supports.

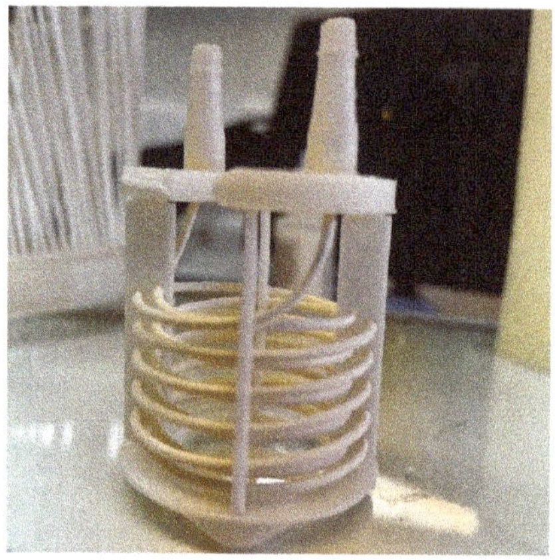

Figure 3. A single hollow fiber anion-exchange membrane module.

2.2. Setup Description

Before use, the membrane module was soaked in water for one hour to allow the membrane to swell, followed by activation. Activation of the membrane module consisted of submerged into the 0.5 mol/L HCl for one hour.

Magnesium hydroxide membrane crystallization was studied using the apparatus depicted in the simplified scheme in Figure 4. The model brine solution located in reactor (No. 2) in which the membrane module was submerged (No. 2a) was continuously stirred throughout the process by overhead stirrer (No. 4). This improved the transfer of OH^- ions across the laminar layer on the brine side, leading to the reaction between OH^- ions and Mg^{2+} ions according to Equation (1).

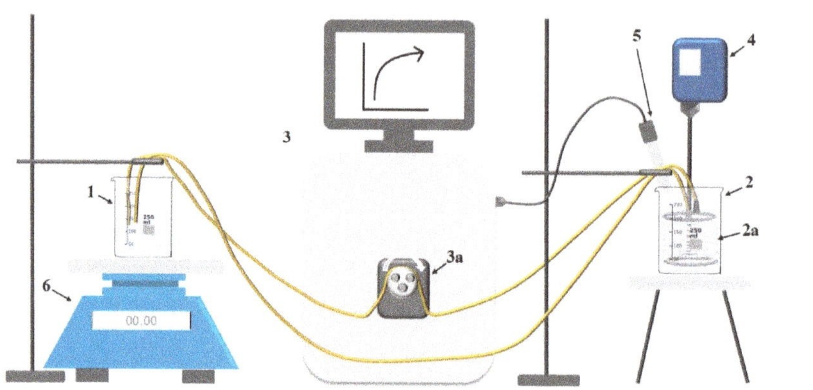

Figure 4. Sketch of the experimental apparatus, 1—reservoir with alkaline solution, 2—reactor with model brine solution, 2a—hollow fiber membrane module, 3—Eppendorf BioFlo® 320 control unit, 3a—peristaltic pump, 4—mixer, 5—pH probe, 6—scales.

The pH of the model brine solution in the reactor was continuously monitored during the process by pH probe (No. 4) as the crystallization of $Mg(OH)_2$ is highly sensitive to pH, occurring within the pH range of 9.8 to approximately 10.4 [28]. After reaching the pH value of 10.4, it was assumed that complete conversion of Mg^{2+} ions occurred [36].

On the other hand, if pH values were higher than 10.4, it could lead to co-precipitation of other components found in real brine, e.g., co-precipitation of Ca^{2+} in the form of

$Ca(OH)_2$. Co-precipitation of other components or ion sorption into the crystal structure could degrade the quality of the product [28].

The weight of the alkaline solution (No. 1) was periodically recorded using scales, No. 6, to verify the functionality of the module and ensure that no damage occurred, which would result in direct contact between the model brine solution and the alkaline solution. If the module was damaged, allowing direct contact between the two solutions, this would be reflected in a significant rise in pH values. Otherwise, the weight of the reservoir changed minimally, within a range of 5 g. These small changes could be attributed either to evaporation of the alkaline solution or the formation of bubbles in the silicone tubes.

2.3. Experimental Procedure

All experiments were conducted with a model solution, consisting only of dissolved $MgCl_2$ and representing the brine, which was placed in the reactor containing the hollow fiber anion-exchange membrane module. The alkaline NaOH solution, serving as the precipitating agent, was held in the reservoir. The 125% excess of OH- ions was selected based on our previous studies, in which a high excess of OH- ions ensured a more stable driving force for mass transfer in batch configuration. Experimental conditions are presented in Table 3.

Table 3. Established operational parameters for membrane crystallization.

Item	Value	Unit
Concentration of $MgCl_2$	9.52	g/L
Concentration of NaOH	20	G/L
Volumetric flow rate	14	mL/min
Agitator speed in the reactor	600	rpm
Agitator speed in the reservoir	100	rpm
Duration of the experiments	300	min

The alkaline solution was pumped from the reservoir into the membrane module through silicone tubing using a peristaltic pump at a flow rate of 14 mL/min. The outgoing precipitating agent was recirculated back to the reservoir. Chosen initial concentrations of $MgCl_2$ were selected based on the published work by Cipollina [26]. The reactor and the reservoir were mechanically stirred at 600 rpm and 100 rpm, respectively. The pH values and flow rates were recorded and controlled using an Eppendorf BioFlo® 320 control unit. Additionally, the mass inside the reservoir was monitored. The duration of each experiment was 5 h.

The regeneration process of the membrane module after each experiment involved submerging the membrane module in 125 mL 0.5 mol/L HCl and circulating it through the system for one hour, followed by multiple rinsing procedures with 125 mL demineralized water. The demineralized water in the rector was exchanged until the conductivity of demineralized water dropped below 1 mS/cm.

2.4. Methodology

The entire process was divided into three stages. The first stage consisted of the membrane crystallization of $Mg(OH)_2$. The process parameters for the first stage remained unchanged throughout all experiments and the parameters are presented in Table 2. The following two stages were introduced in order to study their application for removal of crystals from the membrane surface. The second stage involved the circulation of HCl through the membrane module and the third stage involved the circulation of demineralized water through the membrane module. Variations in the process parameters were made in the second and third stage in order to examine the effect of these process parameters on the crystal yield in the suspension. Throughout all three stages the suspension and the membrane module remained in the reactor. Process parameter changes were made only in one stage in individual experiments in order to evaluate their effects.

Circulation of the concentrated HCl through the hollow fiber causes the repulsive forces of the anion-exchange groups in the membrane to be insufficient to repel cations, which results in the membrane starting to lose its selectivity, therefore allowing these co-ions to pass through the membrane [30]. The introduction of HCl was intended to remove a thin layer of the crystals closest to the membrane surface, according to Equation (2), allowing the upper layer of crystals to be released into the suspension. This would increase the crystal yield in the suspension and simultaneously generate potential crystal nuclei in the suspension that could continue to grow.

$$Mg(OH)_2(s)+2HCl(aq) \rightarrow MgCl_2(aq)+2H_2O(l) \quad (2)$$

The third stage involved the circulation of demineralized water in the membrane module, which aimed to dislodge crystals in the vicinity of the membrane surface into the suspension. Demineralized water was pumped through the fiber placed in the reactor and due to osmosis and hydrodynamic pressure inside the fiber water was transported through the membrane to the reactor.

2.4.1. Modification in the First Stage

As mentioned above, the process parameters of the first stage which were presented in Table 2 were not altered. However, in this set of experiments the effect of 2 mol/L HCl circulation on the crystal yield in the suspension of HCl was investigated if the acid was introduced for a short period of time during the first stage. The introduction of HCl during the first stage occurred at different times from the beginning of the experiment—after 1 h, after 2.5 h, or after 4 h, see Figure 5.

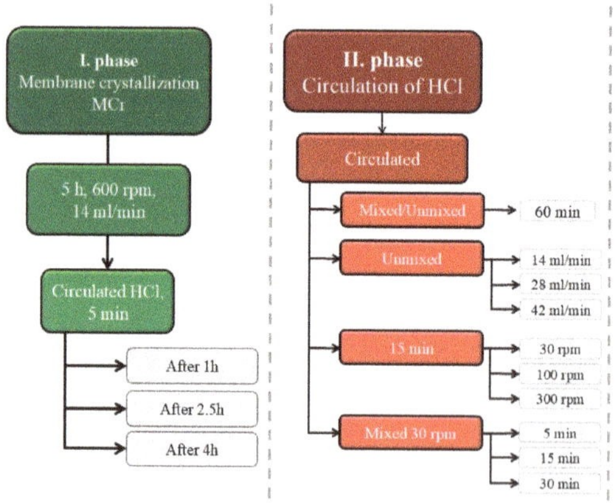

Figure 5. Overview of the experiments performed with the modifications examined at each stage.

2.4.2. Modifications in the Second Stage

In the second stage, four sets of experiments were conducted, see Figure 5. In the first set of experiments, two extremes were examined: stirring the reactor at 600 rpm or not stirring the reactor at all, both for a duration of 60 min. In the second set of experiments, the effect of varying the volumetric flow rate of HCl on the crystal yield in the suspension was investigated. In this set, the suspension was not stirred, and HCl was circulated in the system for 15 min, with the volumetric flow rate of HCl increasing from 14 mL/min to 28 mL/min and then to 42 mL/min. In the third set of experiments, HCl was circulated in the system for 15 min, and the effect of changing the suspension stirring speed inside the reactor on the crystal yield was examined. The stirring speed varied from 30 rpm to 100 rpm, and then to 350 rpm. In the final set of experiments, the suspension was stirred at the lowest possible stirring frequency, 30 rpm, and the effect of varying the circulation time in the system was investigated, with HCl circulating for either 5 min, 15 min, or 30 min. At these low mixing speeds, significant local concentration gradients could have occurred, which might have affected the results after the second stage. The actual outcome of the second stage will only be confirmed in the third stage, which is why the third stage is also displayed in the final graphs.

2.4.3. Modifications in the Third Stage

In the third stage, the effects of the circulation time of demineralized water in the membrane module and the intensity of suspension stirring on the crystal yield in the suspension were investigated, see Figure 5. In the first set of experiments, demineralized water circulated in the system for 20 min, with three tests conducted at different stirring speeds of 600 rpm, 800 rpm, and 1000 rpm. In the second set of experiments, the suspension was stirred at 600 rpm, while demineralized water circulated in the system for either 5 min, 10 min, or 20 min.

2.5. Data Collection and Sampling Procedure

To determine the efficiency of membrane crystallization, it was necessary not only to collect samples but also to gather and record data. The collected data included pH values, which were automatically recorded by the Eppendorf BioFlo® 320 control unit and additionally logged at each sampling point. During the first half of the first stage, samples were collected more frequently—every 30 min. In the second half of the first stage, samples were taken every hour. During the second and third stages, samples were again collected more frequently, every 5 min.

The samples were diluted to fall within the calibration range. Two types of samples were taken every hour. Type I was used for determination of the free dissociated Mg^{2+} ions. This sample was first filtered through a 0.45 µm syringe filter and then diluted with demineralized water. Type II was used for the determination of all Mg^{2+} ions present in the suspension. This sample was diluted with 0.5 mol/L HCl, whose role was to dissolve any formed crystals of $Mg(OH)_2$ and thus convert the magnesium back to the dissociated form.

From the analysis of these two sample types, we were able to determine the concentration of crystals in the suspension according to Equation (3).

$$C_{Mg(OH)_2,suspension} = C_{Mg^{2+},total} - C_{Mg^{2+},(aq)} \tag{3}$$

$C_{Mg(OH)_2,suspension}$ represents the concentration of magnesium hydroxide in the suspension, $C_{Mg^{2+},total}$ is the total concentration of Mg^{2+} ions present in the suspension at a given sample time (sample type II), and $C_{Mg^{2+},(aq)}$ refers to the concentration of free Mg^{2+} ions at a given sample time (sample type I).

The concentration of all $Mg(OH)_2$ crystals (in the suspension and on the membrane) $C_{Mg(OH)_2,total}$ was calculated as the difference between the initial concentration of Mg^{2+} ions in the model solution $C^0_{Mg^{2+}}$ and the concentration of free Mg^{2+} ions in the suspension (sample type I) according to Equation (4).

$$C_{Mg(OH)_2,total} = C^0_{Mg^{2+}} - C_{Mg^{2+},(aq)} \tag{4}$$

The recovery of $Mg(OH)_2$ crystals in the suspension $Y\left(Mg(OH)_{2,suspension}\right)$ was calculated as the difference between the concentration of all Mg^{2+} ions present in the suspension (sample type II) and the concentration of free Mg^{2+} ions (sample type I), divided by the initial concentration of Mg^{2+} ions in the model brine solution as defined by Equation (5). Similarly, the percentage of $Mg(OH)_2$ crystals on the membrane $Y(Mg(OH)_{2,membrane})$ can be determined with the known concentration of $Mg(OH)_2$ crystals in the suspension using Equation (6). Overall conversion to $Mg(OH)_2$, referred to as $Y(Mg(OH)_2)$, was calculated according to Equation (7), using the initial concentration and the concentration of free Mg^{2+} ions (sample type I).

$$Y\left(Mg(OH)_{2,suspension}\right) = \frac{c_{Mg^{2+},total} - c_{Mg^{2+},(aq)}}{c^0_{Mg^{2+}}} \times 100\% \quad (5)$$

$$Y\left(Mg(OH)_{2,membrane}\right) = \frac{c_{Mg(OH)_2,total} - c_{Mg(OH)_2,suspension}}{c^0_{Mg^{2+}}} \times 100\% \quad (6)$$

$$Y(Mg(OH)_2) = \frac{c^0_{Mg^{2+}} - c_{Mg^{2+},(aq)}}{c^0_{Mg^{2+}}} \times 100\% \quad (7)$$

Aqueous samples containing dissociated magnesium were analyzed using an Agilent Infinity 1260 system (Agilent Technologies, Santa Clara, CA, USA) equipped with a cation-exchange column Hamilton PRP-X-800, 4.1 mm × 250 mm, 7 µm, and each sample was analyzed in triplicate. HPLC was employed for indirect absorbance detection at 220 nm with a reference wavelength of 360 nm. The mobile phase consisted of a 1.5 mmol/L $CuSO_4$ aqueous solution with 0.3 mmol/L H_2SO_4, and volumetric flow of mobile phase was 1 mL/min.

3. Results and Discussion

In each experiment, more than 85% magnesium conversion was achieved during the first stage. On average, after the crystallization stage, only 1% of $Mg(OH)_2$ crystals were found in suspension, while the remainder were present on the membrane surface. The results shown below represent the percentages of dissociated magnesium cations, crystals in suspension, and crystals on the membrane, relative to the initial amount of Mg^{2+} ions calculated according to Equations (5) and (6), and are presented as bar charts. To facilitate interpretation, mostly the results from the third (final) stage of one experimental set are compared.

3.1. Evaluation of Single Experiment

The crystallization of magnesium hydroxide depends on the supersaturation of OH^- ions in the reactor, which diffuse through the membrane from the precipitating agent. Once supersaturation is achieved crystals begin to form spontaneously. The concentrations of dissociated Mg^{2+} ions and crystals in the suspension were determined from the samples. Analysis of the samples revealed that the concentration of dissociated magnesium cations decreased during the MCr process, see Figure 6a, which corresponded to the formation of crystals in the solution and on the membrane, see Figure 6b. After the first stage, the concentration of dissociated Mg^{2+} ions was approximately 10 mmol/L, indicating that the remaining Mg^{2+} ions had reacted to form magnesium hydroxide. Figure 6a illustrates the trend in the consumption of dissociated magnesium cations and the simultaneous production of crystals in the solution. The difference between these trends is minimal, already showing that only a small portion of crystals were present in the suspension. Figure 6b directly plots the concentration of crystals in the suspension and it can be seen that their concentration was not excessively high, and most of the crystals remained retained on the membrane, where the supersaturation of OH^- ions in the solution was the highest. Based on these results we can estimate the flux of anions through the membrane assuming that the precipitation is instantaneous and the production of $Mg(OH)_2$ is limited by mass transfer of OH^- ions through the anion-exchange membrane. The average flux of ions through the membrane was 5.58×10^{-5} mol/m^2/s, which is an order of magnitude lower than the flux reported by Vassallo et al. [28], who used flat sheet membranes in their experiments that were thinner than the hollow fiber membranes used in our work.

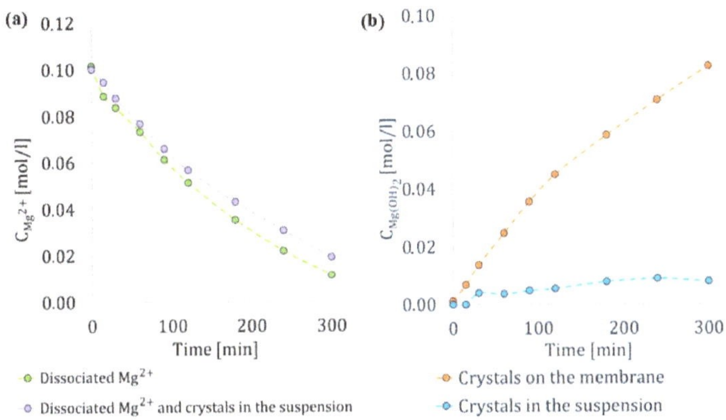

Figure 6. Concentration profiles during the experiment: (**a**) change of Mg^{2+} ions concentration in the reactor through the MCr; (**b**) concentration of $Mg(OH)_2$ in the reactor through all three stages of the experiment.

The production of $Mg(OH)_2$ is also evident from the pH measurements in a typical membrane crystallization experiment conducted using the setup shown in Figure 7, as indicated by the blue line. The diffusion of OH^- ions through the anion-exchange membrane is indicated by an initial sharp increase in pH. This is followed by a gradual decrease as supersaturation is reduced due to $Mg(OH)_2$ formation. Subsequently, the pH rises again as Mg^{2+} ions are progressively consumed, requiring slightly higher OH^- concentrations to sustain further precipitation. To maintain crystallization, the OH^- ion concentration (and thus pH) must continue to rise in order to meet the solubility product requirement.

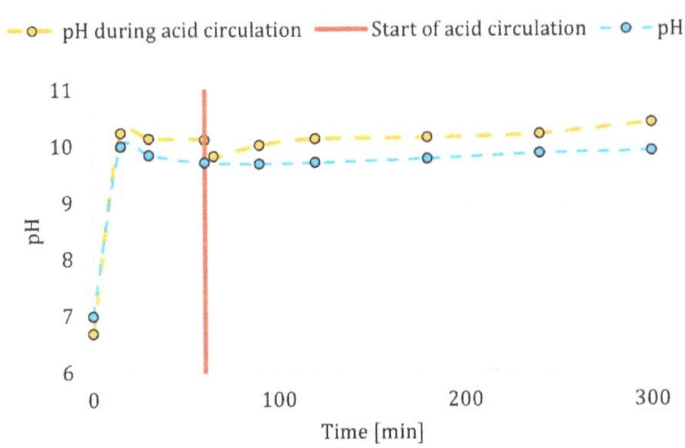

Figure 7. pH profiles in the model experiment and during the first stage with the HCl introduction to the fiber after one hour (red vertical line).

3.2. Circulation of HCl in the 1st Stage

The circulation of HCl in the system is described in more detail in the Section 2.4.1. During the introduction of HCl in the first half of the first stage, that is, after the first hour and after 2.5 h from the start of the experiment, the difference between the percentages of crystals in suspension was not as significant as the percentage difference after the introduction of HCl after the fourth hour of the experiment. It can be concluded that the circulation of HCl in the first half of the first stage had almost the same effect on the percentage of crystals in suspension because a sufficiently thick layer of crystals was not formed on the membrane to be disrupted by HCl and thus capable of detaching. Instead, a small amount of crystals on the membrane surface dissolved and reduced the supersaturation of the solution with OH^- ions in the reactor, which was reflected in the pH values, see orange line in Figure 7. Effectively removing part of the crystals from the membrane surface (Figure 8) would reduce the resistance to OH^- ions transfer, which could also be responsible for the overall higher conversions.

Figure 8. Photo of the experiment during the third stage showing large scales detached from the membrane.

The greatest impact of HCl circulation in the first stage was observed when HCl was circulated in the system after the fourth hour, when the crystal yield was 9.29%. In this case, a sufficiently thick layer of crystals was formed on the membrane, and during the circulation of HCl, the lower layer of crystals on the membrane surface dissolved while the upper layer was released into the suspension. After four hours, a thicker layer of crystals formed on the membrane. Consequently, when HCl was circulated after four hours, a larger portion of the crystals, 9.29% detached into the suspension compared to circulation after one hour, 3.05%. Since a greater portion of the crystals was removed into the suspension, the resistance to the transfer of H^+ ions during HCl supply in the second stage decreased. This led to the dissolution of the crystals, thereby reducing their yield after the third stage. On the other hand, if crystal nuclei had formed on the membrane within the first hour of the experiment, the introduction of HCl after one hour led to their dissolution. However, after completing all three stages of experiments, the final yield of magnesium hydroxide crystals in suspension after the third stage was nearly identical across all three experiments in this set, see Figure 9.

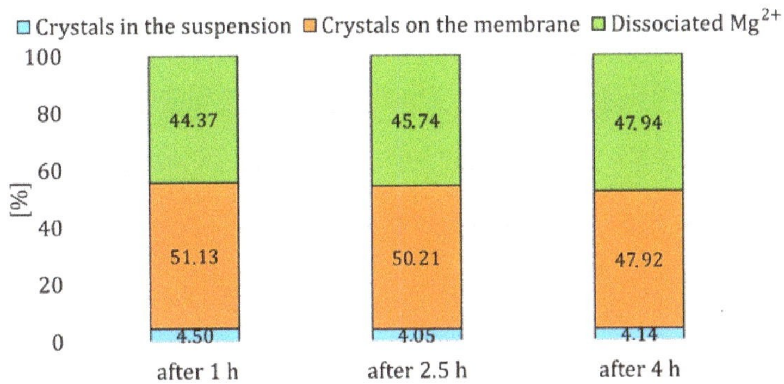

Figure 9. Percentage of Mg^{2+} ions in different forms at the end of the experiments. 1st column—introducing of HCl after the 1st hour; 2nd column—introducing of HCl after the 2.5; 3rd column—introducing of HCl after the 4th hours.

3.3. *Evaluation of Modifications Applied in the 2nd Stage*

The results are described in four chapters, with the graphs showing only the outcomes after the completion of the process. The graphs illustrate the percentage of reacted or dissociated magnesium cations, respectively.

3.3.1. Impact of Mixing Suspension in the Reactor

High stirring speed of 600 rpm improved the mixing of the suspension within the reactor, potentially causing crystals to detach into the suspension. Additionally, vigorous stirring reduced the laminar boundary layer near the membrane surface on the suspension side, enhancing the transfer of H^+ ions into the suspension. This increased ion transfer lowered the solution's supersaturation, leading to the dissolution of crystals. In contrast, when the suspension was not stirred, approximately 12% of the crystals were found in the suspension, Figure 10a. In this set of experiments, a higher yield of crystals was achieved when the suspension was not mixed. Moreover, when comparing the values of dissociated cations and crystals on the membrane, the ratio of these values is almost the same. Since the goal of the process is to dissolve only the bottom layer of crystals and release them into the suspension, it is evident that intensive mixing in the reactor is unsuitable. This is because it facilitates the transfer of protons, which subsequently neutralize all available products, including the portion that has already entered the suspension.

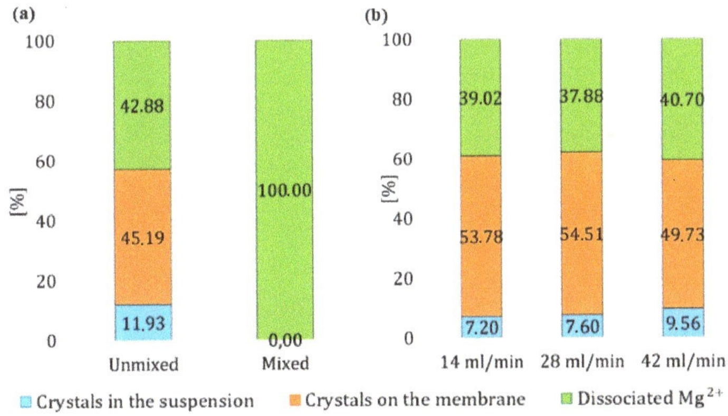

Figure 10. The percentage of reacted and dissociated Mg^{2+} ions after the third experimental stage: (**a**) impact of suspension mixing, (**b**) varying flow rate of acid.

3.3.2. Change in the Acid Flow Rate Within the Fiber

Based on previous results, the suspension has not been stirred in this set of experiments and investigated the effect of the volumetric flow rate of the acid within the fibers on the yield of crystals in the suspension. The fiber utilized is relatively lengthy, and the driving force for proton transport between the module's inlet and outlet can be considerable. This variation in driving force can notably influence the efficiency of proton transfer through the membrane, thereby affecting the overall effectiveness of crystal detachment. An increase in the flow rate of HCl leads to a faster delivery of H^+ ions. Thus, a higher volumetric flow rate of the acid increases the potential for disrupting the lower layer of crystals and promoting their successful detachment from the fibers of the module.

At the end of the third stage, in the first two experiments where the volumetric flow rate was 14 mL/min and 28 mL/min, the crystal yields were almost identical, at 7.20% and 7.60%, respectively, Figure 10b. Ultimately, the yield of crystals in suspension increased with an increase in acid flow rate. The highest yield was achieved 9.56%, if the volumetric flow was 42 mL/min. Moreover, if the percentage values of dissociated magnesium cations are compared, the differences between the values are not as pronounced as in other sets of experiments.

3.3.3. Variation in the Stirring Rate of the Suspension

On the basis of previous experiments, the effect of the intensity of the stirring speed of the suspension was investigated. Although intensive stirring of the suspension negatively affects the proportion of crystals present in the suspension, it also facilitates the transfer of H^+ protons and the detachment of crystals from the membrane surface into the suspension. For this reason, the effect of stirring was studied at three different speeds—30 rpm, 100 rpm, and 350 rpm. Differences only became apparent in the third stage, when the suspension was mixed more intensively, with the final result being shown in Figure 11a. If the suspension was mixed at 30 rpm during the second stage, the highest yield of crystals in suspension was achieved after the third stage in this set of experiments at 9.53%. At this stirring speed in the second stage, the crystals were most probably not detached into the suspension. Instead, only the bottom layer of crystals near the vicinity of the membrane were dissolved, and when mixing was intensified in the third stage, the top layer of crystals was released into the suspension, with only a small percentage of crystals dissolving. In the experiments where the suspension was stirred at higher speeds, the percentage of dissociated magnesium cations in suspension increased to 33.43%. Compared 33.43% and 27.13% of dissociated cations from the experiments with suspension mixing at 350 rpm and the experiment with suspension mixing at 30 rpm, representing an increase in dissociated magnesium cations of more than 6%, which indicates that dissolution of crystals in the suspension occurs with increasing stirring speed of the suspension.

The results from this set of experiments confirmed that more intensive stirring in the second stage facilitates H^+ transport through the membrane leading to dissolution of crystallized magnesium hydroxide throughout the whole solution.

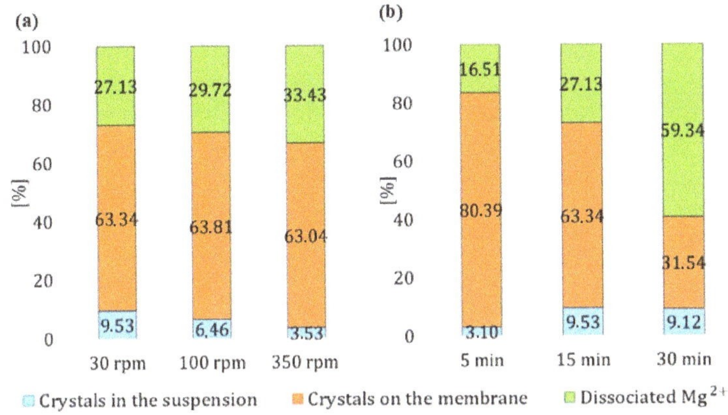

Figure 11. The percentage of reacted and dissociated Mg^{2+} ions in the third experimental stage with varying (**a**) HCl volume flow, (**b**) time duration of HCl circulation.

3.3.4. Variable Time Duration of HCl Circulation

The results from experiments with varying time duration of the second stage show that circulating HCl for 5 min was not sufficient to detach an adequate amount of crystals into the suspension, when only 3.10% of crystals were in the, see Figure 11b. suspension. Conversely, circulating HCl for 30 min increased the portion of crystals in suspension to 9.12%, but this came at the expense of dissociated magnesium cations, whose percentage in the suspension increased more than fivefold to 59.34% compared to 16.51% in the experiment with 5 min second stage. The results suggest that if crystal detachment into the suspension occurred, the acid feeding duration was excessively long, leading to a dissolution of the crystals in the suspension. As a result, crystal dissolution occurred not only on the membrane surface but also within the suspension. If HCl was circulated for 15 min, the highest yield of crystals in suspension in this set was achieved at 9.53%, while the percentage of dissociated cations increased to 27.13% which is 8% higher than in the experiment with a 5 min HCl circulation (16.51%). From this, it can be concluded that crystals were primarily detached from the membrane into the suspension. From this set of experiments, it would be more advantageous to conduct the second stage for a duration of 15 min, because the suspension after the experiment contained twice as few dissociated magnesium cations compared to the experiment with a 30 min second stage and three times more crystals in the suspension compared to the experiment with a 5 min second stage.

3.4. *Evaluation of Modifications Applied in the 3rd Stage*

Conditions for the second stage were determined based on previous sets of experiments. The most suitable conditions were the least intense stirring speeds of the suspension at 30 rpm for 15 min.

3.4.1. Impact of Stirring Speeds of the Reactor

At stirring speeds of 600 rpm, 9.53% of crystals were present in the suspension, at 800 rpm, their portion decreased to 8.59%, and at 1000 rpm, the percentage of crystals in the suspension dropped to 6.69%, as shown in Figure 12a. The portion of the crystals on the membrane remain relatively stable across all stirring speeds tested, ranging between 63.34% and 64.26%. High mixing speeds could not only cause the detachment of crystals in the suspension but also their fragmentation into smaller crystals. Additionally, the water introduced into the module could lead to the additional expulsion of a small amount of H^+

ions present in the membrane pores from previous stage, which then caused the dissolution of the crystals, as indicated by the concentration of dissociated magnesium cations, which slightly increased with stirring speed—from 27.13% to 29.04%. The results of this set show that intensified mixing of the suspension during the third stage caused a decrease in the yield of crystals in the suspension.

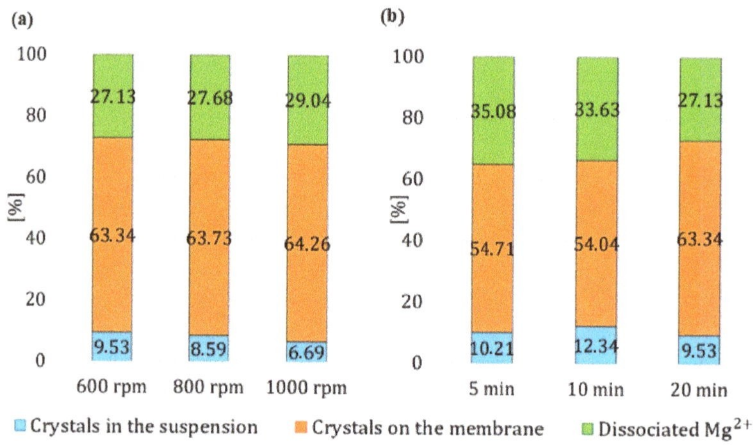

Figure 12. The percentage of reacted and dissociated Mg^{2+} ions in the third experimental stage with varying (**a**) stirring speed, (**b**) time duration of the third stage.

3.4.2. Effect of Time Duration of Circulation During the Third Stage

The data reveals that the proportion of crystals in the suspension increases slightly from 10.21% at 5 min to 12.34% at 10 min, then decreases to 9.53% at 20 min. This trend suggests that moderate circulation time promotes the detachment of crystals into the suspension. The portion of crystals on the membrane (orange segments in Figure 12b remains relatively stable between 5 and 10 min, approximately 54%, but increases significantly to 63.34% after 20 min of water circulation. Simultaneously, the percentage of dissociated Mg^{2+} ions decreases from 35.08% at 5 min to 27.13% at 20 min.

4. Conclusions

The presented study investigated methods to enhance the efficiency of magnesium hydroxide reactive membrane crystallization using a hollow fiber anion-exchange membrane, with the aim of increasing the yield of crystals in the suspension, thereby minimizing membrane fouling and improving product recovery. The membrane crystallization process demonstrated potential for magnesium recovery, achieving a conversion consistently over 85%. However, the majority of produced crystals remained attached on the membrane surface. At the end of the membrane crystallization stage, the suspension contained, on average, only 1% magnesium hydroxide crystals, indicating significant product scaling. To address this, the experiments were designed as batch processes, each consisting of three consecutive stages. The first stage involved 5 h of membrane crystallization under controlled conditions. This was followed by a second stage in which hydrochloric acid was circulated through the membrane to dissolve part of the adhered crystal layer and facilitate detachment. Finally, a third stage involved the circulation of demineralized water to assess further removal and redistribution of the crystals into the suspension. Throughout all three stages, operational parameters such as acid circulation duration and suspension mixing intensity were varied to evaluate their influence on the yield of suspended crystals. This approach allowed for a systematic investigation into in situ methods for scale removal and product recovery, offering practical insight into optimizing membrane crystallization for applications in resource recovery and zero-liquid discharge systems.

The results demonstrated that the timing, duration and intensity of process modifications in each stage significantly influenced the distribution of magnesium species between the membrane, suspension, and its dissociated forms. Introducing HCl in the later stages of the first stage resulted in increased detachment of crystals from the membrane without excessively increasing the concentration of dissociated magnesium ions. In the second stage, moderate conditions for acid circulation of 15 min and low stirring intensity resulted in the highest crystal yield in suspension (9.53%) while minimizing dissolution. Excessively long acid exposure or vigorous mixing promoted undesired dissolution of both membrane-bound and suspended crystals, resulting in more than a twofold increase in the concentration of dissociated magnesium. Similarly, in the third stage, a moderate water circulation time of 10 min at a controlled stirring intensity of 600 rpm achieved the highest overall magnesium hydroxide yield in the suspension: 12.34%. However, a compromise is desired between maximizing suspended crystal recovery and minimizing dissociated magnesium concentration in the solution.

The study confirms that targeted manipulation of process parameters can lead to greater product recovery in the bulk stage. While the demonstrated methods resulted in an approximate 10% increase in the proportion of suspended crystals, it is important to note that the experiments were conducted in batch mode at small scale, using magnesium concentrations typical of desalination brines. As a result, the absolute yield of product was relatively low. However, the findings validate the potential of in situ scale removal strategies and provide a foundation for optimizing reactive membrane crystallization systems—not only for magnesium hydroxide but potentially for other valuable salts in high-salinity streams.

Future work will focus on scaling up the process to continuous operation, using multi-fiber membrane modules. The flux per unit membrane area will be the same. The challenge will be to ensure that all fibers in the module are effectively utilized. In this envisioned setup, a dedicated membrane crystallization stage would be followed by a crystal harvesting stage, enabling sustained operation and higher overall product recovery. Additional strategies, such as integrating gas bubbling into the suspension, will also be explored to further enhance crystal detachment and improve system efficiency at scale.

Author Contributions: Conceptualization, E.K.; Data curation, E.K.; Funding acquisition, I.Č. and J.M.; Investigation, E.K. and L.S.; Methodology, E.K., L.S. and I.Č.; Project administration, J.M.; Resources, I.Č. and J.M.; Software, I.Č.; Validation, E.K. and J.M.; Visualization, E.K.; Writing—original draft, E.K.; Writing—review and editing, I.Č. and J.M. All authors have read and agreed to the published version of the manuscript.

Funding: This work was supported by the Slovak Scientific Agency, Grant No. VEGA 1/0658/24, and the Slovak Research and Development Agency under the contract No. APVV-22-0038.

Institutional Review Board Statement: Not applicable.

Informed Consent Statement: Not applicable.

Data Availability Statement: The data presented in this study are available on request from the corresponding author.

Acknowledgments: The authors gratefully acknowledge MemBrain s.r.o. for providing the hollow fiber anion-exchange membranes and Jakub Fehér for his technical support and assistance during the experimental work and manuscript revision.

Conflicts of Interest: The authors declare no conflicts of interest.

Abbreviations

The following abbreviations are used in this manuscript:

MCr	Membrane crystallization
RO	Reverse osmosis
SEM	Scanning electron microscope

References

1. Roser, M.; Ritchie, H. How Has World Population Growth Changed over Time? Available online: https://ourworldindata.org/population-growth-over-time (accessed on 1 July 2025).
2. Chen, C.; Yang, Y.; Graham, N.J.D.; Li, Z.; Yang, X.; Wang, Z.; Farhat, N.; Vrouwenvelder, J.S.; Hou, L. A Comprehensive Evaluation of the Temporal and Spatial Fouling Characteristics of RO Membranes in a Full-Scale Seawater Desalination Plant. *Water Res.* **2024**, *249*, 120914. [CrossRef] [PubMed]
3. Quist-Jensen, C.A.; Macedonio, F.; Drioli, E. Membrane Crystallization for Salts Recovery from Brine—An Experimental and Theoretical Analysis. *Desalination Water Treat.* **2016**, *57*, 7593–7603. [CrossRef]
4. Cipollina, A.; Misseri, A.; Staiti, G.D.; Galia, A.; Micale, G.; Scialdone, O. Integrated Production of Fresh Water, Sea Salt and Magnesium from Sea Water. *Desalination Water Treat.* **2012**, *49*, 390–403. [CrossRef]
5. Turek, M.; Gnot, W. Precipitation of Magnesium Hydroxide from Brine. *Ind. Eng. Chem. Res.* **1995**, *34*, 244–250. [CrossRef]
6. Vassallo, F.; La Corte, D.; Cancilla, N.; Tamburini, A.; Bevacqua, M.; Cipollina, A.; Micale, G. A Pilot-Plant for the Selective Recovery of Magnesium and Calcium from Waste Brines. *Desalination* **2021**, *517*, 115231. [CrossRef]
7. Van Wyk, S.; Van Der Ham, A.G.J.; Kersten, S.R.A. Potential of Supercritical Water Desalination (SCWD) as Zero Liquid Discharge (ZLD) Technology. *Desalination* **2020**, *495*, 114593. [CrossRef]
8. Arnal, J.M.; Sancho, M.; Iborra, I.; Gozálvez, J.M.; Santafé, A.; Lora, J. Concentration of Brines from RO Desalination Plants by Natural Evaporation. *Desalination* **2005**, *182*, 435–439. [CrossRef]
9. Petersen, K.L.; Paytan, A.; Rahav, E.; Levy, O.; Silverman, J.; Barzel, O.; Potts, D.; Bar-Zeev, E. Impact of Brine and Antiscalants on Reef-Building Corals in the Gulf of Aqaba—Potential Effects from Desalination Plants. *Water Res.* **2018**, *144*, 183–191. [CrossRef] [PubMed]
10. Tsai, J.-H.; Macedonio, F.; Drioli, E.; Giorno, L.; Chou, C.-Y.; Hu, F.-C.; Li, C.-L.; Chuang, C.-J.; Tung, K.-L. Membrane-Based Zero Liquid Discharge: Myth or Reality? *J. Taiwan Inst. Chem. Eng.* **2017**, *80*, 192–202. [CrossRef]
11. Kaplan, R.; Mamrosh, D.; Salih, H.H.; Dastgheib, S.A. Assessment of Desalination Technologies for Treatment of a Highly Saline Brine from a Potential CO2 Storage Site. *Desalination* **2017**, *404*, 87–101. [CrossRef]
12. Ruiz-Aguirre, A.; Andrés-Mañas, J.A.; Fernández-Sevilla, J.M.; Zaragoza, G. Experimental Characterization and Optimization of Multi-Channel Spiral Wound Air Gap Membrane Distillation Modules for Seawater Desalination. *Sep. Purif. Technol.* **2018**, *205*, 212–222. [CrossRef]
13. European Commission. Directorate General for Internal Market, Industry, Entrepreneurship and SMEs. In *Study on the Critical Raw Materials for the EU 2023: Final Report*; Publications Office: Luxembourg, 2023.
14. Morgante, C.; Lopez, J.; Cortina, J.L.; Tamburini, A. New Generation of Commercial Nanofiltration Membranes for Seawater/Brine Mining: Experimental Evaluation and Modelling of Membrane Selectivity for Major and Trace Elements. *Sep. Purif. Technol.* **2024**, *340*, 126758. [CrossRef]
15. Dun, L.; Ouyang, Z.; Sun, Q.; Yue, X.; Wu, G.; Li, B.; Kang, W.; Wang, Y. A Simple and Efficient Magnesium Hydroxide Modification Strategy for Flame-Retardancy Epoxy Resin. *Polymers* **2024**, *16*, 1471. [CrossRef]
16. Sener, A.A.; Demirhan, E. The Investigation of Using Magnesium Hydroxide as a Flame Retardant in the Cable Insulation Material by Cross-Linked Polyethylene. *Mater. Des.* **2008**, *29*, 1376–1379. [CrossRef]
17. Xu, H.; Deng, X. Preparation and Properties of Superfine $Mg(OH)_2$ Flame Retardant. *Trans. Nonferrous Met. Soc. China* **2006**, *16*, 488–492. [CrossRef]
18. Dong, C.; Cairney, J.; Sun, Q.; Maddan, O.L.; He, G.; Deng, Y. Investigation of $Mg(OH)_2$ Nanoparticles as an Antibacterial Agent. *J. Nanoparticle Res.* **2010**, *12*, 2101–2109. [CrossRef]
19. El Bouraie, M.; Masoud, A.A. Adsorption of Phosphate Ions from Aqueous Solution by Modified Bentonite with Magnesium Hydroxide $Mg(OH)_2$. *Appl. Clay Sci.* **2017**, *140*, 157–164. [CrossRef]
20. Li, H.; Liu, S.; Zhao, J.; Feng, N. Removal of Reactive Dyes from Wastewater Assisted with Kaolin Clay by Magnesium Hydroxide Coagulation Process. *Colloids Surf. A Physicochem. Eng. Asp.* **2016**, *494*, 222–227. [CrossRef]
21. Wang, K.; Zhao, J.; Li, H.; Zhang, X.; Shi, H. Removal of Cadmium (II) from Aqueous Solution by Granular Activated Carbon Supported Magnesium Hydroxide. *J. Taiwan Inst. Chem. Eng.* **2016**, *61*, 287–291. [CrossRef]

22. López, J.; Battaglia, G.; Lupo, D.; Fernández De Labastida, M.; Vallès, V.; Luis Cortina, J.; Cipollina, A.; Micale, G. Integration of Ion-Exchange and Crystallisation Processes to Recover Boric Acid and Magnesium Hydroxide from Saltworks Bitterns. *Sep. Purif. Technol.* **2025**, *354*, 129532. [CrossRef]
23. Söhnel, O.; Mareček, J. Precipitation of Magnesium Hydroxide. *Cryst. Res. Technol.* **1978**, *13*, 253–262. [CrossRef]
24. Mullin, J.W.; Murphy, J.D.; Sohnel, O.; Spoors, G. Aging of Precipitated Magnesium Hydroxide. *Ind. Eng. Chem. Res.* **1989**, *28*, 1725–1730. [CrossRef]
25. Alamdari, A.; Rahimpour, M.R.; Esfandiari, N.; Nourafkan, E. Kinetics of Magnesium Hydroxide Precipitation from Sea Bittern. *Chem. Eng. Process. Process Intensif.* **2008**, *47*, 215–221. [CrossRef]
26. Cipollina, A.; Bevacqua, M.; Dolcimascolo, P.; Tamburini, A.; Brucato, A.; Glade, H.; Buether, L.; Micale, G. Reactive Crystallisation Process for Magnesium Recovery from Concentrated Brines. *Desalination Water Treat.* **2015**, *55*, 2377–2388. [CrossRef]
27. La Corte, D.; Vassallo, F.; Cipollina, A.; Turek, M.; Tamburini, A.; Micale, G. A Novel Ionic Exchange Membrane Crystallizer to Recover Magnesium Hydroxide from Seawater and Industrial Brines. *Membranes* **2020**, *10*, 303. [CrossRef]
28. Vassallo, F.; Morgante, C.; Battaglia, G.; La Corte, D.; Micari, M.; Cipollina, A.; Tamburini, A.; Micale, G. A Simulation Tool for Ion Exchange Membrane Crystallization of Magnesium Hydroxide from Waste Brine. *Chem. Eng. Res. Des.* **2021**, *173*, 193–205. [CrossRef]
29. Galama, A.H.; Post, J.W.; Cohen Stuart, M.A.; Biesheuvel, P.M. Validity of the Boltzmann Equation to Describe Donnan Equilibrium at the Membrane–Solution Interface. *J. Membr. Sci.* **2013**, *442*, 131–139. [CrossRef]
30. Tanaka, Y. Ion Exchange Membranes. In *Fundamentals and Applications*; Elsevier: Amsterdam, The Netherlands, 2015; ISBN 978-0-444-63319-4.
31. Chimanlal, I.; Nthunya, L.N.; Quist-Jensen, C.; Richards, H. Membrane Distillation Crystallization for Water and Mineral Recovery: The Occurrence of Fouling and Its Control during Wastewater Treatment. *Front. Chem. Eng.* **2022**, *4*, 1066027. [CrossRef]
32. Alkhatib, A.; Ayari, M.A.; Hawari, A.H. Fouling Mitigation Strategies for Different Foulants in Membrane Distillation. *Chem. Eng. Process. Process Intensif.* **2021**, *167*, 108517. [CrossRef]
33. Hickenbottom, K.L.; Cath, T.Y. Sustainable Operation of Membrane Distillation for Enhancement of Mineral Recovery from Hypersaline Solutions. *J. Membr. Sci.* **2014**, *454*, 426–435. [CrossRef]
34. Li, H.; Lin, H.; Raza, S.; Chen, C.; Yu, W.; Zeng, Q.; Zhao, Z.; Huang, X.; Shen, L. Construction of MXene-MOF Membranes with Photocatalytic Self-Cleaning for Enhanced Oil-Water Emulsion Separation. *J. Membr. Sci.* **2025**, *718*, 123685. [CrossRef]
35. Ouda, A.; Bajón-Fernández, Y.; McAdam, E. Methods to Modify Supersaturation Rate in Membrane Distillation Crystallisation: Control of Nucleation and Crystal Growth Kinetics (Including Scaling). *J. Membr. Sci.* **2024**, *691*, 122249. [CrossRef]
36. Morgante, C.; Vassallo, F.; Battaglia, G.; Cipollina, A.; Vicari, F.; Tamburini, A.; Micale, G. Influence of Operational Strategies for the Recovery of Magnesium Hydroxide from Brines at a Pilot Scale. *Ind. Eng. Chem. Res.* **2022**, *61*, 15355–15368. [CrossRef] [PubMed]

membranes

MDPI

Article

Treatment of Industrial Brine Using a Poly (Vinylidene Fluoride) Membrane Modified with Carbon Nanotubes

Tshifhiwa T. Tshauambea [1,2], Soraya P. Malinga [3] and Patrick G. Ndungu [2,*]

1 Department of Chemistry, Tshwane University of Technology, Private Bag X680, Pretoria 0001, South Africa; tshauambeatt@tut.ac.za
2 Department of Chemistry, University of Pretoria, Private Bag X20, Hartfield 0028, South Africa
3 Department of Chemical Sciences, University of Johannesburg, P.O. Box 17011, Doornfontein, Johannesburg 2028, South Africa; smalinga@uj.ac.za
* Correspondence: patrick.ndungu@up.ac.za; Tel.: + 27-(0)-12-420-3095

Abstract: This study explores using polyvinylidene fluoride (PVDF) membranes modified with multi-walled carbon nanotubes (MWCNTs) to treat simulated and industrial brine from coal power stations. The MWCNTs were acid-treated and characterized using Fourier Transform Infrared Spectroscopy (FTIR), Raman, and nitrogen sorption at 77 K, Thermogravimetric analysis (TGA), and Transmission electron microscopy (TEM). The desired membranes were obtained by casting from a solution of N-Methyl-2-pyrrolidone, PVDF, various weight percentages of MWCNTs, and a small amount of polyvinylpyrrolidone. The acid treatment of the MWCNTs introduced oxygen moieties on the surface, and increased pore volume and surface area while maintaining crystallinity and structural integrity remain preserved. The maximum rejection rate achieved was 41.82% with 1 wt.% of acid-treated MWCNTs in the PVDF membrane. Acid-treated MWCNTs loaded membranes had an improved rejection rate, which was 5× higher than membranes without MWCNTs.

Keywords: membrane technology; polyvinylidene fluoride; multiwalled carbon nanotubes; industrial brine

1. Introduction

One of the greatest difficulties facing the world in the twenty-first century is the provision of safe and affordable water. Furthermore, fast population increases, climate change, and water quality degradation continue to make it challenging for current water supplies to satisfy the ever-increasing demand for safe and affordable water [1,2]. Human activities, particularly industrialization, contribute to rising water shortages in emerging and developed countries by contaminating water resources and putting pressure on associated water resources [1]. People and industries rely on recycling water from different sources and the treatment and reuse of water for different purposes.

Wastewater from industries such as mining, oil and gas, coal-fired power plants, and pharmaceutical processing is difficult to treat in both developed and developing countries, including South Africa, due to its high total dissolved solids content and various pollutants [3,4]. This results in a relatively quick fouling of conventional membranes when using membrane processes and systems to treat such wastewater streams [5–7]. Brine, or the high concentration of salts found in industrial wastewater, has a persistent negative influence on the environment [8]. As a result, future policies and regulations may mandate that industrial wastewater have zero liquid discharge [8]. Salt composition in industrial brines differs depending on the water intake quality of the processes involved in the water treatment from which the final saline effluent is derived [9]. Currently, evaporation ponds are considered the most practical option for disposing of brine solutions from industrial activities, the ponds concentrate the brine, which leads to the salt precipitating into crystals, which are easy to remove from the ponds [9]. Evaporation ponds have disadvantages such as taking up useful land space, overflowing during heavy rain seasons, which leads to pollution in nearby waterbodies, and their functionality is highly dependent on climate conditions [9–11].

Membrane technology is one of several emerging technologies that can have a major impact on the sustainability and cost of wastewater treatment over conventional technologies [12]. This technology has advantages such as high removal capacity and flexible operational methods; it requires less energy and can be very cost-effective [13,14]. Membranes used in water treatment are commonly synthesized using organic (polymers) and inorganic (ceramics) materials [15–17]. In terms of comparison between organic and inorganic membranes, polymeric membranes are extensively utilized for water treatment because they have easy tuneable properties, improved processability, outstanding stability, high surface area for fast decontamination, selectivity to eliminate different pollutants, simple and require fewer steps during installation, and are more cost-effective than inorganic membranes [15,18]. Polymer membranes that have been demonstrated to be useful for water treatment include poly(vinylidene fluoride) (PVDF), polyethersulfone (PES), polyacrylonitrile (PAN), poly(vinyl alcohol) (PVA), poly(vinyl chloride) (PVC), polyethylene (PE), polypropylene (PP), polyamide, and chitosan [13,19]. Despite their outstanding stabilities and flexibility during installation, most polymer membranes are hydrophobic, which may lead to membrane fouling and poor selectivity [15–18]. Fouling results in low permeation flux, reduction or change in selectivity, and short membrane life spans [13,19–21].

To overcome membrane fouling and selectivity, fillers are introduced within the polymer used to make the membranes. Fillers are responsible for improving the mechanical strength, preventing swelling, and improving the chemical and thermal stability of polymeric membranes [13,19]. Nanofillers, when added to polymers within a range of 3–5 wt.% provide similar support properties to polymers that have been modified with 20–30 wt.% of micro-sized fillers. Thus, one of the key advantages nanomaterials have over micromaterials is a distinct weight advantage [19,20]. Nanoscale materials have become increasingly important as fillers due to their high specific surface area, which permits stronger interfacial interactions, in addition to their size [19,20].

Due to its distinct characteristics from its predecessors and their macro- and micro-sized counterparts, carbon nanotechnology has been the subject of extensive research in recent decades [13,20,22–26]. Among carbon nanomaterials, carbon nanotubes (CNTs) have drawn attention in the development of membranes and water treatment due to their inherent adsorption and sieving properties as well as their excellent mechanical and thermal stabilities [20,25–28]. Though they have excellent properties, CNTs are typically hydrophobic and tend to aggregate due to their strong intermolecular interactions, which include van der Waals forces and dipole–dipole interactions [23]. Aggregation properties of CNTs significantly reduces their dispersibility in a majority of solvents. In order to overcome this, CNTs can be functionalised using various chemical or physical techniques, leading to the introduction of functional groups on their surfaces [23,29,30]. The functional groups on CNT surfaces are essential for facilitating good dispersion in a polymer matrix, leading to improved membrane properties, including lower contact angles, enlarged pore size, increased rejection, anti-trade-off between permeability and selectivity, and antifouling properties [12,31,32].

Previous research on polymer membranes containing functionalized-CNTs (f-CNTs) indicated that these membranes could be highly effective in terms of permeability and salt rejection when used in desalination methods [25–28,33].

Sagar et al. synthesized f-CNTs immobilized membrane which was applied in DCMD desalination [25]. A maximum water flux of up to 36.8 kg/m^2 and salt rejection of 99.9% were achieved using a feed of NaCl with a concentration of 10,000 ppm. An electron-spin nanofiber membrane was prepared by incorporating graphene oxide on polyvinylidene fluoride-co-hexafluoropropylene. Application of the membrane in desalination via air gap membrane distillation showed high flux with excellent salt rejection (100%) over 60-h operation using 3.5 wt.% NaCl solution [34].

The objective of this study was to explore and compare PVDF membranes modified with MWCNTs and f-MWCNTs and their application in treating real industrial brine solutions. Several researchers have studied the desalination of brine from simulated brackish and seawater [28–31]. Our focus in this investigation shifts towards the treatment of real industrial brine from a coal-fired power plant. This type of brine has varying compositions, higher levels of organic and inorganic pollutant concentrations, and increased variability. This study provides insights into the potential of polymer membranes to treat industrial brine wastewater with a mixture of inorganic and organic pollutants and to process water volumes with potentially lower energy usage and reduced pressure requirements.

2. Materials and Methods

No further modification and processing were done on all the reagents used. Multi-walled CNTs were bought from Timesnano, NMP solvent was purchased from Srichem, and diethyl ether was purchased from Merck. PVDF (Poly (vinylidene fluoride), with an average molecular weight of ~180,000 determined by gel permeation chromatography) and PVP (polyvinylpyrrolidone with an average molecular weight of 10,000) were purchased from Sigma Aldrich (Johannesburg, South Africa). Nitric acid (55%), sulfuric acid (98%), ethanol, and hydrochloric acid (32%) were purchased from accredited local suppliers. The industrial brine used in this work originated from a local coal-fired power plant (Eskom South Africa).

2.1. Methods

2.1.1. Functionalization of Pristine-MWCNTs (p-MWCNTs)

MWCNTs were functionalized using methods adapted from Wand et al. [35]. In a round-bottom flask, 1.008 g of the purchased MWCNTs was mixed with 20 mL of HNO_3 (55%) and 60 mL of H_2SO_4 (98%). The mixture was stirred for 90 min at 90 °C. The solution was cooled for a few minutes, and then diluted in 800 mL of deionized water and left standing for 12 h to allow separation to occur. The oxidized MWCNTs were washed several times with double-distilled water, filtered, and then dried in a vacuum oven overnight at 60 °C.

2.1.2. Synthesis of Mixed Matrix Membranes with Carbon Nanomaterials

Membranes with different weight percentages (0.2%, 0.5%, 1%, and 2%) based on the weight of the polymer PVDF and of carbon nanomaterials (p-MWCNTs, and f-MWCNTs) were synthesized [24]. Firstly 0.08 g (0.5%) of p-MWCNTs was dissolved in 85 mL of NMP by sonication for 2 h. Then 14 g of PVDF followed by 1 g of PVP was added into the mixture, and the solution was stirred with an overhead stirrer for 24 h with the temperature kept at 60 °C. The prepared solution was removed from the stirring set-up, and it was left standing for another 24 h to remove all the air bubbles before casting it into a membrane. Before casting, the mixture was visibly inspected to ensure the CNTs had not settled out of the solution. To cast the membrane, the mixture was poured onto a glass plate and the casting knife set at 180 μm was used to cast the membrane. The glass plate was immersed in a water bath where the sheet peeled off into a membrane. The membrane was left in a water bath for 24 h to remove the residual solvent. After 24 h, the membranes were dried at room temperature.

2.2. Membrane Testing

For contact angle measurements, the analysis was done by dropping a few drops of ultrapure water using a needle at 5 different spots on each membrane (Sessile drop method). For the water intake capacity (*WIC*) measurement, the dry membranes were measured and thereafter the membranes were immersed in deionized water for 24 h and dapped with a paper towel before measuring the wet weight on the weighing balance. The following equation was used to calculate *WIC*%:

$$WIC = \frac{W_w - W_d}{W_w} \times 100 \tag{1}$$

where W_w is wet weight, W_d is dry weight.

Each membrane sample (0.00126 m^2) was compacted using deionized water for an hour at 200 kPa until a stable flux was obtained. This was done before taking measurements for both water flux and solute flux. The water flux and solute flux were calculated using Equation (2) where J_w is replaced by J_s for solute flux.

$$J_s/J_w = \frac{Q}{A \cdot \Delta t} \tag{2}$$

where:

J_w is the water flux (L/m^2·h)
J_s is the solute flux (L/m^2·h)
Q is the volume of the permeated water (L)
A is the effective area of the membrane (m^2)
Δt is the permeation time (h)

Synthesized membranes were further tested if they were influential in removing salts from industrial brine and simulated brine. The simulated brine solution was prepared by mixing Calcium (Ca) = 601.31 mg/L, sodium (Na) = 3028.99 mg/L, Magnesium (Mg) = 102.13 mg/L, and chlorine (Cl) = 1584 mg/L (prepared from calcium chloride, magnesium chloride and sodium chloride) with the concentration of ions chosen based on the concentration before on real industrial brine solution [9].

Brine Testing

The rejection tests were done in terms of the removal of salts from the brine and the removal of total organic carbons using a TOC analyser. There was no pre-treatment of the industrial brine before analysis. A conductivity meter was used to measure conductivity in mS and TDS in mg/L. The conductivity and TDS of the simulated brine and industrial brine were measured before treating (feed concentration) both the solution with the membranes and after treatment (permeate concentration) [36–38]. The rejection of salts was calculated using,

$$R(\%) = \left(1 - C_p/C_f\right) \times 100 \tag{3}$$

where:

R is rejection (%)
C_p is the permeate concentration (mg/L) measured TDS.
C_f is the feed concentration (mg/L) measured TDS.

Membrane samples, with each loading, were synthesized in duplicate, and each sample was tested.

The studies on the removal of total organic carbons were done only on the industrial brine.

3. Results and Discussion

3.1. Characterization of Nanomaterials

3.1.1. FTIR Analysis

Figure 1 shows the FTIR spectra of MWCNTs and f-MWCNTs, which provides some insight on the functional groups on the surfaces of the MWCNTs, as well as certain modifications after treating the MWCNTs samples with strong acids (H_2SO_4 and HNO_3) at 90 °C. The FTIR spectra of MWCNTs and f-MWCNTs show similarities in terms of functional groups identified, though, from the MWCNTs spectra, the intensity of the C-H at 1383 cm^{-1} is higher than that of the f-MWCNTs and this may indicate additional functional groups on the f-MWCNTs. Both the MWCNTs and f-MWCNTs have the C=C bond (stretch) at around 1630 cm^{-1}, and only the f-MWCNTs samples had a small shoulder peak at 1715 cm^{-1} caused by vibrations of C=O bonds. The appearance of the C=O vibrations on the f-MWCNTs indicate the acid treatment successfully introduced new surface groups to the carbon nanomaterials. The symmetric and asymmetric stretching vibrations of the methylene group are assigned to the peaks observed at 2975 cm^{-1} and 2921 cm^{-1} on the MWCNTs. On the spectrum of the f-MWCNTs, the methylene group peaks appear to have shifted to 2966 cm^{-1} and 2925 cm^{-1} with less intensity. The C-OH stretching vibration is assigned to the 3445 cm^{-1} for MWCNTs and the 3427 cm^{-1} for the f-MWCNTs. The shoulder peak (see insert on Figure 1) on the f-MWCNTs and peak shifts accompanied by lowering of intensities gives some idea that the addition of oxygen functional groups has occurred on the surface [39,40]. The presence of hydrophilic oxygen functional groups in membrane synthesis speeds up the solvent-non-solvent exchange process [41,42]. This results in membranes with large pores, improving the water flux even at low operating pressures [43].

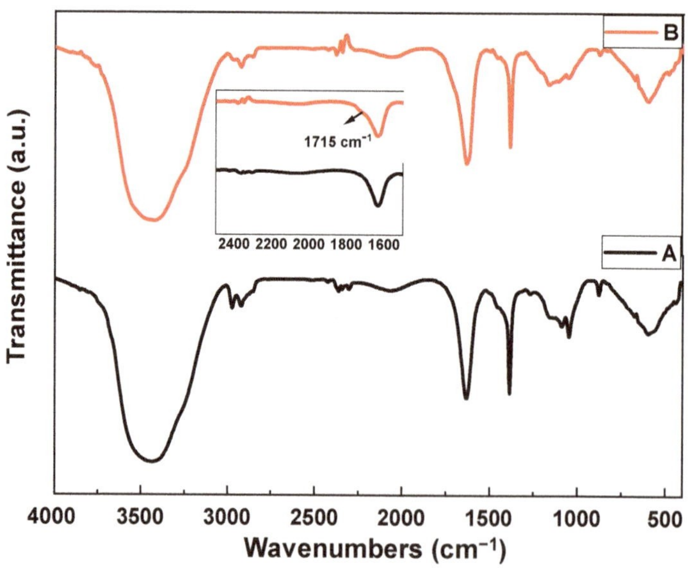

Figure 1. FTIR spectra of (A) MWCNTs and (B) f-MWCNTs.

3.1.2. XRD Results

The XRD patterns of MWCNTs show diffraction peaks at $2\theta = 25.52°$, $43.77°$, and $78°$ corresponding to (002), (100), and (110) reflection planes (Figure 2), and these are patterns typically associated with graphitic carbon. The f-MWCNTs have two diffraction peaks at $2\theta = 25.31°$ and $43.01°$, and a relatively low intensity and broad peak around the (110) reflections plane. The (002) diffraction peak can be used to provide some information about the crystalline structure of CNTs. The increase in the FWHM and the peak shift of the (002) diffraction peak from 25.52° to 25.31° (see Table 1) indicates a decrease in the crystallinity of the MWCNTs after acid treatment (002) [44]. The 90-min acid treatment used in this study indicates samples were functionalized with minimal degradation of the crystallinity of the MWCNTs [45,46]. The preserved crystallinity is beneficial for interaction at polymer/f-MWCNTs interfaces, which may enhance the mechanical strength of membranes modified with MWCNTs [47].

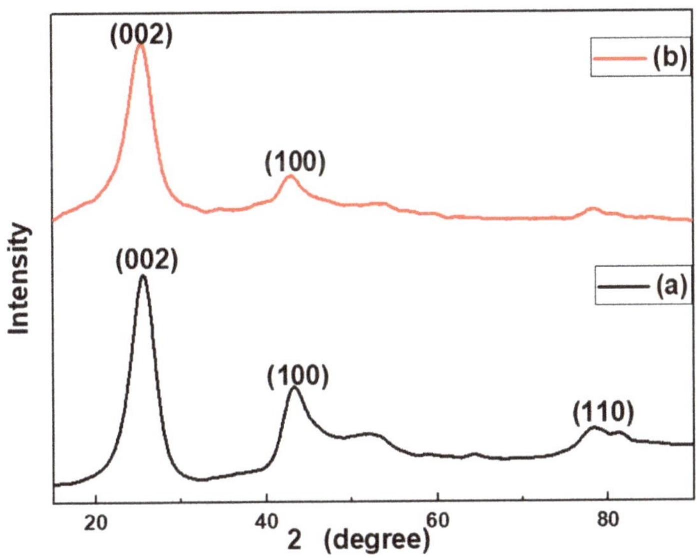

Figure 2. XRD patterns of (a) MWCNTs and (b) f-MWCNTs.

Table 1. Comparison of the interspacing layer of the MWCNTs and f-MWCNTs.

Sample	$2\theta/°$	FWHM/°	Interplanar Spacing $d_{(002)}$/nm
MWCNTs	25.52	2.958	0.3486
f-MWCNTs	25.31	4.034	0.3515

Further investigation using Bragg's law ($n\lambda = 2d\sin\theta$) shows that there is an increase in the interspacing distance from 0.3486 nm to 0.3515 nm for f-MWCNTs. Although the values are close, the small difference can be used as an indication of some degree of functionalization, and this can be attributed to the addition of oxygen functional groups (as per FTIR findings).

3.1.3. Raman Analysis

The results of the Raman analysis are presented in Figure 3. From Figure 3, the MWCNTs exhibited peaks with the D band at 1449 cm^{-1}, and the G band at 1601 cm^{-1} Similarly, the f-MWCNTs exhibited peaks with the introduction of defects during oxidation resulting in the shifting of the peaks [48], thus the D band appears at 1347 cm^{-1}, the G band at 1597 cm^{-1}, and 2D at 2414 cm^{-1}, the overtones appears at 1515 and 1738 cm^{-1} respectively. The D band is due to double resonance scattering caused by the structural defects during the attachment of different chemical groups on f-MWCNTs and is typically attributed to sp^3 carbon [45,49]. The vibration of the graphitic carbon (sp^2) along the nanotube axis is responsible for the G band [50,51]. The functionalization of MWCNTs results in changes in the intensity and the increase of the FWHM of G and D bands, as well as shifts in the peak positions of the G and the D bands [52,53]. Intensified structural defects caused by strong acid oxidation result in shifts in the position of the Raman peaks [54] and a decrease in the intensity of the G and D bands, the results obtained here indicate that MWCNTs underwent acid oxidation with minimum structural defects. The G′ on the f-MWCNTs is the result of two phonons scattering around the Brillouin zone's K point [55]; this band is the defect-induced D-band's second overtone [56]. The results from the Raman analysis correlate with the FTIR observations; specifically, the increase in the FWHM for both the D and G bands indicating increased defect structures due to the modification, and the presence of new functional groups (FTIR analysis) on the surfaces of the f-MWCNTs. The crystallinity of f-MWCNTs is confirmed by the G band which represents the graphitic structure of carbon (sp^2), this is in line with the XRD results.

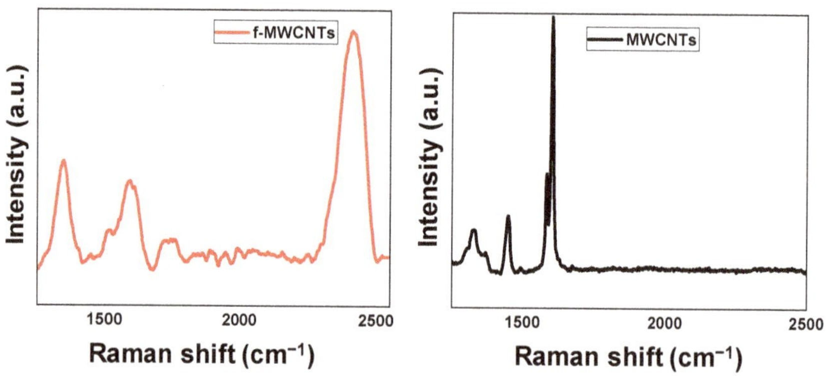

Figure 3. Raman spectrum of MWCNTs and f-MWCNTs showing their D and G bands.

The introduction of defects, via chemical oxidation in this case, on the MWCNTs surfaces can be inferred from the I_D/I_G ratio. Results in Table 2 show that the I_D/I_G ratio is higher for the f-MWCNTs (0.85) than the I_D/I_G of the MWCNTs (0.26). This corresponds with the addition of the oxygen-containing functional groups on the MWCNTs surface, as seen in FTIR results. The increase in I_D/I_G value for f-MWCNTs not only confirms the defects on the carbon surface, but also some change in the degree of graphitization [48]. The defects on the carbon surfaces and introduction of new chemical functional groups are crucial in membrane applications, as they improve the dispersal properties of the CNTs in the polymer matrix, promote hydrophilicity, and play a role in the reduction of membrane fouling [57].

Table 2. Comparison of the spectral parameters in the Raman of MWCNTs with f-MWCNTs.

Sample	D		G		
	Position	FWHM	Position	FWHM	I_D/I_G
MWCNTs	1449.33	22.015	1601	11.31	0.26
f-MWCNTs	1347.61	58.38	1591.07	85.48	0.85

The G and 2D prime band of f-MWCNTs shows a clear separation and this characteristic has been seen in other CNTs where the 2D has been associated with the overtone mode of the D band [58,59]. The increase in the FWHM on the f-MWCNTs is attributed to the oxidation process at higher temperatures for a relatively long period of time. The results obtained here are comparable to what has been reported in other studies where the oxidation of the MWCNTs is confirmed by the shifting of the D and G bands [48,60].

3.1.4. Thermal Stability (TGA Analysis)

Thermal stability studies, presented in Figure 4, show that MWCNTs are stable, with decomposition starting at a temperature of 600 °C, which is similar to previous reports [48,61]. The f-MWCNTs, on the other hand, decompose in three steps, with the first step starting at 40 °C and reaching a maximum of 56 °C representing the removal of water molecules adsorbed by the samples [62], this accounts for the loss of weight of 3.6%. The second decomposition, which accounts for the weight loss of 8.31%, is caused by the loss of the hydroxyl and carbonyl groups attached to f-MWCNTs surface starting at 180 °C and reaching maximum decomposition at 218 °C [48]. The loss of weight of 8.31% on the f-MWCNTs is a further indication of the functionalization of the MWCNTs with various oxygen moieties attached to the surface, which corroborates the FTIR and Raman results. The last step of decomposition is at 620 °C, which is due to the oxidation of the remaining carbon frameworks [63].

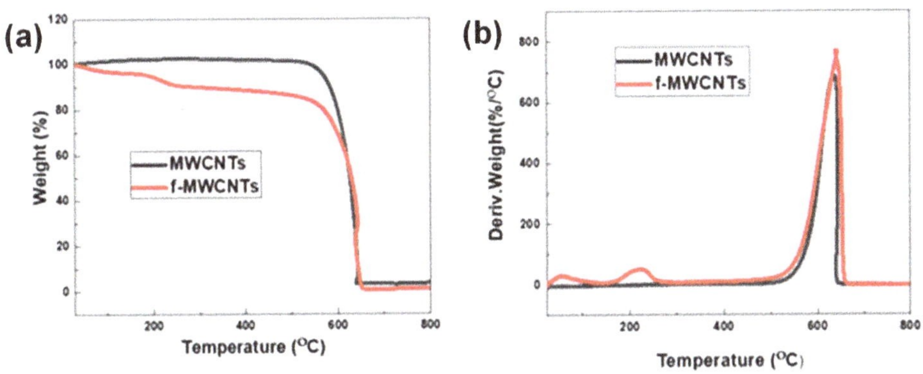

Figure 4. Thermal stabilities of MWCNTs and f-MWCNTs as represented by their (**a**) TGA curves and (**b**) derivative weight loss curves.

The FTIR spectrum of f-MWCNTs did provide some indication that the treatment had introduced various oxygen functional groups. Further analysis with XRD and Raman indicated and supported the FTIR analysis in terms of the functionalization of the f-MWCNTs. The TGA decomposition peak at 218 °C on the f-MWCNTs, and the lack of a similar peak on the MWCNTs further confirms that there are various oxygen moieties on the f-MWCNTs surface. TGA provided insight into the thermal stability of the nanotubes before and after functionalization. The 8.3% weight (%) loss observed with the f-MWCNTs is a semi-quantitative measure of the degree of functionalization of the nanotubes. In addition, the derivative weight loss peak for the MWCNTs and f-MWCNTs are similar, which shows the oxidative treatment was not only successful, but it did not significantly damage or degrade the nanotube properties. From the results obtained, the f-MWCNTs are likely to disperse and incorporate into a polymer matrix and the resulting membranes will likely have improved physical properties due to the addition of the nanotubes.

3.1.5. TEM Analysis

Transmission electron microscopy (TEM) analysis of the MWCNTs and the f-MWCNTs are presented in Figure 5. Figure 5 shows some minor damage to the wall structure of the MWCNTs. Most of the outer and some of the inner walls of the f-MWCNTs were distorted, with some indication of a change in the roughness of the wall surface (highlighted in red circles Figure 5D). Figure 5A shows MWCNTs that are highly entangled, caused by the strong van der Waals forces [57]. The acid treatment at high temperature led to the cleavage of some nanotubes, opening of nanotubes caps, and ingress of the treatment into the inside pores as evidenced by roughening of the inner walls (Figure 5B).

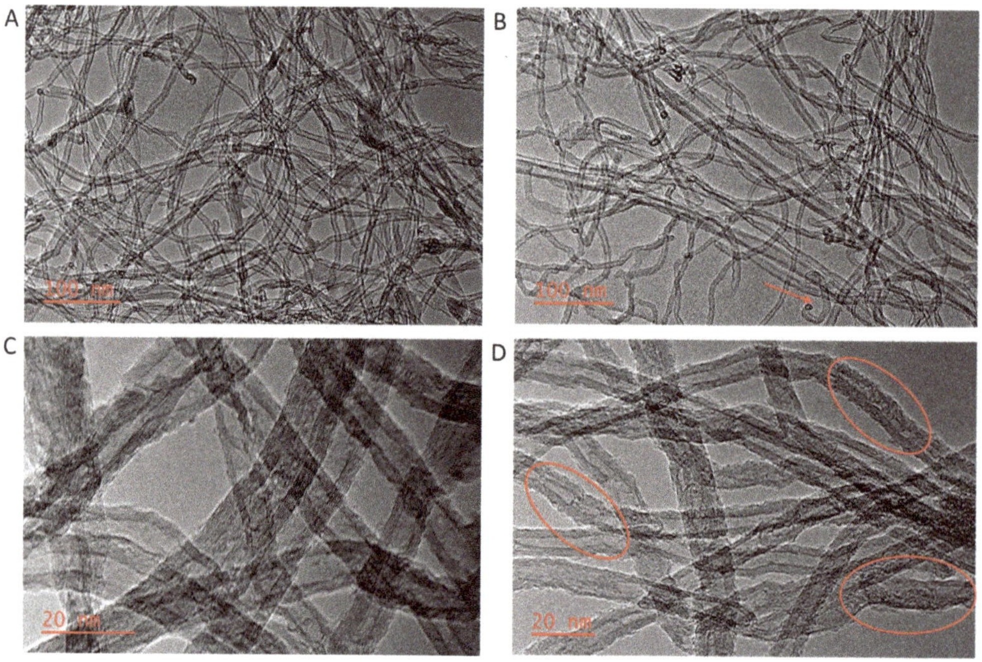

Figure 5. TEM images of (**A**) MWCNTs, (**B**) f-MWCNTs, (**C**) higher magnification image of MWCNTs, and (**D**) higher magnification image of f-MWCNTs. Arrow on micrograph (**B**) indicates nanotube end, and circles in (**D**) highlight walls of the nanotubes.

The changes on the surface of the f-MWCNTs show that oxidation under acidic conditions does not only bring chemical change but also affects the morphology, which is an indication of the disordered carbon on the surface of MWCNTs. The I_D/I_G ratio from Raman increased from MWCNTs to f-MWCNTs. This agrees with defects in the outer and inner walls of f-MWCNTs as seen on TEM images. This characteristic is an indication of disordered carbon on the surface. Similar characteristics to what was observed in this work were reported where small damages on the surface of the MWCNTs before acid treatment were observed and this is associated to the purification by the manufacturer to remove amorphous carbon and catalytic metal [64]. While the defects on the surface of the nanotubes may be detrimental to the mechanical properties of the CNTs, they are beneficial in terms of the dispersion properties of the nanotubes, which facilitates the interfacial bonding with the polymer matrix. Also, the defect sites may also contain chemical moieties that enhance ion-exchange processes or absorption and rejection mechanisms, resulting in the overall improvement of the membrane performance.

The method of functionalization has a significant impact on the degree of surface functionalization of the nanotubes. A study done on the duration of the treatment of the MWCNTs revealed that the weight percent of oxygen functional groups on MWCNTs increases with increasing oxidation time [46,48]. This was achieved when medium-concentration mixtures of HNO_3/H_2SO_4 were used, as reflux at high concentrations may result in the structural damage of the MWCNTs [46,64]. In contrast, the 90-min reflux time used in this study, with concentrated acid mixtures did result in alteration of the surface chemistry of the samples as well as the morphology, but the thermal stability remained similar to the untreated samples.

3.1.6. Textural Characteristics of the MWCNTs

The surface area of MWCNTs is highly affected by modifications such as surface functionalization. The theoretical studies reported that the surface area of CNTs as a function of the number of walls and diameters is in the range between 50–1315 m^2/g, as reported by Piegney et al. [65]. The experimental surface area of MWCNTs ranges between 10–50 m^2/g, and it is mostly the function of several walls. From Table 3, the surface area of f-MWCNT is 244.5 m^2/g which is an increase from 202.3 m^2/g of MWCNTs.

Table 3. BET textural properties of MWCNTs and f-MWCNTs.

Sample	Surface Area (m^2/g)	Pore Volume (m^3/g)	Pore Size (nm)
MWCNTs	202.31	1.12	22.26
f-MWCNTs	244.51	1.95	31.96

It is expected that functionalizing MWCNTs results in higher surface area, due to such factors as roughening of the walls of the nanotubes (see Figure 5B), opening the ends of the nanotubes, or detangling the CNTs. The BET theory assumes uniform adsorption over the surface, whereas the adsorption sites on MWCNTs differ due to aggregation and defects, as demonstrated in TEM pictures of f-MWCNTs [65]. The BET surface area of f-MWCNTs (244.51 m^2/g) reported in this work is similar to the 244.86 m^2/g (MWNTs) which was reported by Shabaan et al. [66]. The increase in the BET surface area of f-MWCNTs confirms that acid treatment not only added oxygen functional groups successfully on the surface, but also detangled the nanotubes, roughened the walls of the CNTs, and opened up some ends of the MWCNTs [67]. The sorption isotherms of both MWCNTs and f-MWCNTs display a type IV isotherm with a hysteresis loop, which is a feature of a mesoporous material [68–72], as shown in Figure 6.

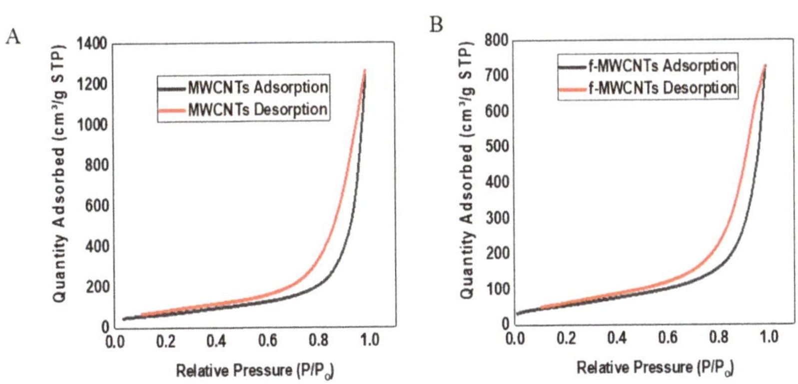

Figure 6. Nitrogen adsorption isotherms of (**A**) MWCNTs and (**B**) f-MWCNTs.

Functionalization of MWCNTs under acidic medium resulted in an increase in the pore volume and a similar observation was seen when MWCNT-COOH functionalised using 3:1 H_2SO_4 and HNO_3 had a total pore volume of 1.410 cm^3/g as compared to the 0.570 cm^3/g of the MWCNTs [73].

3.2. *Membrane Characterizations*

3.2.1. ATR-FTIR

The FTIR spectra of the PVDF/PVP membrane and membranes doped with varying wt.% of p-MWCNTs, and f-MWCNTs were recorded using an ATR accessory, and all the membranes had similar characteristics peaks (Figure 7). Peaks at 1402 cm^{-1} and 1173 cm^{-1} are due to C-C's stretching and deformation vibrations on the aromatic ring and C-F from the PVDF [74]. The PVP main characteristics peaks are the 1674 cm^{-1} and 1272 cm^{-1} due to the stretching and bending vibrations of the C=O and C-N [24,75–77]. Weak peaks at 2977 and 3021cm^{-1} are because of the stretching vibrations of C-H. The peak arising from 3020 cm^{-1} and 2971 cm^{-1} is caused by stretching vibrations from the CH_2 from polymer PVDF [74,78].

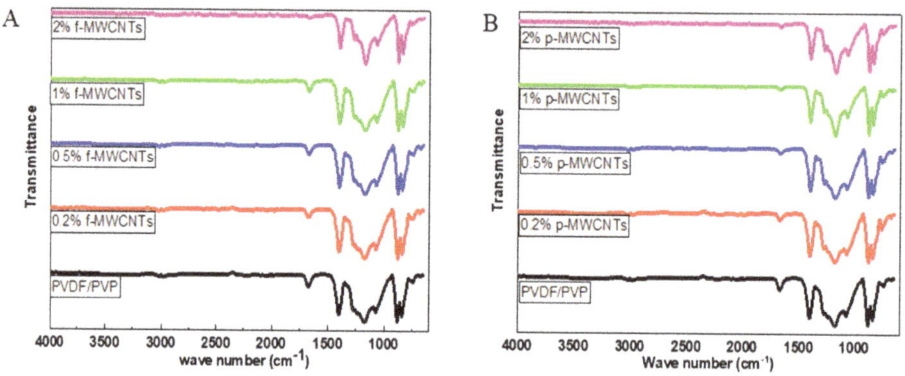

Figure 7. ATR-FTIR spectrum of PVDF/PVP membranes loaded with different wt.% of (**A**) f-MWCNTs and (**B**) p-MWCNTs.

The relatively low weight percentages of the MWCNTs and f-MWCNTs in the polymer membrane did not result in the observation of any key peaks from the nanomaterials within the nanocomposites. In addition, some of the key peaks occur in the same region as the polymer peaks and could have been obscured by the more intense polymer vibrational peaks.

3.2.2. XRD Analysis

The crystallinity of nanocomposite membranes was evaluated by making use of XRD. All the synthesized membranes in Figure 8 show a sharp peak corresponding to the (110) diffraction peak. The (110) diffraction peak appears at 2θ region of 20.28–20.43° respectively. In prior studies, it was discovered that β-phase of PVDF occurs at 2θ region of 20.2–20.6° [79]. The β-phase in this study was caused by stretching, elongation forces, and pulling during the casting of the membrane [80]. A weak peak (211) diffraction plane is observed at 2θ = 40.1° on the bare PVDF/PVP membrane. This peak has previously been related to the γ-phase of the PVDF [79]. From Figure 8 0.2 and 0.5% f-MWCNTs shows a diffraction peak at 2θ = 25° and this peak is due to the f-MWCNTs as it was seen in the previously on X-ray diffraction pattern of f-MWCNTs. The outcomes acquired in this investigation exhibit similarity to prior observations documented in the literature, wherein the β-phase of PVDF/P[MMA-IL]-MWCNTs manifested at 2θ = 20.6°. In the present study, the β-phase is discerned at 2θ = 20.2°, aligning with the established findings in the field [81].

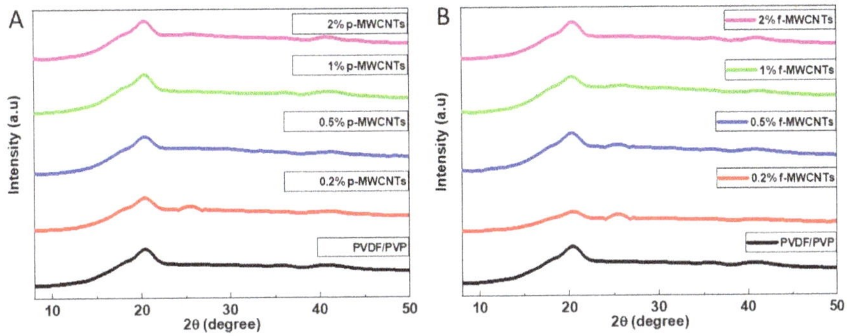

Figure 8. X-ray diffraction patterns of synthesized membranes loaded with 0.2–2% of the (**A**) f-MWCNTs and (**B**) p-MWCNTs.

3.2.3. SEM (Surface and Cross-Sectional)

Studies and comparisons of the membrane surface morphology were done using SEM. The diameter of the membrane pores was measured using ImageJ (version 1.53v), the image of the membrane captured from SEM was loaded on ImageJ and the scale was set to 10 μm (from the scale on the image), for each membrane 200 pores were measured by drawing a line across each pore. A distribution curve was plotted using the Origin software program, and each membrane's average diameter was obtained. The diameter measurements of the pores are shown by their distribution curves underneath each membrane's SEM image.

From Figure 9, all of the synthesized membranes generally exhibited a relatively large number of pores on the membrane surface, and incorporation of the MWCNTs and f-MWCNTs had an influence on the overall pore size distribution. The PVDF/PVP membrane pore distribution curve was centred at 0.5 μm with similar count for the various pore sizes. The 1 wt.% PVDF/PVP/MWCNTs pore size distribution curve was centred at 0.32 μm with a higher count for the smaller pore sizes. The 1 wt.% PVDF/PVP/f-MWCNT's pore size distribution curve was centred at 0.35 μm with counts distributed over a narrower range over smaller diameters. Addition of the MWCNT's or f-MWCNT's resulted in a higher count of pores with smaller diameters, indicating the nanotubes do affect the formation of the surface pores. PVP is well known as a pore former in PVDF membranes, and it's addition to the casting solution enhances the kinetics and phase separation processes during casting and results in greater number of pores [82,83], and in our system wide distribution of large pores on the surface. The addition of the nanotubes affected the pore formation process, possibly due to some favourable interactions between the PVP and the nanotubes, and resulted in smaller pores and narrower distributions. Our results are similar to other reports in the literature, were carbon nanotubes have been observed to result in the formation of smaller surface pores [47]. Prior reports in the literature have attributed the fast exchange between solvent and non-solvent during the phase inversion process to the hydrophilic groups on the surface of the nanotubes [84–86]. The cross-sectional image of the PVDF/PVP membrane (Figure 9d) showed the expected asymmetric morphology and structure with finger-like macro-voids [87,88]. The addition of the nanotubes had a minimal effect on the formation of the macro-void structures; however, the root-like structures at the bottom of the membrane observed with the PVDF/PVP membranes were not prominent on the PVDF/PVP/ 1 wt.% MWCNTs and the PVDF/PVP/ 1 wt.% f-MWCNTs.

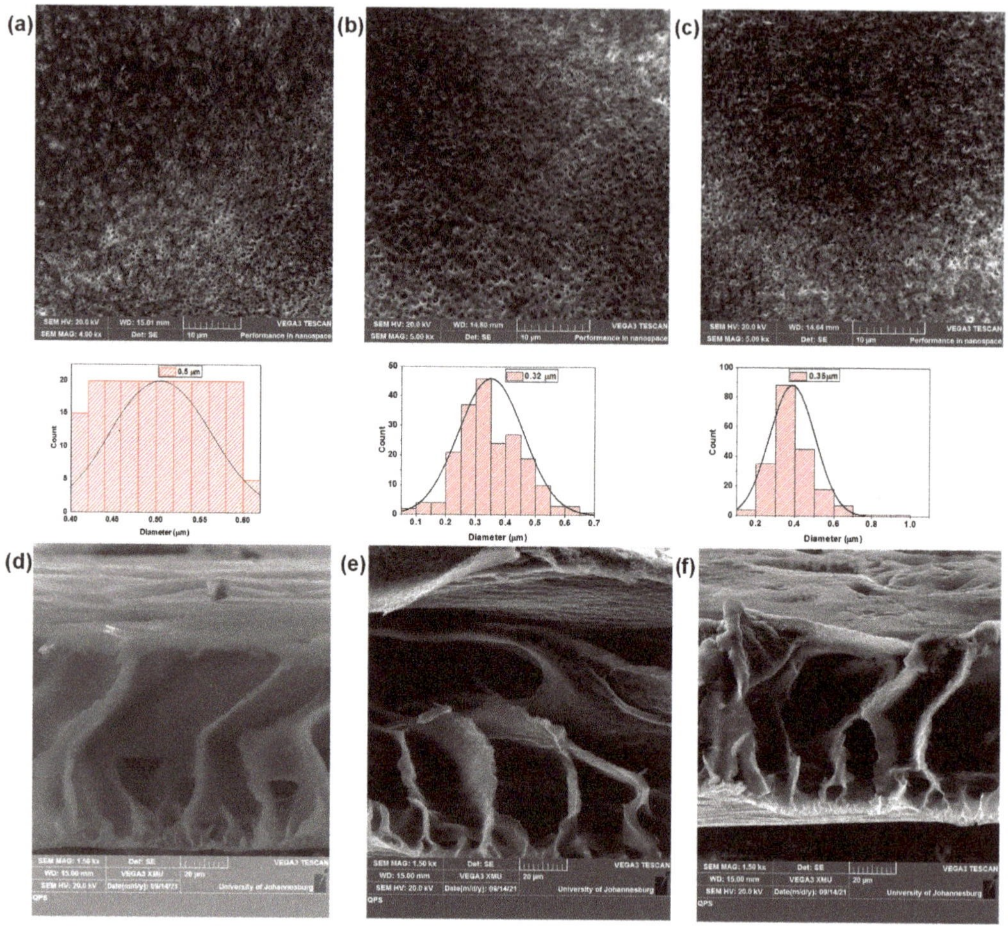

Figure 9. SEM images showing surface of (**a**) PVDF/PVP with corresponding pore size distribution below the micrograph, (**b**) PVDF/PVP/ 1 wt.% MWCNTs with corresponding pore size distribution below the micrograph, and (**c**) PVDF/PVP/ 1 wt.% f-MWCNTs with corresponding pore size distribution below the micrograph, and the cross-section micrographs of (**d**) PVDF/PVP, (**e**) PVDF/PVP/ 1wt %p-MWCNTs, and (**f**) PVDF/PVP/ 1 wt.% f-MWCNTs.

3.2.4. Contact Angle and Water Intake Measurements

Results of the WIC % and contact angle measurements (an average from measurements of five different spots on each membrane) are represented in Table 4.

The membranes with smaller contact angles (<90°) resemble a more hydrophilic nature than membranes with higher contact angles (>90°) [89,90]. The bare membrane (PVDF/PVP) has a contact angle of 85.45°, which is not much larger than the value of 83.8° reported by Guangyong et al. This high contact angle is related to the intrinsic hydrophobic nature of PVDF [91]. Recent studies show that PVDF without the modification by PVP has a contact angle of 141°; when comparing it to the contact angle of 85.45° obtained in this work, it can be confirmed that adding PVP improves the hydrophilicity of the PVDF [92].

Incorporating the carbon nanomaterials in the membranes further enhances the membrane's hydrophilicity, and hydrophilicity increases with increasing the wt.% of nanomaterials. As the membranes' hydrophilicity and pore size improve, their water intake capacity (WIC) % increases. Previously, enhanced hydrophilicity was linked to nanoparticle migration towards the membrane/water interface, resulting in a decrease in interface energy during the phase inversion process [93]. From Table 4 the contact angle membranes decrease in the following order: MWCNTs 79.41–72.63° > f-MWCNTs 72.57–68.25°. This order shows that the membranes with nanomaterials rich in oxygen functional groups are more hydrophilic than those with fewer oxygen functional groups. From the results obtained in this study, it can be confirmed that the hydrophilicity of a membrane varies according to the chemical composition of the additive, which in this study are PVP, MWCNTs, and f-MWCNTs [94]. The improved hydrophilicity enables the formation of a hydrated layer, thereby enhancing the water intake capacity of the membranes, which has been reported in previous studies, and shown to improve the water flux and anti-fouling performance [95,96].

Table 4. Contact angle measurements and WIC for the membrane and nanocomposite membranes.

Sample	Contact Angle	Water Intake Capacity (WIC)%
PVDF/PVP	85.45	75.51
PVDF/PVP/0.2% MWCNTs	79.41	79.52
PVDF/PVP/0.5% MWCNTs	77.43	73.72
PVDF/PVP/1% MWCNTs	75.76	77.06
PVDF/PVP/2% MWCNTs	72.63	76.77
PVDF/PVP/0.2% f-MWCNTs	72.57	78.39
PVDF/PVP/0.5% f-MWCNTs	71.46	73.80
PVDF/PVP/1% f-MWCNTs	70.55	78.71
PVDF/PVP/2% f-MWCNTs	68.26	78.10

3.3. *Performance Tests on the Membranes*

3.3.1. Water Flux and Permeate Flux Tests

Water and permeate flux were done at a pressure of 100 kPa and time t = 10 min. As seen from Figure 10, the bare membrane (PVDF/PVP) water flux was increased upon the addition of 0.2 wt.% p-MWCNT nanomaterial; this is due to the water transport promoted by the nano-corridors, which forms when polymers interact with the nanomaterial's functional groups during phase inversion process [97]. Previous studies have shown that increasing nanomaterial loading in the polymer matrix may result in agglomeration of nanomaterials, causing pore blockage, and this may be the reason for the decrease in water flux with increasing nanomaterial loading [98]. The pure water flux of the membranes decreased when wt.% of the nanomaterials was increased from 0.5 wt.%, increasing the nanomaterials in the casting solution results in blocking the pore connection, and this hinders water transportation [99,100].

The permeation flux tests were done using the industrial brine solution as well as simulated brine solution, and the results from Figure 10 shows that all the synthesized membranes have permeate flux which is much lower than the water flux. A decrease in the membrane permeate flux indicates fouling that occurred on the surface of the membrane. The SEM images of the membranes obtained after treatment of the brine solution (Figure 11) show that each membrane has a layer deposited with crystals and particles blocking pores on the membrane surface. The decrease in the permeate flux of the simulated brine is caused by friction between the membrane matrix with friction between water and ions as it controls water flow through the membrane [101,102]. The large decrease in the MWCNTs loaded membrane is associated with the smaller pores as was seen previously on the SEM images showing the smaller diameter of the membrane pores.

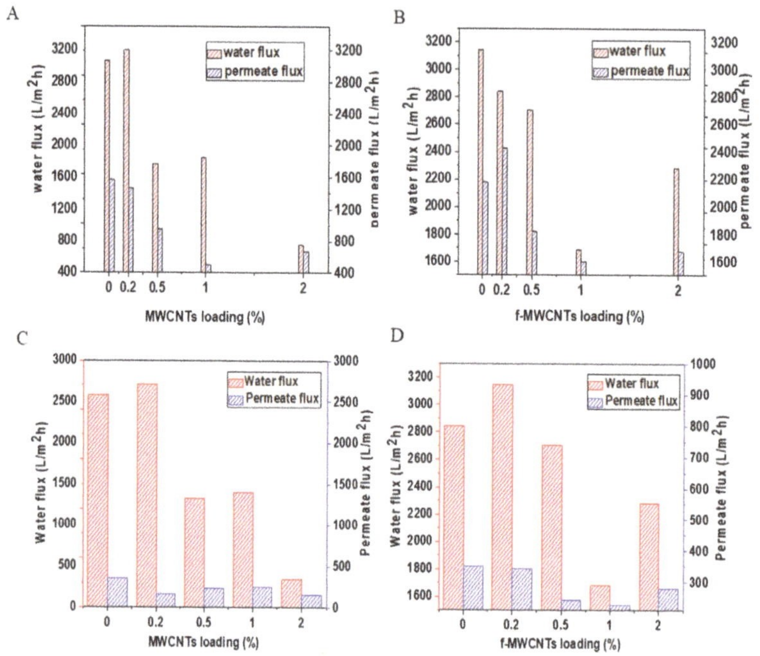

Figure 10. Water and permeate flux of PVDF/PVP membranes loaded with (**A**) p-MWCNTs and (**B**) f-MWCNTs (simulated brines) and (**C**) p-MWCNTs and (**D**) f-MWCNTs (industrial brines).

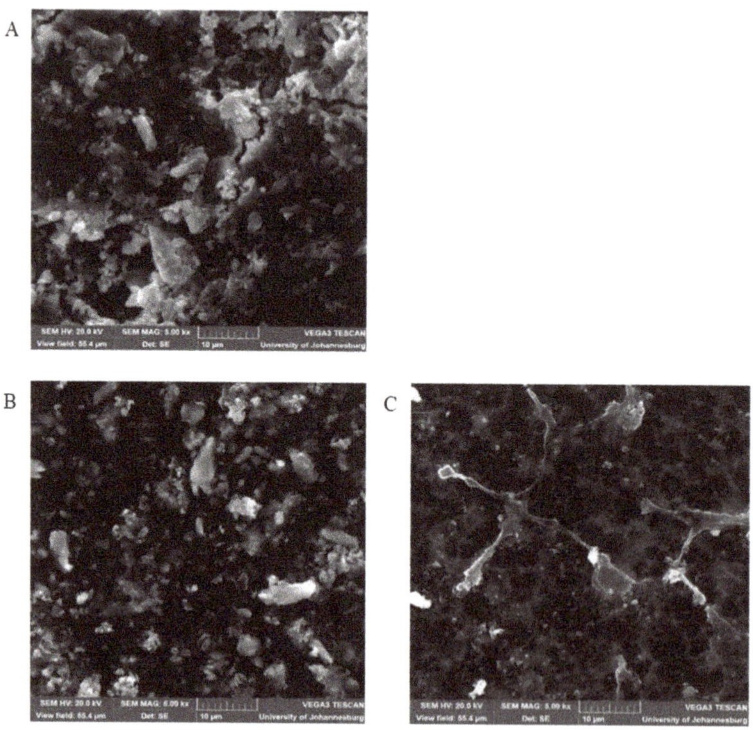

Figure 11. Shows accumulation of solutes and particles on membrane surfaces of (**A**) PVDF/PVP, (**B**) PVDF/PVP/1 wt % p-MWCNTs, and (**C**) PVDF/PVP/ 1 wt % f-MWCNTs.

3.3.2. Rejection Tests: Effects of Nanomaterial Loading

The effects of nanomaterials dosage were done to determine the optimum wt.% loading of the nanomaterials (MWCNTs, and f-MWCNTs) that resulted in good membrane performance. The membranes without MWCNTs had a rejection rate of 7.23% when using simulated brine. With the addition of 0.2 wt.% loading of MWCNTs, the rejection rate increased to 8.69%. After adding 0.5 wt.% of MWCNTs to the membrane, the simulated brine rejection rate dropped to 5.93%, which is related to poor distribution of the nanotubes in the polymer matrix. However, the rejection rate increased to 7.33% with the 1 wt.% loaded membrane. Another decline in the rejection rate to 5.65% was observed, after loading the membranes with 2 wt.% of MWCNTs, and is likely associated with the agglomeration of the MWCNTs resulting in defects in the polymer matrix, which may affect the membranes performance [92]. The performance of the MWCNTs loaded membranes on industrial brines yielded results that are different from those of the simulated brine with a rejection rate up to 40.13%. The bare membrane has a low rejection rate at 7.15%, which is similar to the simulated brine. When the membrane was loaded with 0.2 wt.% MWCNTs, the rejection rate increased to 12.08%, but dropped to 6.1% with 0.5 wt.% loading, and then further increases were observed with the 1 (38.1 5 rejection rate) and 2 wt.% (40.1% rejection rate) MWCNTs loaded membranes. The membranes with f-MWCNTs incorporated in their matrix have a slightly higher rejection rate on simulated brine compared to the rejection rate of membranes with MWCNTs. The rejection rate increased from 9.99% when 0.2 wt.% of f-MWCNTs was introduced and increased at 0.5 wt.% to 10.92%, further increased to 14% at 1 wt.%, and then decreased to 10.61% at 2 wt.%. For industrial brine, the f-MWCNTs membrane loaded membranes showed rejection rates of 7, 5, 6, 44, and 12% for the 0, 0.2, 0.5, 1, and 2 wt.% loaded membranes. From Figure 12, the membranes loaded with 1 wt.% of the f-MWCNTs performed better than the other membrane loadings for both the simulated and industrial brines. The 1 wt.% loading may have resulted in enough oxygen moieties required in the membrane matrix for ion exchange, another factor could be the homogenous distribution during the preparation of the casting solution [103]. Increasing the wt.% to 2 resulted in a decrease in the rejection rate with both simulated and industrial brines, this may be due to the poor distribution of the high-content nanomaterials in the casting solution [104]. Overall, the 1 wt.% f-MWCNTs membrane has a higher rejection rate than the MWCNTs, and this confirms that introducing functional groups on MWCNTs assists in opening pores at the tips leading to access of the walls inside the nanotubes, introducing charged oxygen containing groups on the surface of the CNTs which assist with the rejection of ions in the brine [104]. Comparing permeate flux and rejection rates, it can be seen that membranes with low permeate flux have higher rejection rates when compared to membranes with high permeate flux. The low permeate flux is an added advantage as it permits more interaction between the contaminants and the nanomaterials within the membrane's matrix.

The overall rejection obtained in this work was ~41.8%, which is low in comparison to the rejection rates that have been reported in the literature. This is due to the large pore size on the membrane and the use of simulated and real industrial brine solutions, containing multiple salts at high concentration. In contrast, the membranes reported in Table 5 were tested on synthetic brine containing a single salt, which may result in higher rejections under more controlled conditions. Moreover, the membranes synthesized in this study exhibited high water and permeate flux at a low operating pressure of 1 bar, whereas the membranes reported in Table 5 required significantly higher pressures (10–16 bar) to achieve comparatively lower flux values. The use of low pressure (100 kPa) makes the membrane synthesized in this study essential for saving energy and may be applicable in desalinating larger volumes of wastewater [105]. Rejection rates of up to

35% of 0.06 M NaCl have been reported before, where the temperature-induced phase separation technique was used to synthesize the PVDF-CNT [105].

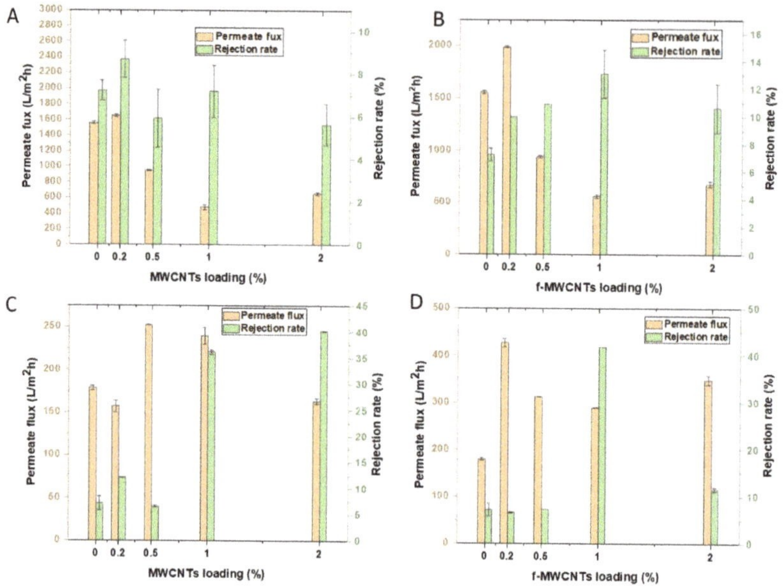

Figure 12. Effects of nanomaterials loading on the PVDF/PVP membranes (**A**) simulated brine on MWCNTs loaded membranes, (**B**) simulated brine of f-MWCNTs loaded membranes, (**C**) industrial brine on MWCNTs loaded membranes, and (**D**) industrial brine on f-MWCNTs loaded membranes.

Table 5. Comparison of rejection rates between different desalination membranes and this work.

Membrane	Salt Rejection	Pressure (kPa)	Target Source	Concentration (ppm)	Ref
PVDF/PVP/1%MWNCTs	41.8%	100	Real industrial brine	1080–14,000	This work
GO/polymer (PES)	98.5%	1000	Synthetic NaCl	2000	[106]
NMPS/PA	98.7%	1600	Synthetic NaCl	2000	[107]
TEMPO/CFNs/PA	96.2%	1500	Synthetic NaCl	2000	[108]
TFN-f-nTiO_2/PA	98.4	1500	Synthetic NaCl	2000	[109]
	54.8	1500	Synthetic Boric acid	5	
PSf-TiO_2	72.8%	250	Synthetic NaCl	1160	[86]

3.3.3. Effects of Contact Time

The effect of contact time on the removal of salts from the industrial brine was studied to determine how often the membrane will require to be changed. From Figure 13, the membranes show a high rejection rate during the first 5 min; the rejection rate decreases with increasing time. The continuous deposition of solutes and particles clogging the pores on the surface of the membranes is responsible for the low rejection as the contact time is increased [88]. The clogging of the pores has been confirmed by the low permeate flux (see Figure 10) compared to water flux. Several distinct mechanisms of negative rejections have been identified in studies, each characterized by different dependencies on membrane properties, feed solution composition, and operational conditions [110,111]. It has been reported that an imbalance in ion concentrations, with one or more ions present at higher levels than others, can lead to the high permeation of ions present in lower concentrations in the solution [110]. The rejection test was conducted using an industrial brine solution containing a mixture of salts at various concentrations. In this study, negative rejections were observed due to an increased concentration of salts in the membrane phase and the difference in the concentration of ions in the industrial brine solution.

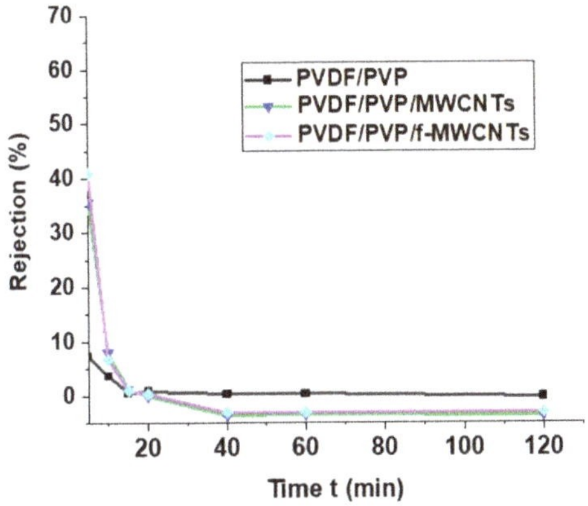

Figure 13. Effects of contact time on the removal of salts from industrial brine.

3.4. TOC Removal Analysis

Total organic carbon (TOC) in the brine solution was examined using a TOC analyser using KHP as the standard. The MWCNTs and f-MWCNTs (Figure 14) loaded membranes results showed fluctuations with different loadings. It is crucial to note that the industrial brine solution contains different organic pollutants with varying physical-chemical properties, and these can affect the organic removal performance [112].

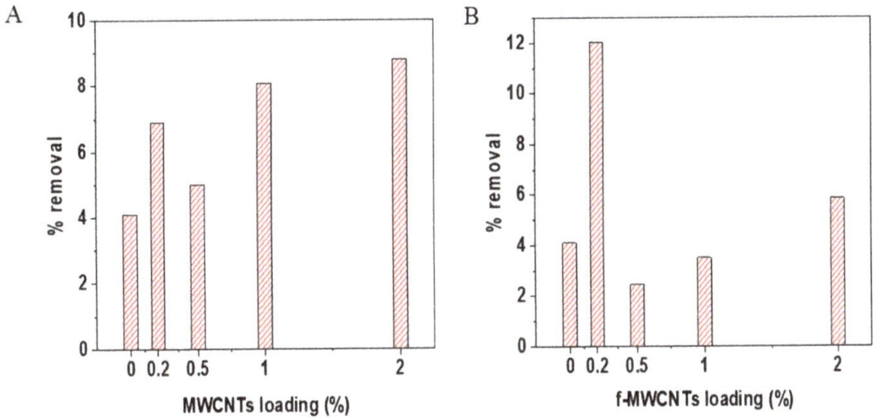

Figure 14. Shows effects of nanomaterials loading on the removal of TOC content in the industrial brine solution. Results for the membranes synthesized with MWCNTS are presented in (**A**), and membranes synthesized with f-MWCNTs are presented in (**B**).

Overall, the membranes with 0.2 wt.% f-MWCNTs had the highest removal percentage of 12% in terms of TOC. In contrast, the 2 wt.% MWCNTs were able to remove 8.5%. Thus, functionalizing the MWCNTs allows for a lower wt.% to be incorporated into the membrane for the removal of total organic carbon. However, the low 0.2 wt.% of f-MWCNTs is not ideal for the removal of key brine constituents (see Figure 12). Thus, these results may indicate that there are competitive removal processes between the organic carbon compounds present in the brine and the inorganic ions, which favour the inorganic over the organic. In addition, in terms of the TOC levels in the industrial brine tested, the initial

TOC values (20 mg/L) are within permissible limits for discharge under various global regulations, but are also above acceptable concentrations for discharge in other locales. Also, when considering future trends and possible changes in policies such values may likely be considered high in the near future [113]. These factors highlight that further investigations are required using more sensitive chromatographic techniques (such as gas or liquid chromatography coupled to suitable mass spectrometry detectors) to ascertain which moieties are favoured in the removal processes, as TOC does not offer insight into the specific types of organic compounds present in the pollutant waste stream, which is important in terms of understanding the toxicity of the effluent.

The scope of this study was on the initial application of the synthesized nanocomposite membranes. Further work is needed in the investigation of usage of these membranes in systems and processes that can clean and regenerate the membranes, optimisation for long term operation, and investigating recycling and other sustainability options for these membranes.

4. Conclusions

Over the years, carbon nanotubes (CNTs) have attracted noteworthy attention in membrane technology due to their extraordinary electrical conductivity and mechanical strength. Functionalization of CNTs enhances their interaction with polar water molecules, improving water flux and enabling membrane operation at lower pressures. This, in turn, contributes to reduced energy consumption in water treatment processes. The purchased MWCNTs were functionalized with strong acids (sulphuric and nitric) to produce f-MWCNTs. The XRD of f-MWCNTs showed that they were oxidized without destroying the graphitic carbon, and this is validated by the presence of the (002) peak on f-MWCNTs. An increase in I_D/I_G was observed on the Raman spectrum between the MWCNTs and f-MWCNTs, indicating the acid treatment was successful in the oxidation of the nanotubes. Functionalization of MWCNTs requires special attention in terms of the temperature used, the reagents used for the oxidation, and the time taken. The membrane's hydrophilicity was improved by the addition of MWCNTs and f-MWCNTs. The high water and permeate flux obtained in this work is an indication that the membranes synthesized has the potential for application in large volumes of water. Furthermore, the use of low pressure is an added advantage of reduced energy consumption in terms of pressure. The rejection rate was improved from 6.89% on the bare membranes to 41.85% when using membranes modified with acid-treated MWCNTs. This indicates that the nanomaterials in the membranes had a significant impact on the treatment of highly saline wastewater. The scope of this study was on the initial application of the synthesized membranes, however it is important to highlight the importance of studying the re-usability and membrane cleaning properties as this is crucial for determining operational stability and sustainability. The use of dead-end cells in this study affected the rejection, due to low rotational speed causing low mass transfer, and high concentration on the membrane. An alternative that could be used is the crossflow because membranes work longer on crossflow and the feed flows parallel as opposed to the dead-end cell where the feed flows perpendicularly. Based on the rejection rate obtained in this study, it is imperative to take into consideration the pore size of the membrane in relation to the molecular weight of the salts in the brine solution to obtain higher rejection rates. In future work, we intend to investigate the integration of PDVF-based membranes with diverse nanomaterials beyond carbon-based variants. Our focus will be on comprehensively exploring their applicability in resource recovery from wastewater, aligning with the global trend towards a circular economy.

Author Contributions: Conceptualization, P.G.N. and S.P.M.; formal analysis, T.T.T., P.G.N. and S.P.M.; funding acquisition, P.G.N.; investigation, T.T.T.; methodology, T.T.T., P.G.N. and S.P.M.; project administration, P.G.N. and S.P.M.; resources P.G.N. and S.P.M.; supervision, P.G.N. and S.P.M.; validation, T.T.T., P.G.N. and S.P.M.; visualization, T.T.T., P.G.N. and S.P.M.; writing—original draft, T.T.T.; writing—review and editing, T.T.T., P.G.N. and S.P.M. All authors have read and agreed to the published version of the manuscript.

Funding: This research was partially funded by the National Nanoscience Postgraduate Teaching & Training platform, University of Johannesburg, and the University of Pretoria.

Institutional Review Board Statement: Not applicable.

Data Availability Statement: The raw data supporting the conclusions of this article will be made available by the authors on request.

Acknowledgments: Acknowledgement is extended for the financial support and training facilitated by the national nanoscience postgraduate teaching and training platform, which significantly contributed to the successful execution of this project. Furthermore, sincere appreciation is conveyed to the instrument technician team within the Department of Chemical Sciences at the University of Johannesburg for their valuable assistance. Finally, the authors acknowledge research support from the University of Pretoria and the University of Johannesburg.

Conflicts of Interest: The authors declare no conflicts of interest.

Abbreviations

The following abbreviations are used in this manuscript:

PVDF	Poly (vinylidene) fluoride
PVP	Polyvinylpyrrolidone
MWCNTs	Multi-walled carbon nanotubes
f-MWCNTS	Functionalized multi-walled carbon nanotubes
p-MWCNTs	Pristine multi-walled carbon nanotubes
FTIR	Fourier Transform Infrared Spectroscopy
TEM	Transmission electron microscopy
TGA	Thermogravimetric analysis
SEM	Scanning electron microscopy
TOC	Total organic carbon
WIC	water intake capacity
KHP	Potassium hydrogen phthalate

References

1. Qu, X.; Alvarez, P.J.; Li, Q. Applications of nanotechnology in water and wastewater treatment. *Water Res.* **2013**, *47*, 3931–3946. [CrossRef] [PubMed]
2. Khan, N.A.; Khan, S.U.; Ahmed, S.; Farooqi, I.H.; Dhingra, A.; Hussain, A.; Changani, F. Applications of nanotechnology in water and wastewater treatment: A review. *Asian J. Water Environ. Pollut.* **2019**, *16*, 81–86. [CrossRef]
3. Guo, L.; Xie, Y.; Sun, W.; Xu, Y.; Sun, Y. Research progress of high-salinity wastewater treatment technology. *Water* **2023**, *15*, 684. [CrossRef]
4. Panagopoulos, A. Study and evaluation of the characteristics of saline wastewater (brine) produced by desalination and industrial plants. *Environ. Sci. Pollut. Res.* **2021**, *29*, 23736–23749. [CrossRef] [PubMed]
5. Vidic, R.D.; Brantley, S.L.; Vandenbossche, J.M.; Yoxtheimer, D.; Abad, J.D. Impact of shale gas development on regional water quality. *Science* **2013**, *340*, 1235009. [CrossRef] [PubMed]
6. Shaffer, D.L.; Arias Chavez, L.H.; Ben-Sasson, M.; Romero-Vargas Castrillón, S.; Yip, N.Y.; Elimelech, M. Desalination and reuse of high-salinity shale gas produced water: Drivers, technologies, and future directions. *Environ. Sci. Technol.* **2013**, *47*, 9569–9583. [CrossRef] [PubMed]
7. Gregory, K.B.; Vidic, R.D.; Dzombak, D.A. Water management challenges associated with the production of shale gas by hydraulic fracturing. *Elements* **2011**, *7*, 181–186. [CrossRef]

8. Panagopoulos, A.; Michailidis, P. Membrane technologies for sustainable wastewater treatment: Advances, challenges, and applications in zero liquid discharge (zld) and minimal liquid discharge (mld) systems. *Membranes* **2025**, *15*, 64. [CrossRef] [PubMed]
9. Petrik, L.; Fatoba, O.; Fey, M.; Ndlela, R.; Ndlovu, N.; Omoniyi, E.; Bent, D.; Nell, P. *Industrial Brine Minimization: Determining the Physical Chemical Parameters That Affect Evaporation Rates on Multi-Component Hyper-Saline Effluents: Report to the Water Research Commission*; Water Research Commission: Pretoria, South Africa, 2015.
10. Panagopoulos, A.; Haralambous, K.-J.; Loizidou, M. Desalination brine disposal methods and treatment technologies—A review *Sci. Total Environ.* **2019**, *693*, 133545. [CrossRef] [PubMed]
11. El Zayat, H.; Nasr, P.; Sewilam, H. Investigating sustainable management of desalination brine through concentration using forward osmosis. *Environ. Sci. Pollut. Res.* **2021**, *28*, 39938–39951. [CrossRef] [PubMed]
12. Ma, L.; Dong, X.; Chen, M.; Zhu, L.; Wang, C.; Yang, F.; Dong, Y. Fabrication and water treatment application of carbon nanotubes (CNTs)-based composite membranes: A review. *Membranes* **2017**, *7*, 16. [CrossRef] [PubMed]
13. Jhaveri, J.H.; Murthy, Z. A comprehensive review on anti-fouling nanocomposite membranes for pressure driven membrane separation processes. *Desalination* **2016**, *379*, 137–154. [CrossRef]
14. Criscuoli, A. Water–Energy Nexus: Membrane Engineering Towards a Sustainable Development. *Membranes* **2025**, *15*, 98. [CrossRef] [PubMed]
15. Ng, L.Y.; Mohammad, A.W.; Leo, C.P.; Hilal, N. Polymeric membranes incorporated with metal/metal oxide nanoparticles: A comprehensive review. *Desalination* **2013**, *308*, 15–33. [CrossRef]
16. Jarvis, P.; Carra, I.; Jafari, M.; Judd, S.J. Ceramic vs polymeric membrane implementation for potable water treatment. *Water Res.* **2022**, *215*, 118269. [CrossRef] [PubMed]
17. Kotobuki, M.; Gu, Q.; Zhang, L.; Wang, J. Ceramic-polymer composite membranes for water and wastewater treatment: Bridging the big gap between ceramics and polymers. *Molecules* **2021**, *26*, 3331. [CrossRef] [PubMed]
18. Berber, M.R. Current advances of polymer composites for water treatment and desalination. *J. Chem.* **2020**, *2020*, 7608423. [CrossRef]
19. Arundhathi, B.; Pabba, M.; Raj, S.S.; Sahu, N.; Sridhar, S. Advancements in Mixed-Matrix Membranes for Various Separation Applications: State of the Art and Future Prospects. *Membranes* **2024**, *14*, 224. [CrossRef] [PubMed]
20. Bhattacharya, M. Polymer nanocomposites—A comparison between carbon nanotubes, graphene, and clay as nanofillers. *Materials* **2016**, *9*, 262. [CrossRef] [PubMed]
21. Li, H.; Chen, V. Membrane fouling and cleaning in food and bioprocessing. In *Membrane Technology*; Elsevier: Amsterdam, The Netherlands, 2010; pp. 213–254.
22. Corry, B. Designing carbon nanotube membranes for efficient water desalination. *J. Phys. Chem. B* **2008**, *112*, 1427–1434. [CrossRef] [PubMed]
23. Punetha, V.D.; Rana, S.; Yoo, H.J.; Chaurasia, A.; McLeskey, J.T., Jr.; Ramasamy, M.S.; Sahoo, N.G.; Cho, J.W. Functionalization of carbon nanomaterials for advanced polymer nanocomposites: A comparison study between CNT and graphene. *Prog. Polym. Sci.* **2017**, *67*, 1–47. [CrossRef]
24. Van Tran, T.T.; Kumar, S.R.; Nguyen, C.H.; Lee, J.W.; Tsai, H.-A.; Hsieh, C.-H.; Lue, S.J. High-permeability graphene oxide and poly (vinyl pyrrolidone) blended poly (vinylidene fluoride) membranes: Roles of additives and their cumulative effects. *J. Membr. Sci.* **2021**, *619*, 118773. [CrossRef]
25. Ali, S.; Rehman, S.A.U.; Luan, H.-Y.; Farid, M.U.; Huang, H. Challenges and opportunities in functional carbon nanotubes for membrane-based water treatment and desalination. *Sci. Total Environ.* **2019**, *646*, 1126–1139. [CrossRef] [PubMed]
26. Regmi, C.; Kshetri, Y.K.; Wickramasinghe, S.R. Carbon-Based Nanocomposite Membranes for Membrane Distillation: Progress, Problems and Future Prospects. *Membranes* **2024**, *14*, 160. [CrossRef] [PubMed]
27. Roy, K.; Mukherjee, A.; Maddela, N.R.; Chakraborty, S.; Shen, B.; Li, M.; Du, D.; Peng, Y.; Lu, F.; García Cruzatty, L.C. Outlook on the bottleneck of carbon nanotube in desalination and membrane-based water treatment—A review. *J. Environ. Chem. Eng.* **2020**, *8*, 103572. [CrossRef]
28. Li, Q.; Yang, D.; Guan, M.; Zhang, H.; Xu, X.; Shi, H.; Liu, Q. Effects of modification groups and defects on the desalination performance of multi-walled carbon nanotube (MWNT) membranes. *Sep. Purif. Technol.* **2022**, *297*, 121444. [CrossRef]
29. Lavagna, L.; Nisticò, R.; Musso, S.; Pavese, M. Functionalization as a way to enhance dispersion of carbon nanotubes in matrices: A review. *Mater. Today Chem.* **2021**, *20*, 100477. [CrossRef]
30. Alosime, E.M. A review on surface functionalization of carbon nanotubes: Methods and applications. *Discov. Nano* **2023**, *18*, 12. [CrossRef] [PubMed]
31. Sianipar, M.; Kim, S.H.; Iskandar, F.; Wenten, I.G. Functionalized carbon nanotube (CNT) membrane: Progress and challenges. *RSC Adv.* **2017**, *7*, 51175–51198. [CrossRef]
32. Das, R.; Ali, M.E.; Abd Hamid, S.B.; Ramakrishna, S.; Chowdhury, Z.Z. Carbon nanotube membranes for water purification: A bright future in water desalination. *Desalination* **2014**, *336*, 97–109. [CrossRef]

33. Subramani, A.; Jacangelo, J.G. Emerging desalination technologies for water treatment: A critical review. *Water Res.* **2015**, *75*, 164–187. [CrossRef] [PubMed]
34. Woo, Y.C.; Tijing, L.D.; Shim, W.-G.; Choi, J.-S.; Kim, S.-H.; He, T.; Drioli, E.; Shon, H.K. Water desalination using graphene-enhanced electrospun nanofiber membrane via air gap membrane distillation. *J. Membr. Sci.* **2016**, *520*, 99–110. [CrossRef]
35. Wang, L.; Qiu, H.; Liang, C.; Song, P.; Han, Y.; Han, Y.; Gu, J.; Kong, J.; Pan, D.; Guo, Z. Electromagnetic interference shielding MWCNT-Fe_3O_4@ Ag/epoxy nanocomposites with satisfactory thermal conductivity and high thermal stability. *Carbon* **2019**, *141*, 506–514. [CrossRef]
36. Berkessa, Y.W.; Lang, Q.; Yan, B.; Kuang, S.; Mao, D.; Shu, L.; Zhang, Y. Anion exchange membrane organic fouling and mitigation in salt valorization process from high salinity textile wastewater by bipolar membrane electrodialysis. *Desalination* **2019**, *465*, 94–103. [CrossRef]
37. Chai, Y.; Qin, P.; Wu, Z.; Bai, M.; Li, W.; Pan, J.; Cao, R.; Chen, A.; Jin, D.; Peng, C. A coupled system of flow-through electro-Fenton and electrosorption processes for the efficient treatment of high-salinity organic wastewater. *Sep. Purif. Technol.* **2021**, *267*, 118683. [CrossRef]
38. Zhang, X.; Guo, Z.; Zhang, C.; Luan, J. Exploration and optimization of two-stage vacuum membrane distillation process for the treatment of saline wastewater produced by natural gas exploitation. *Desalination* **2016**, *385*, 117–125. [CrossRef]
39. Avid, A.; Jafari, S.H.; Khonakdar, H.A.; Ghaffari, M.; Krause, B.; Pötschke, P. Surface modification of MWCNT and its influence on properties of paraffin/MWCNT nanocomposites as phase change material. *J. Appl. Polym. Sci.* **2020**, *137*, 48428. [CrossRef]
40. Alwan, S.H.; Alshamsi, H.A. In situ synthesis NiO/F-MWCNTs nanocomposite for adsorption of malachite green dye from polluted water. *Carbon Lett.* **2022**, *32*, 1073–1084. [CrossRef]
41. Yadav, A.; Singh, K.; Shahi, V.K. Side-chain grafted functional groups poly (vinylidene fluoride-hexafluoropropylene) anti-fouling fluorinated polymer membrane with tuneable hydrophobicity for distillation. *Desalination* **2022**, *525*, 115501. [CrossRef]
42. Jaber, L.; Almanassra, I.W.; Backer, S.N.; Kochkodan, V.; Shanableh, A.; Atieh, M.A. A comparative analysis of the effect of carbonaceous nanoparticles on the physicochemical properties of hybrid polyethersulfone ultrafiltration membranes. *Membranes* **2022**, *12*, 1143. [CrossRef] [PubMed]
43. Dong, X.; Lu, D.; Harris, T.A.; Escobar, I.C. Polymers and solvents used in membrane fabrication: A review focusing on sustainable membrane development. *Membranes* **2021**, *11*, 309. [CrossRef] [PubMed]
44. Yaragalla, S.; Anilkumar, G.; Kalarikkal, N.; Thomas, S. Structural and optical properties of functionalized multi-walled carbon nanotubes. *Mater. Sci. Semicond. Process.* **2016**, *41*, 491–496. [CrossRef]
45. Ramoraswi, N.O.; Ndungu, P.G. Photo-catalytic properties of TiO_2 supported on MWCNTs, SBA-15 and silica-coated MWCNTs nanocomposites. *Nanoscale Res. Lett.* **2015**, *10*, 427. [CrossRef] [PubMed]
46. Chiang, Y.-C.; Lin, W.-H.; Chang, Y.-C. The influence of treatment duration on multi-walled carbon nanotubes functionalized by H_2SO_4/HNO_3 oxidation. *Appl. Surf. Sci.* **2011**, *257*, 2401–2410. [CrossRef]
47. Zhou, R.; Rana, D.; Matsuura, T.; Lan, C.Q. Effects of multi-walled carbon nanotubes (MWCNTs) and integrated MWCNTs/SiO_2 nano-additives on PVDF polymeric membranes for vacuum membrane distillation. *Sep. Purif. Technol.* **2019**, *217*, 154–163. [CrossRef]
48. Hoa, L.T.M. Characterization of multi-walled carbon nanotubes functionalized by a mixture of HNO_3/H_2SO_4. *Diam. Relat. Mater* **2018**, *89*, 43–51.
49. Dhall, S.; Jaggi, N.; Nathawat, R. Functionalized multiwalled carbon nanotubes based hydrogen gas sensor. *Sens. Actuators A Phys.* **2013**, *201*, 321–327. [CrossRef]
50. Dresselhaus, M.S.; Dresselhaus, G.; Saito, R.; Jorio, A. Raman spectroscopy of carbon nanotubes. *Phys. Rep.* **2005**, *409*, 47–99. [CrossRef]
51. Gómez, S.; Rendtorff, N.M.; Aglietti, E.F.; Sakka, Y.; Suárez, G. Surface modification of multiwall carbon nanotubes by sulfonitric treatment. *Appl. Surf. Sci.* **2016**, *379*, 264–269. [CrossRef]
52. Theodore, M.; Hosur, M.; Thomas, J.; Jeelani, S. Influence of functionalization on properties of MWCNT–epoxy nanocomposites. *Mater. Sci. Eng. A* **2011**, *528*, 1192–1200. [CrossRef]
53. Li, Z.; Deng, L.; Kinloch, I.A.; Young, R.J. Raman spectroscopy of carbon materials and their composites: Graphene, nanotubes and fibres. *Prog. Mater. Sci.* **2023**, *135*, 101089. [CrossRef]
54. Chang, C.-C.; Hsu, I.-K.; Aykol, M.; Hung, W.-H.; Chen, C.-C.; Cronin, S.B. A new lower limit for the ultimate breaking strain of carbon nanotubes. *Acs Nano* **2010**, *4*, 5095–5100. [CrossRef] [PubMed]
55. Gatabi, M.P.; Moghaddam, H.M.; Ghorbani, M. Point of zero charge of maghemite decorated multiwalled carbon nanotubes fabricated by chemical precipitation method. *J. Mol. Liq.* **2016**, *216*, 117–125. [CrossRef]
56. Bourchak, M.; Juhany, K.; Salah, N.; Ajaj, R.; Algarni, A.; Scarpa, F. Determining the tensile properties and dispersion characterization of CNTs in epoxy using Tem and Raman spectroscopy. *Mech. Compos. Mater.* **2020**, *56*, 215–226. [CrossRef]
57. Yuan, X.-T.; Xu, C.-X.; Geng, H.-Z.; Ji, Q.; Wang, L.; He, B.; Jiang, Y.; Kong, J.; Li, J. Multifunctional PVDF/CNT/GO mixed matrix membranes for ultrafiltration and fouling detection. *J. Hazard. Mater.* **2020**, *384*, 120978. [CrossRef] [PubMed]

58. Bobenko, N.; Bolotov, V.; Egorushkin, V.; Korusenko, P.; Melnikova, N.; Nesov, S.; Ponomarev, A.; Povoroznyuk, S. Experimental and theoretical study of electronic structure of disordered MWCNTs. *Carbon* **2019**, *153*, 40–51. [CrossRef]
59. Sharma, M.; Rani, S.; Pathak, D.K.; Bhatia, R.; Kumar, R.; Sameera, I. Manifestation of anharmonicities in terms of phonon modes' energy and lifetime in multiwall carbon nanotubes. *Carbon* **2021**, *171*, 568–574. [CrossRef]
60. Santangelo, S.; Messina, G.; Faggio, G.; Abdul Rahim, S.; Milone, C. Effect of sulphuric–nitric acid mixture composition on surface chemistry and structural evolution of liquid-phase oxidised carbon nanotubes. *J. Raman Spectrosc.* **2012**, *43*, 1432–1442. [CrossRef]
61. Hanbali, G.; Jodeh, S.; Hamed, O.; Bol, R.; Khalaf, B.; Qdemat, A.; Samhan, S. Enhanced ibuprofen adsorption and desorption on synthesized functionalized magnetic multiwall carbon nanotubes from aqueous solution. *Materials* **2020**, *13*, 3329. [CrossRef] [PubMed]
62. Kumar, D.; Singh, K.; Verma, V.; Bhatti, H. Low-temperature hydrothermal synthesis and functionalization of multiwalled carbon nanotubes. *Indian J. Phys.* **2016**, *90*, 139–148. [CrossRef]
63. Kaskun, S.; Akinay, Y.; Kayfeci, M. Improved hydrogen adsorption of ZnO doped multi-walled carbon nanotubes. *Int. J. Hydrogen Energy* **2020**, *45*, 34949–34955. [CrossRef]
64. Sezer, N.; Koç, M. Oxidative acid treatment of carbon nanotubes. *Surf. Interfaces* **2019**, *14*, 1–8. [CrossRef]
65. Lehman, J.H.; Terrones, M.; Mansfield, E.; Hurst, K.E.; Meunier, V. Evaluating the characteristics of multiwall carbon nanotubes. *Carbon* **2011**, *49*, 2581–2602. [CrossRef]
66. Shabaan, O.A.; Jahin, H.S.; Mohamed, G.G. Removal of anionic and cationic dyes from wastewater by adsorption using multiwall carbon nanotubes. *Arab. J. Chem.* **2020**, *13*, 4797–4810. [CrossRef]
67. Abbas, A.; Yi, Y.M.; Saleem, F.; Jin, Z.; Veksha, A.; Yan, Q.; Lisak, G.; Lim, T.M. Multiwall carbon nanotubes derived from plastic packaging waste as a high-performance electrode material for supercapacitors. *Int. J. Energy Res.* **2021**, *45*, 19611–19622. [CrossRef]
68. Ma, C.-Y.; Huang, S.-C.; Chou, P.-H.; Den, W.; Hou, C.-H. Application of a multiwalled carbon nanotube-chitosan composite as an electrode in the electrosorption process for water purification. *Chemosphere* **2016**, *146*, 113–120. [CrossRef] [PubMed]
69. Hu, C.; Grant, D.; Hou, X.; Xu, F. High rhodamine B and methyl orange removal performance of graphene oxide/carbon nanotube nanostructures. *Mater. Today Proc.* **2021**, *34*, 184–193. [CrossRef]
70. Zhang, K.; Gao, X.; Zhang, Q.; Li, T.; Chen, H.; Chen, X. Preparation and microwave absorption properties of asphalt carbon coated reduced graphene oxide/magnetic CoFe2O4 hollow particles modified multi-wall carbon nanotube composites. *J. Alloys Compd.* **2017**, *723*, 912–921. [CrossRef]
71. Shukla, A.K.; Alam, J.; Ansari, M.A.; Alhoshan, M.; Alam, M.; Kaushik, A. Selective ion removal and antibacterial activity of silver-doped multi-walled carbon nanotube/polyphenylsulfone nanocomposite membranes. *Mater. Chem. Phys.* **2019**, *233*, 102–112. [CrossRef]
72. Duman, O.; Tunç, S.; Polat, T.G.; Bozoğlan, B.K. Synthesis of magnetic oxidized multiwalled carbon nanotube-κ-carrageenan-Fe_3O_4 nanocomposite adsorbent and its application in cationic Methylene Blue dye adsorption. *Carbohydr. Polym.* **2016**, *147*, 79–88. [CrossRef] [PubMed]
73. Doğan, M.; Selek, A.; Turhan, O.; Kızılduman, B.K.; Bicil, Z. Different functional groups functionalized hexagonal boron nitride (h-BN) nanoparticles and multi-walled carbon nanotubes (MWCNT) for hydrogen storage. *Fuel* **2021**, *303*, 121335. [CrossRef]
74. Wang, Z.; Yu, H.; Xia, J.; Zhang, F.; Li, F.; Xia, Y.; Li, Y. Novel GO-blended PVDF ultrafiltration membranes. *Desalination* **2012**, *299*, 50–54. [CrossRef]
75. Beygmohammdi, F.; Kazerouni, H.N.; Jafarzadeh, Y.; Hazrati, H.; Yegani, R. Preparation and characterization of PVDF/PVP-GO membranes to be used in MBR system. *Chem. Eng. Res. Des.* **2020**, *154*, 232–240. [CrossRef]
76. Chang, X.; Wang, Z.; Quan, S.; Xu, Y.; Jiang, Z.; Shao, L. Exploring the synergetic effects of graphene oxide (GO) and polyvinylpyrrodione (PVP) on poly (vinylylidenefluoride)(PVDF) ultrafiltration membrane performance. *Appl. Surf. Sci.* **2014**, *316*, 537–548. [CrossRef]
77. Guo, Z.; Xu, X.; Xiang, Y.; Lu, S.; Jiang, S.P. New anhydrous proton exchange membranes for high-temperature fuel cells based on PVDF–PVP blended polymers. *J. Mater. Chem. A* **2015**, *3*, 148–155. [CrossRef]
78. Safarpour, M.; Khataee, A.; Vatanpour, V. Effect of reduced graphene oxide/TiO_2 nanocomposite with different molar ratios on the performance of PVDF ultrafiltration membranes. *Sep. Purif. Technol.* **2015**, *140*, 32–42. [CrossRef]
79. Begum, S.; Ullah, H.; Kausar, A.; Siddiq, M.; Aleem, M.A. Fabrication of epoxy functionalized MWCNTs reinforced PVDF nanocomposites with high dielectric permittivity, low dielectric loss and high electrical conductivity. *Compos. Sci. Technol.* **2018**, *167*, 497–506. [CrossRef]
80. Yu, H.; Huang, T.; Lu, M.; Mao, M.; Zhang, Q.; Wang, H. Enhanced power output of an electrospun PVDF. *Nanotechnology* **2013**, *24*, 405401. [CrossRef] [PubMed]
81. Li, Z.; Kang, W.; Han, Z.; Yan, J.; Cheng, B.; Liu, Y. Hierarchical MnOx@ PVDF/MWCNTs tree-like nanofiber membrane with high catalytic oxidation activity. *J. Alloys Compd.* **2019**, *780*, 805–815. [CrossRef]

82. Hou, C.; Pang, Z.; Xie, S.; Yang, Z.; Wong, N.H.; Sunarso, J.; Peng, Y. Dual PVP roles for preparing PVDF hollow fiber membranes with bicontinuous structures via the complex thermally induced phase separation (c-TIPS). *Sep. Purif. Technol.* **2024**, *332*, 125766. [CrossRef]
83. Kartohardjono, S.; Salsabila, G.M.K.; Ramadhani, A.; Purnawan, I.; Lau, W.J. Preparation of PVDF-PVP Composite Membranes for Oily Wastewater Treatment. *Membranes* **2023**, *13*, 611. [CrossRef] [PubMed]
84. Rahimpour, A.; Jahanshahi, M.; Khalili, S.; Mollahosseini, A.; Zirepour, A.; Rajaeian, B. Novel functionalized carbon nanotubes for improving the surface properties and performance of polyethersulfone (PES) membrane. *Desalination* **2012**, *286*, 99–107. [CrossRef]
85. Zhang, J.; Xu, Z.; Shan, M.; Zhou, B.; Li, Y.; Li, B.; Niu, J.; Qian, X. Synergetic effects of oxidized carbon nanotubes and graphene oxide on fouling control and anti-fouling mechanism of polyvinylidene fluoride ultrafiltration membranes. *J. Membr. Sci.* **2013**, *448*, 81–92. [CrossRef]
86. Emadzadeh, D.; Lau, W.J.; Matsuura, T.; Rahbari-Sisakht, M.; Ismail, A.F. A novel thin film composite forward osmosis membrane prepared from PSf–TiO_2 nanocomposite substrate for water desalination. *Chem. Eng. J.* **2014**, *237*, 70–80. [CrossRef]
87. Chen, W.; Xiao, T.; Yang, X. Facile pore structure control of poly(vinylidene fluoride) membrane for oil/water separation. *Sep. Purif. Technol.* **2020**, *251*, 117305. [CrossRef]
88. Yang, S.; Fang, P.; Li, X.; Li, Q.; Zhou, Y.; Fan, E. Preparation and properties of patterned P-MWCNTs/PVDF functional membranes. *J. Polym. Res.* **2025**, *32*, 134. [CrossRef]
89. Ma, Y.; Cao, X.; Feng, X.; Ma, Y.; Zou, H. Fabrication of super-hydrophobic film from PMMA with intrinsic water contact angle below 90. *Polymer* **2007**, *48*, 7455–7460. [CrossRef]
90. Zhenyu, S.; Zhanqiang, L.; Hao, S.; Xianzhi, Z. Prediction of contact angle for hydrophobic surface fabricated with micro-machining based on minimum Gibbs free energy. *Appl. Surf. Sci.* **2016**, *364*, 597–603. [CrossRef]
91. Zeng, G.; He, Y.; Zhan, Y.; Zhang, L.; Pan, Y.; Zhang, C.; Yu, Z. Novel polyvinylidene fluoride nanofiltration membrane blended with functionalized halloysite nanotubes for dye and heavy metal ions removal. *J. Hazard. Mater.* **2016**, *317*, 60–72. [CrossRef] [PubMed]
92. Bakhsh, N.; Ahmed, Z.; Mahar, R.B.; Khatri, Z. Development and application of electrospun modified polyvinylidene fluoride (PVDF) nanofibers membrane for biofouling control in membrane bioreactor. *Desalin. Water Treat.* **2021**, *217*, 74–82. [CrossRef]
93. Xu, Z.; Wu, T.; Shi, J.; Teng, K.; Wang, W.; Ma, M.; Li, J.; Qian, X.; Li, C.; Fan, J. Photocatalytic antifouling PVDF ultrafiltration membranes based on synergy of graphene oxide and TiO_2 for water treatment. *J. Membr. Sci.* **2016**, *520*, 281–293. [CrossRef]
94. Silva, T.L.; Morales-Torres, S.; Figueiredo, J.L.; Silva, A.M. Multi-walled carbon nanotube/PVDF blended membranes with sponge-and finger-like pores for direct contact membrane distillation. *Desalination* **2015**, *357*, 233–245. [CrossRef]
95. Bilal, A.; Yasin, M.; Akhtar, F.H.; Gilani, M.A.; Almohamadi, H.; Younas, M.; Mushtaq, A.; Aslam, M.; Hassan, M.; Nawaz, R. Enhancing water purification by integrating titanium dioxide nanotubes into polyethersulfone membranes for improved hydrophilicity and anti-fouling performance. *Membranes* **2024**, *14*, 116. [CrossRef] [PubMed]
96. Sun, Z.K.; Zhou, Y.; Jiao, Y.; Cheng, X.Q.; Zhang, Y.; Wang, P.; Liang, H.; Yang, X.; Drioli, E.; Figoli, A. Multi-hydrophilic functional network enables porous membranes excellent anti-fouling performance for highly efficient water remediation. *J. Membr. Sci.* **2020**, *608*, 118191. [CrossRef]
97. Shan, M.; Kang, H.; Xu, Z.; Li, N.; Jing, M.; Hu, Y.; Teng, K.; Qian, X.; Shi, J.; Liu, L. Decreased cross-linking in interfacial polymerization and heteromorphic support between nanoparticles: Towards high-water and low-solute flux of hybrid forward osmosis membrane. *J. Colloid Interface Sci.* **2019**, *548*, 170–183. [CrossRef] [PubMed]
98. Nasrollahi, N.; Aber, S.; Vatanpour, V.; Mahmoodi, N.M. Development of hydrophilic microporous PES ultrafiltration membrane containing CuO nanoparticles with improved antifouling and separation performance. *Mater. Chem. Phys.* **2019**, *222*, 338–350. [CrossRef]
99. Athanasekou, C.; Sapalidis, A.; Katris, I.; Savopoulou, E.; Beltsios, K.; Tsoufis, T.; Kaltzoglou, A.; Falaras, P.; Bounos, G.; Antoniou, M. Mixed matrix PVDF/graphene and composite-skin PVDF/graphene oxide membranes applied in membrane distillation. *Polym. Eng. Sci.* **2019**, *59*, E262–E278. [CrossRef]
100. Yang, M.; Zhao, C.; Zhang, S.; Li, P.; Hou, D. Preparation of graphene oxide modified poly (m-phenylene isophthalamide) nanofiltration membrane with improved water flux and antifouling property. *Appl. Surf. Sci.* **2017**, *394*, 149–159. [CrossRef]
101. Wang, L.; Cao, T.; Dykstra, J.E.; Porada, S.; Biesheuvel, P.; Elimelech, M. Salt and water transport in reverse osmosis membranes: Beyond the solution-diffusion model. *Environ. Sci. Technol.* **2021**, *55*, 16665–16675. [CrossRef] [PubMed]
102. Geise, G.M.; Park, H.B.; Sagle, A.C.; Freeman, B.D.; McGrath, J.E. Water permeability and water/salt selectivity tradeoff in polymers for desalination. *J. Membr. Sci.* **2011**, *369*, 130–138. [CrossRef]
103. Yang, G.; Xie, Z.; Cran, M.; Ng, D.; Gray, S. Enhanced desalination performance of poly (vinyl alcohol)/carbon nanotube composite pervaporation membranes via interfacial engineering. *J. Membr. Sci.* **2019**, *579*, 40–51. [CrossRef]
104. Fornasiero, F.; In, J.B.; Kim, S.; Park, H.G.; Wang, Y.; Grigoropoulos, C.P.; Noy, A.; Bakajin, O. pH-tunable ion selectivity in carbon nanotube pores. *Langmuir* **2010**, *26*, 14848–14853. [CrossRef] [PubMed]

105. Dhand, V.; Hong, S.K.; Li, L.; Kim, J.-M.; Kim, S.H.; Rhee, K.Y.; Lee, H.W. Fabrication of robust, ultrathin and light weight, hydrophilic, PVDF-CNT membrane composite for salt rejection. *Compos. Part B Eng.* **2019**, *160*, 632–643. [CrossRef]
106. Kim, S.; Ou, R.; Hu, Y.; Li, X.; Zhang, H.; Simon, G.P.; Wang, H. Non-swelling graphene oxide-polymer nanocomposite membrane for reverse osmosis desalination. *J. Membr. Sci.* **2018**, *562*, 47–55. [CrossRef]
107. Song, X.; Wang, Y.; Jiao, C.; Huang, M.; Wang, G.-H.; Jiang, H. Microstructure regulation of polyamide nanocomposite membrane by functional mesoporous polymer for high-efficiency desalination. *J. Membr. Sci.* **2020**, *597*, 117783. [CrossRef]
108. Liu, S.; Low, Z.-X.; Hegab, H.M.; Xie, Z.; Ou, R.; Yang, G.; Simon, G.P.; Zhang, X.; Zhang, L.; Wang, H. Enhancement of desalination performance of thin-film nanocomposite membrane by cellulose nanofibers. *J. Membr. Sci.* **2019**, *592*, 117363. [CrossRef]
109. Chong, C.-Y.; Lau, W.-J.; Yusof, N.; Lai, G.-S.; Ismail, A.F. Roles of nanomaterial structure and surface coating on thin film nanocomposite membranes for enhanced desalination. *Compos. Part B Eng.* **2019**, *160*, 471–479. [CrossRef]
110. Epsztein, R.; Cheng, W.; Shaulsky, E.; Dizge, N.; Elimelech, M. Elucidating the mechanisms underlying the difference between chloride and nitrate rejection in nanofiltration. *J. Membr. Sci.* **2018**, *548*, 694–701. [CrossRef]
111. Yaroshchuk, A.E. Negative rejection of ions in pressure-driven membrane processes. *Adv. Colloid Interface Sci.* **2008**, *139*, 150–173. [CrossRef] [PubMed]
112. Zhao, D.; Lee, L.Y.; Ong, S.L.; Chowdhury, P.; Siah, K.B.; Ng, H.Y. Electrodialysis reversal for industrial reverse osmosis brine treatment. *Sep. Purif. Technol.* **2019**, *213*, 339–347. [CrossRef]
113. Park, J.W.; Kim, S.Y.; Noh, J.H.; Bae, Y.H.; Lee, J.W.; Maeng, S.K. A shift from chemical oxygen demand to total organic carbon for stringent industrial wastewater regulations: Utilization of organic matter characteristics. *J. Environ. Manag.* **2022**, *305*, 114412. [CrossRef] [PubMed]

membranes

Article

High-Performance Thin Film Composite Nanofiltration (NF) Membrane Constructed on Modified Polyvinylidene Fluoride (PVDF) Substrate

Junliang Dong, Qianzhi Sun, Xiaolin Feng and Ruijun Zhang *

School of Civil and Transportation Engineering, Hebei University of Technology, Tianjin 300401, China
* Correspondence: zrj@hebut.edu.cn; Tel.: +86-(18)-822231917

Abstract: The inherent hydrophobic nature of PVDF material renders it challenging to establish a stable aqueous hydration layer, thereby limiting its suitability as a substrate for the preparation of nanofiltration (NF) membranes. In this study, we developed a novel modification approach that effectively enhances the hydrophilicity of PVDF substrates through the incorporation of sulfonic acid-doped polyaniline (SPANI) and hyperbranched polyester (HPE) into the PVDF casting solution, followed by cross-linking with trimesoyl chloride (TMC). The introduction of SPANI and HPE, which contain reactive polar amino and hydroxyl groups, improved the hydrophilicity of the substrate, while the subsequent cross-linking with TMC effectively anchored these components within the substrate through the covalent linking between TMC and the reactive sites. Additionally, the hydrolysis of TMC yielded non-reactive carboxyl groups, which further enhanced the hydrophilicity of the substrate. As a result, the modified PVDF substrate exhibited improved hydrophilicity, facilitating the construction of an intact polyamide layer. In addition, the fabricated TFC NF membrane demonstrated excellent performance in the advanced treatment of tap water, achieving a total dissolved solid removal rate of 57.9% and a total organic carbon removal rate of 85.3%. This work provides a facile and effective route to modify PVDF substrates for NF membrane fabrication.

Keywords: nanofiltration membrane; substrate; PVDF; hydrophilicity; drinking water treatment

1. Introduction

As a typical pressure-driven membrane technology ranging between RO and ultrafiltration (UF), nanofiltration (NF) has emerged as a promising method for effectively purifying drinking water, profiting from its simplicity, energy efficiency, affordability, and environmental friendliness [1–3]. Compared to porous UF membranes and non-porous RO membranes, NF membranes' relatively dense selective layer and charged polymer matrix endow them with selectivity for contaminant removal; heavy metals, antibiotics, and other harmful organic substances can be effectively removed while retaining certain minerals that are beneficial to human health [4–7]. As the cornerstone of NF water purification technology, NF membranes significantly influence both purification efficiency and operational costs [8]. Consequently, many researchers are dedicated to the design and modification of high-performance NF membranes from various perspectives.

Currently, thin film composite (TFC) NF membranes, which feature an ultrathin selective layer formed through interfacial polymerization (IP) on a porous substrate, continue to dominate research interest and industrial applications [9,10]. During the IP process, the substrate is first impregnated with an aqueous solution containing an amine monomer such as piperazine (PIP) or m-phenylenediamine (MPD) [11]. It is then brought into contact with an immiscible organic phase containing acyl chloride (usually trimesoyl chloride, TMC), leading to the formation of a cross-linked polyamide (PA) layer. Most of the studies aimed at enhancing membrane performance primarily concentrated on the regulation of the PA selective layer. A variety of strategies have been explored including the screening of novel monomers [12], the optimization of IP reaction parameters [13], blending with nanomaterials [14], and post-treatment techniques [15]. However, a major obstacle that limits the practical application of these modification strategies is their negative impact on the integrity of the resulting selective layer [16]. In recent years, the design and modification of substrates have emerged as another promising strategy for enhancing the performance of TFC NF membranes [17,18]. However, most of the porous substrates in both reported and commercial TFC NF membranes are composed of polysulfone (PSF) and polyethersulfone (PES), rather than other polymers [19–21]. This is primarily due to the challenge of finding cost-effective alternatives that possess the necessary mechanical strength, hydrophilicity, and chemical stability required for the synthesis of TFC NF membranes.

Polyvinylidene fluoride (PVDF) has been widely utilized in the construction of microfiltration (MF) and UF membranes because of its strong thermal stability, mechanical strength, and chemical resistance [22]. However, to the best of our knowledge, no commercially viable TFC NF membranes currently utilize PVDF substrates—a gap attributed to their inherent hydrophobicity. This property critically precludes the formation of a stable aqueous hydration layer, which is essential for synthesizing a defect-free PA selective layer. To address this challenge, several pioneering researchers have proposed various strategies which can be classified into two routes. The first one involves innovating the IP procedure while maintaining the use of a hydrophobic PVDF substrate. For instance, ethanol/acetone can be mixed into the aqueous phase to increase the wettability of PVDF [23–25], but the permselectivity of the resultant NF membranes is not desirable. Taking the optimized NF sample prepared by Tang et al. as an example, its water permeance was only 8.1 L m^{-2} h^{-1} bar^{-1} [24]. Different from this idea, Lee at al. proposed to wet the PVDF substrate via the pressurization of the aqueous phase or flowing the organic solution at a specific velocity [26,27]. The second route involves directly regulating the physicochemical properties of the PVDF substrate. Based on Jurin's law, which states that the pressure exerted by the membrane surface to push water away is inversely proportional to the radius of the membrane pore, a hydrophobic PVDF membrane with a large pore size is expected to become wet by the aqueous phase. Accordingly, Wu et al. successfully prepared a TFC NF membrane on a PVDF substrate with a large pore size up to 100 nm [28]. The resultant NF membrane presented a high water permeance up to 27.0 L m^{-2} h^{-1} bar^{-1}, but its Na_2SO_4 rejection was only 96.2%. In addition, low-temperature plasmas were employed by Kim et al. to modify a commercial PVDF MF membrane, thus improving its hydrophilicity [29]. However, the oxidation of plasmas might decrease the mechanical strength of the PVDF membrane.

By comparing the two technical routes mentioned above, it is evident that modifying the characteristics of the PVDF substrate is more convenient and feasible, allowing for better compatibility with the existing industrial production processes of TFC NF membranes. To shorten the modification process, hydrophilic fragments can be directly introduced into the PVDF polymer matrix during the non-solvent-induced phase separation (NIPS) process. However, additional measures must be implemented to ensure the stable incorporation of these hydrophilic fragments into the resulting PVDF substrate. Only through such regulatory measures can we obtain a high-performance PVDF-supported TFC NF membrane.

To achieve this vision, in this work, sulfonic acid-doped polyaniline (SPANI) and hyperbranched polyester (HPE), which possess numerous hydrophilic groups and reactive sites, were blended into a PVDF casting solution and subsequently cross-linked with TMC. Subsequently, an intact PA layer can be formed on the modified PVDF substrate through a conventional IP strategy between PIP and TMC. This process can result in the synthesis of a high-performance PVDF-supported TFC NF membrane, which can be used to purify drinking water by further reducing dissolved organic and inorganic contaminants that conventional water treatment technologies cannot remove. Compared to previously reported studies, our strategy is both facile and effective, making it easily adoptable by existing TFC membrane manufacturers.

2. Materials and Methods

2.1. Materials

The polyvinylidene fluoride (PVDF, Solef® 6020, with a density of 1750~1800 mg/cm^3) used for membrane preparation was acquired from Solvay. The PET nonwoven fabric used as a support for PVDF membrane formation was obtained from Suka New Material Technology Co., Ltd (Shanghai, China). Hyperbranched polyester (HPE, H30), featuring a molecular weight of 3387.4 g/mol, was generously supplied by Weihai Chenyuan Dendrimer Technology Co., Ltd. (Weihai, China). Sulfonic acid-doped polyaniline (SPANI, 98%) was purchased from Shanghai Maclin Biochemical Technology Co., Ltd. (Shanghai, China). N-methylpyrrolidone (NMP, 99%), which served as the solvent for preparing the casting solution, was also purchased from Shanghai Maclin Biochemical Technology Co., Ltd. Trimesoyl chloride (TMC, >98.0%) was obtained from Tokyo Chemical Industry Co., Ltd. (Shanghai, China). Piperazine (PIP, >99.0%), n-hexane, sodium sulfate (Na_2SO_4, >99.0%), sodium chloride (NaCl, >99.0%), and Bovine Serum Albumin (BSA) were all sourced from Aladdin Industrial Corporation Co., Ltd. (Shanghai, China). Unless otherwise specified, all solutions were prepared using ultrapure water produced by a reverse osmosis and ion exchange purification system (RUPF-30, Xinrui, China).

2.2. Preparation of Modified PVDF Substrate and PVDF-Supported TFC NF Membrane

PVDF powder, NMP, HPE, and SPANI were used to prepare the casting solution. The PVDF powder intended for membrane preparation should be vacuum-dried for 24 h at a pressure of −0.08 MPa and a temperature of 60 °C. During the preparation of the casting solution, HPE (4 wt%) or SPANI (0.3 wt%) was first added to the NMP solvent, followed by stirring with a magnetic stirrer for 15 min and ultrasound dispersion for 30 min. Afterward, PVDF powder (19 wt%) was added to the mixture and mechanically mixed for 24 h to obtain a homogeneous casting solution. The casting solution should be placed in a vacuum oven, with the pressure set between −0.08 MPa and −0.06 MPa, and then degassed at 25 °C for 24 h to achieve a uniform and stable casting liquid. Then it can be cast onto the PET nonwoven fabric fixed on the glass sheet using a casting knife. The thickness of the casting solution was fixed at 200 μm. The membrane obtained after casting was subsequently immersed in a coagulation bath (pure water at 25 °C) for 24 h to facilitate phase inversion.

For the cross-linking modification of the HPE/SPANI doping PVDF membrane, two strategies were proposed. The first approach involved directly adding TMC (0.2 wt%) into the casting solution prior to phase inversion. The resultant casting solution containing PVDF/NMP/HPE/SPANI/TMC was used to prepare the modified PVDF membrane. This modified membrane was denoted as the HCDM-PVDF substrate. The second approach entailed cross-linking after phase inversion. For this approach, the casting solution containing PVDF/NMP/HPE/SPANI was used to prepare the modified PVDF membrane. After phase inversion, the resultant membrane was further soaked in a TMC hexane solution (0.2 wt%) for 1 min. Then the modified PVDF membrane was denoted as the HCAM-PVDF substrate.

As shown in Figure 1, the TFC NF membrane was fabricated using the conventional IP method on the HCDM-PVDF or HCAM-PVDF substrate. The aqueous phase solution was prepared by dissolving 0.6 $w/v\%$ PIP in pure water, while the organic phase solution was composed of 0.1 $w/v\%$ TMC dissolved in n-hexane. The substrate membrane was first coated with the PIP solution for 5 min, after which excess aqueous solution was removed using a rubber scraper. Next, the TMC n-hexane solution was gently poured onto the surface of the substrate to initiate the IP reaction for 20 s, followed by rinsing to remove the residual TMC with pure n-hexane. Finally, the membrane sample was heated in an oven at 50 °C for 3 min to stabilize the membrane structure. The resulting TFC NF membrane was stored in ultrapure water at 4 °C until testing.

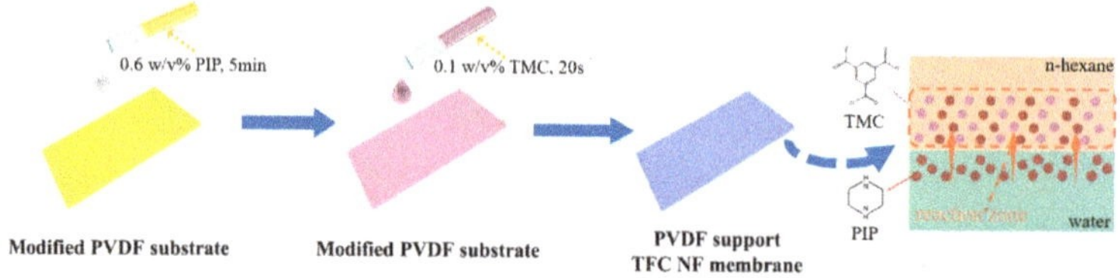

Figure 1. An illustrative diagram of the NF membrane preparation procedure on the modified PVDF substrate.

2.3. Membrane Characterization

The selected membrane sample for characterization was naturally dried in a well-ventilated environment for 24 h, followed by further drying in a vacuum oven at 30 °C and a pressure of −0.06 MPa for an additional 24 h to remove residual moisture. The membrane's surface and cross-sectional morphology were characterized using a scanning electron microscope (SEM, TESCAN MIRA LMS, Brno, Czech Republic). The hydrophilicity of the membrane surface was assessed by measuring the water contact angle (WCA) using a drop shape analyzer (Dataphysics OCAI5EC, Stuttgart, Germany). The reported WCA value represents the average of five measurements taken at different locations on each membrane sample, with a water droplet volume of 0.2 µL used for the WCA assessment.

2.4. Evaluation of Membrane Separation Performance

Membrane filtration experiments were conducted using a self-made cross-flow membrane filtration system (as illustrated in Figure 2), with the membrane cell sourced from Sterlitech Corporation (CF016P, Auburn, WA, USA). The effective membrane area in this cell was 20.6 cm^2. For the modified PVDF substrate, the filtration test commenced by pre-pressurizing the membrane with pure water at 2 bar until a steady flux was achieved. Measurements were then conducted under a pressure of 1 bar at 25 °C. In the case of the PVDF-supported TFC NF membrane, the filtration test began with pre-pressurization using pure water at 6 bar until a steady flux was attained. Subsequently, measurements were performed at 5 bar and 25 °C, with a cross-flow velocity of 0.45 m/s to reduce concentration polarization. The pure water permeance (PWP, $L \cdot m^{-2} \cdot h^{-1} \cdot bar^{-1}$) was calculated according to Equation (1). In this equation, A is the effective membrane area (m^2), Δt is the time interval (h), ΔP is the trans-membrane pressure (bar), and V is the permeate volume (L).

$$PWP = \frac{V}{A \times \Delta t \times \Delta P} \quad (1)$$

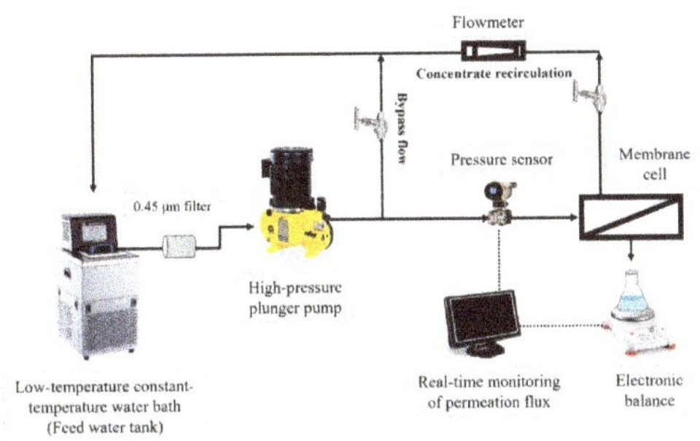

Figure 2. A diagram of the self-made cross-flow membrane filtration system.

BSA is a large (~66.5 kDa), monomeric, globular protein which is highly soluble in water and buffers over a wide pH range. It has a hydrodynamic diameter of approximately 7 nm and a globular structure, making it an ideal molecular probe for evaluating pore sizes in ultrafiltration (UF) membranes. This is particularly relevant for UF membranes with a molecular weight cut-off (MWCO) of 50~100 kDa, as their pore sizes align well with BSA's dimensions. Notably, UF membranes in this MWCO range are optimal substrates for TFC NF membrane preparation. Therefore, in this work, 200 mg/L BSA was used to assess the selectivity of modified PVDF substrates. Subsequently, 2000 mg/L Na_2SO_4 was employed to evaluate the salt rejection of the PVDF-supported TFC NF membrane. The observed solute rejection (R, %) was calculated according to Equation (2). The BSA concentration was determined using ultraviolet spectrophotometry, while the Na_2SO_4 concentration was assessed by measuring the conductivities of both the feed and permeate solutions with a conductivity meter (Ohaus ST3100C/F, Pine Brook, NJ, USA). In the following equation, Cf is the solute concentration in feed, and Cp is the solute concentration in the permeate.

$$R = \left(1 - \frac{C_p}{C_f}\right) \times 100\% \tag{2}$$

The PVDF-supported TFC NF membrane was further utilized to treat tap water sourced from the municipal network in the Hongqiao District of Tianjin, China. The main quality indicators of this tap water are presented in Table 1. The pH of the water sample was measured using a calibrated pH meter (Lei-ci, DZS-706A, Shanghai, China), while the concentrations of various ions were analyzed using ion chromatography (Metrohm 883 Basic IC plus, Herisau, Switzerland). The total organic carbon (TOC) content was determined using a TOC analyzer (TOC-L CPH, Shimadzu, Kyoto, Japan). It can guarantee a 100% oxidation recovery rate for all carbon in the sample through catalytic combustion at a fixed temperature of 680 °C. A high-sensitivity NDIR (non-dispersive infrared detector), combined with a large sample volume combustion system, enables the detection of TOC at levels as low as 10 μg/L.

Table 1. Water quality indicators of tap water.

Index	Content	Unit
pH	7.4~7.7	-
TOC	2.2	mg/L
TDS	155.2	mg/L
Na^+	9.32	mg/L
K^+	2.19	mg/L
Ca^{2+}	45.33	mg/L

Table 1. *Cont.*

Index	Content	Unit
Mg^{2+}	7.45	mg/L
Cl^{-}	8.99	mg/L
SO_4^{2-}	29.34	mg/L

3. Results and Discussion

3.1. Influence of SPANI/HPE Doping Modification on PVDF Substrate

To identify the influence of SPANI/HPE doping modification on the PVDF substrate, four kinds of substrates were prepared, as illustrated in Figure 3a. Then the separation performance and physicochemical properties of the resultant membranes were characterized. As shown in Figure 3b, the pure PVDF substrate without any dopant exhibited a PWP around 185 L m^{-2} h^{-1} bar^{-1} and a poor BSA rejection of ~62%. This inferior separation performance can be attributed to the presence of macro-porous defects observed in the SEM image of Figure 3d, which facilitates the passage of water as well as BSA molecules. Additionally, a remarkable high water contact angle (WCA) of nearly 90° can be observed in Figure 3c for the pure PVDF substrate, indicating the inherent hydrophobicity of PVDF material.

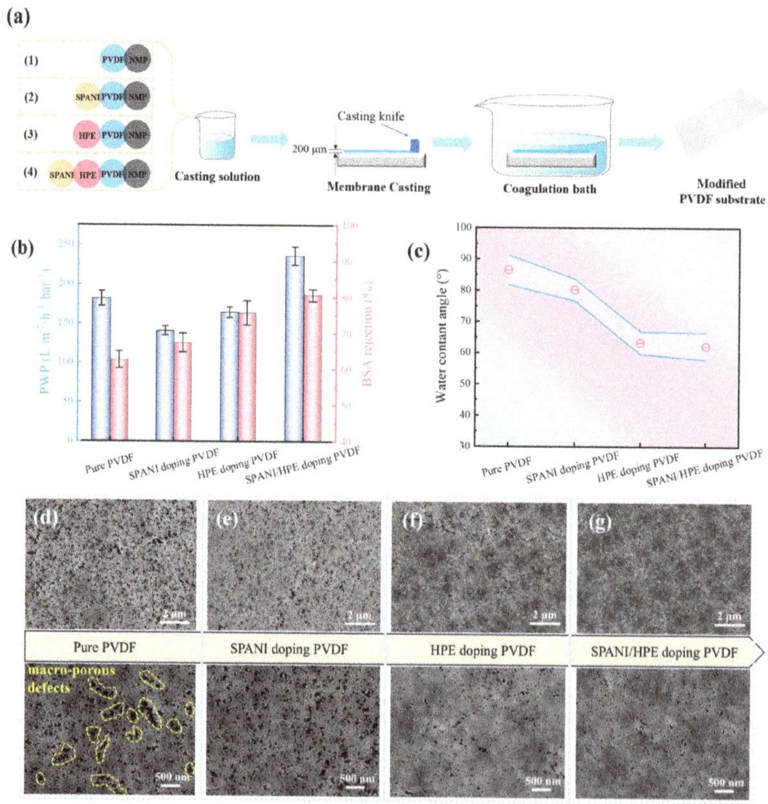

Figure 3. Property evaluation of PVDF-based substrate. (**a**) Illustrative diagram of preparation procedure. (**b**) Separation performance (**c**) water contact angle and (**d–g**) SEM surface morphology of modified PVDF substrates.

After the addition of SPANI into the PVDF casting solution, the membrane's PWP decreased, while its BSA rejection level increased. However, the changes in surface WCA (Figure 3c) and SEM morphology (Figure 3e) were not significant. This suggests that the doping modification with SPANI did not affect the surface properties of the membrane. The observed changes in separation performance are likely due to the accumulation of SPANI particles within the inner pore structures of PVDF. Unlike SPANI, HPE doping could simultaneously alter both the separation performance and surface characteristics of the membrane. As shown in Figure 3b, the BSA rejection of HPE doping PVDF membranes obviously improved, accompanied by a slight decrease in PWP. Meanwhile, the water contact angle sharply decreased to 63.3°. In addition, as shown in Figure 3f, the large pore defects were eliminated, and membrane surface porosity was reduced.

To harness the modification effects of both SPANI and HPE, a substrate denoted as SPANI/HPE doping PVDF was prepared and tested. Interestingly, there seems to be some synergistic modification effect between SPANI and HPE. In comparison to HPE doping PVDF, SPANI/HPE doping PVDF presented a similar water contact angle and more even pore size distribution. Consequently, a significant enhancement in both PWP (~230 L m^{-2} h^{-1} bar^{-1}) and BSA rejection (~81%) was produced, as shown in Figure 3b. However, the BSA rejection rate of 81% was still inadequate, and the water contact angle of 62.1° was not sufficiently low. According to previous research [30–32], the water contact angle of a substrate for TFC NF membrane preparation should be lower than 50°.

3.2. Effect of TMC Cross-Linking on Modification of PVDF Substrate

To further optimize the properties of the PVDF substrate, two types of TMC cross-linking methods were employed for subsequent modifications. As depicted in Figure 4a,b, the first method involves directly incorporating TMC into the PVDF casting solution prior to phase inversion, referred to as HCDM-PVDF. The second method entails immersing the PVDF substrate into the TMC/hexane solution after phase inversion, termed HCAM-PVDF.

The separation performance and physicochemical properties of the TMC-modified PVDF substrate were characterized. As presented in Figure 4c, compared to the uncross-linked PVDF substrate, both the HCDM-PVDF and HCAM-PVDF substrates exhibited slightly enhanced PWP and BSA rejection. Meanwhile, Figure 4d demonstrates a substantial decline in WCA (~45°) for the TMC cross-linked substrate. This decline can be attributed to the susceptibility of the acyl chloride functional groups in the TMC molecule to hydrolysis and their conversion to carboxyl groups, which substantially increases the hydrophilicity of the PVDF substrate. In addition, SEM images (Figure 4e–g) revealed a notable increase in pore structures on the substrate following TMC modification. Notably, the BSA rejection level (~88%) of the HCDM-PVDF substrate and its hydrophilicity (44.3°) were slightly greater than those of the HCAM-PVDF substrate (~87%, 45.7°). Additionally, The SEM morphology of the HCDM-PVDF substrate revealed a higher density of pore structures compared to the HCAM-PVDF substrate. These findings indicate that the direct incorporation of TMC into the PVDF casting solution for the HCDM-PVDF substrate could bring a better modification effect. Moreover, the first cross-linking modification strategy, which involves directly adding TMC to the PVDF casting solution before phase inversion, is more convenient and cost-effective.

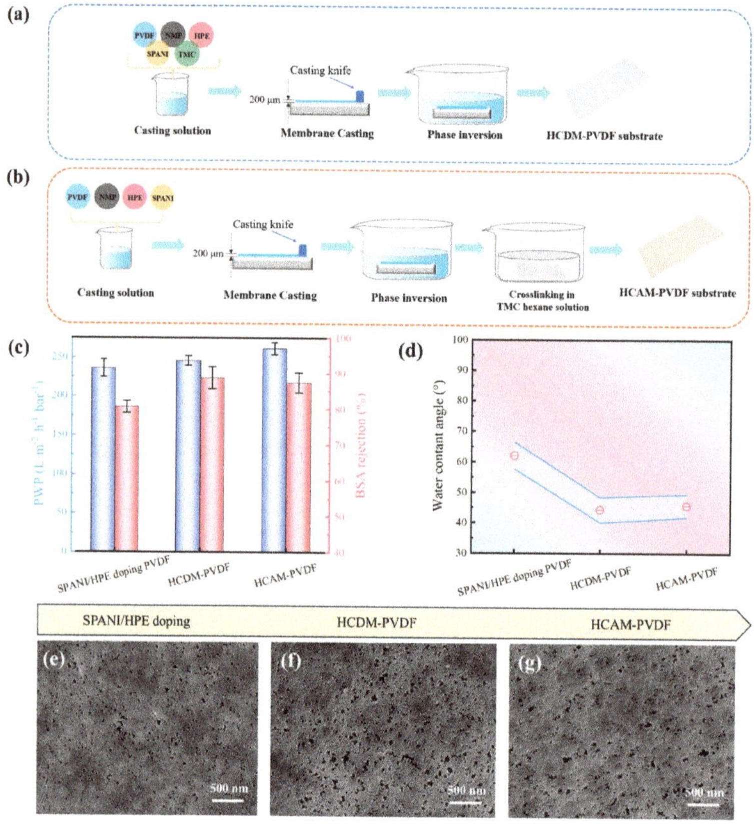

Figure 4. Property evaluation of TMC cross-linking PVDF substrates. Illustrative diagram of preparation procedure of (**a**) HCDM-PVDF substrate and (**b**) HVAM-PVDF substrate. (**c**) Separation performance, (**d**) water contact angle, and (**e–g**) SEM surface morphology of TMC cross-linking PVDF substrates.

3.3. Impact of PVDF Substrate on Performance of Resultant TFC NF Membranes

The preceding sections confirmed the effectiveness of SPANI/HPE doping followed by TMC modification in enhancing the hydrophilicity and optimizing the pore structure of PVDF substrates. To assess the impact of these modification processes on interfacial polymerization, six types of PVDF-based substrates were utilized to prepare the TFC NF membrane. The separation performance of six resultant NF membranes is shown in Figure 5a. The pure PVDF substrate, due to its inherent hydrophobicity, yielded an NF membrane with a low PWP of 13.1 L m^{-2} h^{-1} bar^{-1} and an inferior Na_2SO_4 rejection of only 38.2%. Doping SPANI or HPE into the PVDF substrate led to a typical "trade-off" effect, where a significant enhancement in the Na_2SO_4 rejection (~95%) of the fabricated NF membrane was achieved at the expense of reduced PWP (8 L m^{-2} h^{-1} bar^{-1}). Interestingly, the alliance of SPANI and HPE further slightly increased the PWP of the fabricated NF membrane, likely due to the higher PWP of the SPANI/HPE doping PVDF substrate. Notably, the cross-linking modification step involving TMC breaks through the "trade-off" limitation, wherein both PWP and Na_2SO_4 rejection were concurrently enhanced. During the two TMC modification methods, the HCDM-PVDF substrate, where TMC was directly incorporated into the casting solution, demonstrated better NF performance based on its enhanced modification effect. Specifically, the NF membrane fabricated with the HCDM-PVDF substrate exhibited a high PWP of 15.7 L m^{-2} h^{-1} bar^{-1} alongside a remarkable Na_2SO_4 rejection up to 98.8%. By comparison, the PVDF-supported TFC NF membrane reported by Tang et al. exhibited a lower PWP of 8.1 L m^{-2} h^{-1} bar^{-1} [24]. Similarly, while Wu et al. achieved a higher PWP of 27.0 L m^{-2} h^{-1} bar^{-1} using a PVDF substrate with larger pore size, their Na_2SO_4 rejection was limited to 96.2% [28]. Therefore, it can be concluded that our modification strategy achieved a good balance between water permeability and selectivity.

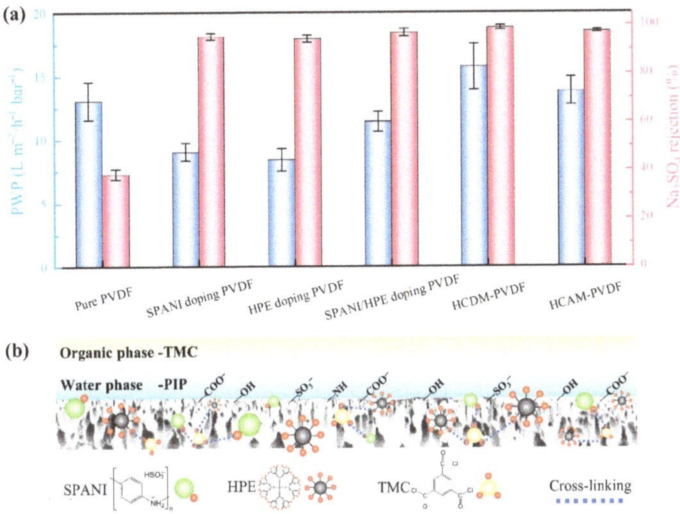

Figure 5. (a) Separation performance of fabricated NF membrane. (b) Illustrative diagram of impact of substrate on resultant TFC NF membrane.

As the TFC NF membrane constructed on the HCDM-PVDF substrate presented the best separation performance, its modification mechanism was highlighted and is illustrated in Figure 5b to elucidate the impact of the substrate on NF membrane formation. The typical NF membrane formation process involves the sequential immersion of the substrate in aqueous PIP and organic TMC solutions, driving interfacial polymerization (IP) between PIP and TMC on the stable hydration layer of the substrate to form an intact selective layer. In this study, the substrate was modified with SPANI, HPE, and TMC to enhance its properties. The introduction of SPANI and HPE, which contain reactive polar amino and hydroxyl groups, respectively, enhanced the hydrophilicity of the substrate and created additional reactive sites for TMC coordination, as the functional groups of SPANI and HPE readily react with the acyl chlorine groups of TMC. Additionally, the PVDF substrate was cross-linked with TMC, which covalently bonded with the reactive sites of SPANI and HPE, effectively anchoring these components within the substrate. In addition, TMC underwent hydrolysis, yielding non-reactive carboxyl groups that further enhanced the substrate's hydrophilicity. Consequently, the improved hydrophilicity of the PVDF substrate facilitated the formation of a stable hydration layer on its surface, which in turn contributed to the creation of an intact selective layer. Moreover, the reactive sites introduced by SPANI and HPE promoted additional reactions that replenished defects, thereby promoting the formation of a high-performance NF membrane.

3.4. Performance Evaluation of PVDF-Supported TFC NF Membrane

The demand for high-quality drinking water has catalyzed the focus on advanced treatment technologies for municipal tap water production [33]. Taking the group Health Drinking Water Quality Standard (T/BJWA 001-2021) [34] in China as example, it imposes more stringent requirements, specifying TDS levels of 50~300 mg/L and TOC $\leq$ 1.0 mg/L. In the context of this trend, NF technology has emerged as a promising candidate due to its high-quality effluent [35]. To assess the viability of the NF membrane fabricated on the HCDM-PVDF substrate, it was used to treat the tap water sourced from the municipal network in the Hongqiao District of Tianjin, China. The main quality indicators of this tap water are presented in Table 1. As depicted in Figure 6a, this NF membrane exhibited a notable tap water permeance of 15 L m^{-2} h^{-1} bar^{-1}, alongside a moderate total dissolved solid (TDS) removal rate of 57.9%. Notably, the rejection rates of bivalent ions ($SO_4{}^{2-}$ (86.9%), Mg^{2+} (42.9%), and Ca^{2+} (32.6%)) were significantly higher than those of monovalent ions (Na^+ (18.7%), K^+ (17.3%), and Cl^- (18.9%)). This discrepancy can be attributed to the larger hydration diameter and electric charge of the bivalent ions. Additionally, the organic matter removal efficiency of the NF membrane was assessed through TOC rejection, which revealed an excellent removal effect for TOC up to 85.3%. As shown in Table 1, the TOC of raw tap water was approximately 2.2 mg/L. After the NF treatment, this level was reduced to 0.32 mg/L, effectively meeting the high-quality drinking water standard of TOC < 1 mg/L. This proficiency in organic matter removal is conducive to ensuring the safety of drinking water. In addition, Excitation Emission Matrix (EEM) Spectra were employed to quantify the organic matter present within tap water before and after the NF treatment. In general, the fluorescence response regions of the EEM can be categorized as M, P, H, and F, corresponding to microbial products, proteins, humic acids, and fulvic acids, respectively [36]. As demonstrated in Figure 6b, only small amounts of fulvic acid (F region) and humic acid (H region) were detected in the raw tap water, reflecting its relatively clean quality. It is well known that fulvic acid and humic acid are types of natural organic matter (NOM). Despite their diversity and complexity, they typically possess large molecular weights. Therefore, the NF membrane is expected to effectively remove these substances with >96% rejection [37,38]. Consequently, as shown in Figure 6c, these fractions were detected at nearly negligible levels in the NF permeate water. These results demonstrated that the NF membrane constructed on the HCDM-PVDF substrate can significantly reduce the content of organic impurities in tap water, underscoring its exceptional ability to produce high-quality drinking water.

To evaluate the long-term stability of the PVDF-supported TFC NF membrane, we conducted a continuous 100 h tap water filtration experiment at 5 bar and 25 °C. Throughout this extended operation, water permeation flux, TDS removal, and TOC removal were monitored at 10 h intervals. The results in Figure 6d demonstrate exceptional operational stability: permeation flux decreased marginally from 74.3 to 71.7 $L{\cdot}m^{-2}{\cdot}h^{-1}$, while TDS and TOC rejection remained consistently high at approximately 58.3% and 85.9%, respectively. This robust performance confirms the system's long-term stability potential.

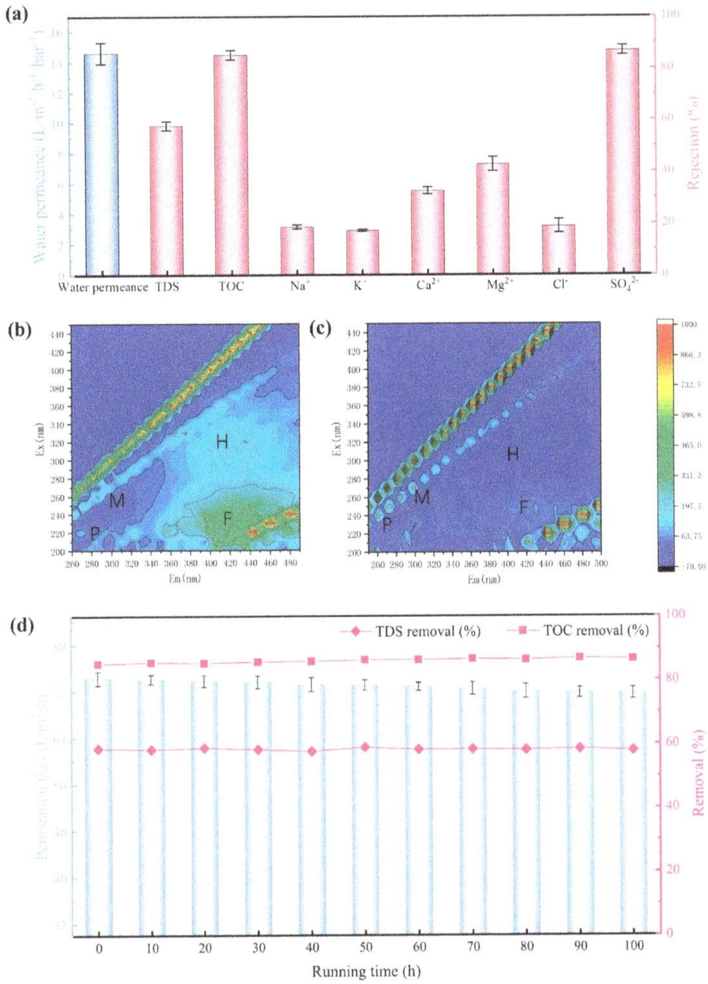

Figure 6. (**a**) Tap water treatment performance. (**b**) Tap water and (**c**) NF membrane effluent measured by three-dimensional fluorescence. (**d**) Long-term tap water filtration performance of PVDF-supported TFC NF membrane tested at 5 bar and 25 °C.

4. Conclusions

In this study, we successfully enhanced the hydrophilicity of the PVDF substrate by incorporating SPANI and HPE into the casting solution, followed by cross-linking with TMC. This modification rendered the substrate highly suitable for constructing high-performance TFC NF membranes. The presence of reactive polar groups (amino from SPANI and hydroxyl from HPE) significantly decreased the WCA from 86.5° to 62.1°. The subsequent TMC cross-linking effectively anchored these hydrophilic components within the substrate matrix and introduced non-reactive carboxyl groups via hydrolysis, further reducing the WCA to 44.3°. Crucially, this enhanced hydrophilicity promoted the formation of a dense and intact PA selective layer. The resulting TFC NF membrane exhibited desirable performance in advanced tap water treatment, achieving proper removal rates of 57.9% for TDS and 85.3% for TOC. This enables the permeate (TDS: ~65.3 mg/L; TOC: ~0.32 mg/L) to meet the stringent requirements of the Chinese group Health Drinking Water Quality

Standard (T/BJWA 001-2021). Beyond offering a straightforward and efficient method for PVDF substrate hydrophilization, the proposed approach in this work could be adapted to modify other hydrophobic polymer substrates in future studies. Additionally, broader feed water testing, such as wastewater containing specific organic pollutants and brackish water with higher ionic strength, is also necessary to validate its robustness and to identify suitable application niches.

Author Contributions: Conceptualization, J.D. and R.Z.; methodology, J.D. and X.F.; software, J.D. and Q.S.; validation, J.D. and X.F.; formal analysis, J.D. and Q.S.; investigation, J.D. and X.F.; writing—original draft preparation, J.D.; writing—review and editing, R.Z.; visualization, J.D. and Q.S.; supervision, R.Z.; project administration, R.Z.; funding acquisition, R.Z. All authors have read and agreed to the published version of the manuscript.

Funding: This work was financially supported by Central Guiding Fund for Local Science and Technology Development Project (236Z3605G).

Institutional Review Board Statement: Not applicable.

Data Availability Statement: Data is unavailable due to privacy or ethical restrictions.

Conflicts of Interest: The authors declare no conflicts of interest.

References

1. Yang, Z.; Wu, C.; Tang, C.Y. Making waves: Why do we need ultra-permeable nanofiltration membranes for water treatment? *Water Res. X* **2023**, *19*, 100172. [CrossRef] [PubMed]
2. Zhang, R.; Sun, Q.; Tian, J.; Van der Bruggen, B.; Zhu, J. Non-polyamide nanofiltration (NPA-NF) membrane: A non-mainstream but indispensable member of the "membrane family". *Desalination* **2023**, *564*, 116772. [CrossRef]
3. Shen, Q.; Xu, S.-J.; Dong, Z.-Q.; Zhang, H.-Z.; Xu, Z.-L.; Tang, C.Y. Polyethyleneimine modified carbohydrate doped thin film composite nanofiltration membrane for purification of drinking water. *J. Membr. Sci.* **2020**, *610*, 118220. [CrossRef]
4. Tian, J.; Chang, H.; Gao, S.; Zhang, R. How to fabricate a negatively charged NF membrane for heavy metal removal via the interfacial polymerization between PIP and TMC? *Desalination* **2020**, *491*, 114499. [CrossRef]
5. Abedi, F.; Dubé, M.A.; Kruczek, B. Adsorption of heavy metals on the surface of nanofiltration membranes: "A curse or blessing"? *J. Membr. Sci.* **2023**, *685*, 121988. [CrossRef]
6. Wu, B.; Wang, N.; Shen, Y.; Jin, C.-G.; An, Q.-F. Inorganic salt regulated zwitterionic nanofiltration membranes for antibiotic/monovalent salt separation. *J. Membr. Sci.* **2023**, *666*, 121144. [CrossRef]
7. Liu, Y.; Du, J.; Wu, H.; Cong, C.; Zhang, H.; Wang, J.; Wang, Z. Antifouling streptomycin-based nanofiltration membrane with high permselectivity for dye/salt separation. *Sep. Purif. Technol.* **2022**, *297*, 121443. [CrossRef]
8. Wang, K.; Wang, X.; Januszewski, B.; Liu, Y.; Li, D.; Fu, R.; Elimelech, M.; Huang, X. Tailored design of nanofiltration membranes for water treatment based on synthesis–property–performance relationships. *Chem. Soc. Rev.* **2022**, *51*, 672–719. [CrossRef] [PubMed]
9. Li, X.; Wang, Z.; Han, X.; Liu, Y.; Wang, C.; Yan, F.; Wang, J. Regulating the interfacial polymerization process toward high-performance polyamide thin-film composite reverse osmosis and nanofiltration membranes: A review. *J. Membr. Sci.* **2021**, *640*, 119765. [CrossRef]
10. An, X.C.; Ming, S.J.; Xiang, Y.X.; He, J.S.; Zhang, K.; Luo, H.B.; Fan, L.Q.; Chen, W.; Ma, D.D.; Cheng, L. Hierarchical graphitic carbon nitride modified interface supported ultrathin polyamide layer for dye/salt separation. *Appl. Surf. Sci.* **2024**, *671*, 160732. [CrossRef]
11. Lu, X.; Elimelech, M. Fabrication of desalination membranes by interfacial polymerization: History, current efforts, and future directions. *Chem. Soc. Rev.* **2021**, *50*, 6290–6307. [CrossRef] [PubMed]
12. Ormanci-Acar, T.; Tas, C.E.; Keskin, B.; Ozbulut, E.B.S.; Turken, T.; Imer, D.; Tufekci, N.; Menceloglu, Y.Z.; Unal, S.; Koyuncu, I. Thin-film composite nanofiltration membranes with high flux and dye rejection fabricated from disulfonated diamine monomer. *J. Membr. Sci.* **2020**, *608*, 118172. [CrossRef]
13. Zhan, Z.-M.; Xu, Z.-L.; Zhu, K.-K.; Tang, Y.-J. How to understand the effects of heat curing conditions on the morphology and performance of polypiperazine-amide NF membrane. *J. Membr. Sci.* **2020**, *597*, 117640. [CrossRef]

14. Johnson, D.J.; Hilal, N. Nanocomposite nanofiltration membranes: State of play and recent advances. *Desalination* **2022**, *524*, 115480. [CrossRef]
15. Guo, S.; Chen, X.; Wan, Y.; Feng, S.; Luo, J. Custom-tailoring loose nanofiltration membrane for precise biomolecule fractionation: New insight into post-treatment mechanisms. *ACS Appl. Mater. Interfaces* **2020**, *12*, 13327–13337. [CrossRef] [PubMed]
16. Lau, W.-J.; Lai, G.-S.; Li, J.; Gray, S.; Hu, Y.; Misdan, N.; Goh, P.-S.; Matsuura, T.; Azelee, I.W.; Ismail, A.F. Development of microporous substrates of polyamide thin film composite membranes for pressure-driven and osmotically-driven membrane processes: A review. *J. Ind. Eng. Chem.* **2019**, *77*, 25–59. [CrossRef]
17. Lim, Y.J.; Goh, K.; Lai, G.S.; Zhao, Y.; Torres, J.; Wang, R. Unraveling the role of support membrane chemistry and pore properties on the formation of thin-film composite polyamide membranes. *J. Membr. Sci.* **2021**, *640*, 119805. [CrossRef]
18. Li, P.; Xie, H.X.; Bi, Y.F.; Miao, C.C.; Chen, K.; Xie, T.T.; Zhao, S.C.; Sun, H.X.; Yang, X.J.; Hou, Y.F.; et al. Preparation of high flux organic solvent nanofiltration membrane based on polyimide/Noria composite ultrafiltration membrane. *Appl. Surf. Sci.* **2023**, *618*, 156650. [CrossRef]
19. Peng, L.E.; Yang, Z.; Long, L.; Zhou, S.; Guo, H.; Tang, C.Y. A critical review on porous substrates of TFC polyamide membranes: Mechanisms, membrane performances, and future perspectives. *J. Membr. Sci.* **2022**, *641*, 119871. [CrossRef]
20. Tian, J.; Chang, H.; Gao, S.; Zong, Y.; Van der Bruggen, B.; Zhang, R. Direct generation of an ultrathin (8.5 nm) polyamide film with ultrahigh water permeance via in-situ interfacial polymerization on commercial substrate membrane. *J. Membr. Sci.* **2021**, *634*, 119450. [CrossRef]
21. He, M.; Li, T.; Hu, M.; Chen, C.; Liu, B.; Crittenden, J.; Chu, L.-Y.; Ng, H.Y. Performance improvement for thin-film composite nanofiltration membranes prepared on PSf/PSf-g-PEG blended substrates. *Sep. Purif. Technol.* **2020**, *230*, 115855. [CrossRef]
22. Zou, D.; Lee, Y.M. Design strategy of poly (vinylidene fluoride) membranes for water treatment. *Prog. Polym. Sci.* **2022**, *128*, 101535. [CrossRef]
23. Zhang, H.; Zheng, J.; Zhao, Z.; Han, C.C. Role of wettability in interfacial polymerization based on PVDF electrospun nanofibrous scaffolds. *J. Membr. Sci.* **2013**, *442*, 124–130. [CrossRef]
24. Tang, Y.; Zhu, J.; Yu, H.; Zhang, F.; Hu, S.; Chen, H.; Zhang, C.; Wu, H.; Yu, L.; Wang, X. Novel PA/PVDF hollow fiber nanofiltration membrane with high permeability and Ca^{2+}/antibiotics selectivity for drinking water purification. *Adv. Membr.* **2024**, *4*, 100102. [CrossRef]
25. Marquez, J.A.D.; Ang, M.B.M.Y.; Doma, B.T., Jr.; Huang, S.-H.; Tsai, H.-A.; Lee, K.-R.; Lai, J.-Y. Application of cosolvent-assisted interfacial polymerization technique to fabricate thin-film composite polyamide pervaporation membranes with PVDF hollow fiber as support. *J. Membr. Sci.* **2018**, *564*, 722–731. [CrossRef]
26. Lee, J.S.; Lee, H.H.; Seo, J.A.; Park, H.S.; Park, J.; Min, B.R. Interfacial polymerization on hydrophobic PVDF UF membranes surface: Membrane wetting through pressurization. *Appl. Surf. Sci.* **2015**, *356*, 1207–1213. [CrossRef]
27. Lee, J.S.; Seo, J.A.; Lee, H.H.; Min, B.R. Salt rejection characterization of PA/PVDF composite membranes prepared by pressurization method. *Chem. Lett.* **2015**, *44*, 1404–1406. [CrossRef]
28. Wu, J.; Xia, M.; Li, Z.; Shen, L.; Li, R.; Zhang, M.; Jiao, Y.; Xu, Y.; Lin, H. Facile preparation of polyvinylidene fluoride substrate supported thin film composite polyamide nanofiltration: Effect of substrate pore size. *J. Membr. Sci.* **2021**, *638*, 119699. [CrossRef]
29. Kim, E.-S.; Kim, Y.J.; Yu, Q.; Deng, B. Preparation and characterization of polyamide thin-film composite (TFC) membranes on plasma-modified polyvinylidene fluoride (PVDF). *J. Membr. Sci.* **2009**, *344*, 71–81. [CrossRef]
30. Dai, R.B.; Yang, Z.; Qiu, Z.W.; Long, L.; Tang, C.Y.; Wang, Z.W. Distinct impact of substrate hydrophilicity on performance and structure of TFC NF and RO polyamide membranes. *J. Membr. Sci.* **2022**, *662*, 120966. [CrossRef]
31. Sun, Q.Z.; Chai, Z.X.; An, Z.X.; Dong, J.L.; Zhang, R.J. Construction of a high-performance hollow fiber composite NF membrane via the simultaneous modification of substrate and selective layer. *Sep. Purif. Technol.* **2024**, *341*, 126908. [CrossRef]
32. Geng, C.B.; Zhao, F.B.; Niu, H.Y.; Zhang, J.M.; Dong, H.X.; Li, Z.G.; Chen, H.X. Enhancing the permeability, anti-biofouling performance and long-term stability of TFC nanofiltration membrane by imidazole-modified carboxylated graphene oxide/polyethersulfone substrate. *J. Membr. Sci.* **2022**, *664*, 121099. [CrossRef]
33. Miller, J.D.; Staddon, C.; Salzberg, A.; Lucks, J.B.; de Bruin, W.B.; Young, S.L. Self-reported anticipated harm from drinking water across 141 countries. *Nat. Commun.* **2024**, *15*, 7320. [CrossRef] [PubMed]
34. *T/BJWA 001-2021*; Health Drinking Water Quality Standard. Comprehensive Testing Center of Chinese Academy of Inspection and Quarantine: Beijing, China, 2021.
35. Liu, L.; Lan, H.C.; Cui, Y.Q.; Tang, Q.W.; Bai, J.Q.; An, X.Q.; Sun, M.; Liu, H.J.; Qu, J.H. A Janus membrane with electro-induced multi-affinity interfaces for high-efficiency water purification. *Sci. Adv.* **2024**, *10*, 8696. [CrossRef] [PubMed]
36. Jeon, I.; Ryberg, E.C.; Alvarez, P.J.J.; Kim, J.H. Technology assessment of solar disinfection for drinking water treatment. *Nat. Sustain.* **2022**, *5*, 801–808. [CrossRef]

37. Ruohomäki, K.; Väisänen, P.; Metsämuuronen, S.; Kulovaara, M.; Nyström, M. Characterization and removal of humic substances in ultra-and nanofiltration. *Desalination* **1998**, *118*, 273–283. [CrossRef]
38. Geng, C.; Zhao, F.; Niu, H.; Zhang, J.; Liu, Y.; Liu, L.; Dong, H.; Li, Z.; Chen, H. Zwitterionic loose polyamide nanofiltration membrane with good antifouling properties for efficient removal of humic acid and dyes. *J. Environ. Chem. Eng.* **2023**, *11*, 110285. [CrossRef]

membranes

MDPI

Article

Oil and Water Recovery from Palm Oil Mill Effluent: A Comparative Study of PVDF and α-Al_2O_3 Ultrafiltration Membranes

Saqr A. A. Al-Muraisy [1,*], Jiamin Wu [1], Mingliang Chen [1], Begüm Tanis [1], Sebastiaan G. J. Heijman [1], Shahrul bin Ismail [2], Jules B. van Lier [1] and Ralph E. F. Lindeboom [1]

1 Sanitary Engineering Section, Department of Water Management, Faculty of Civil Engineering and Geosciences, Delft University of Technology, 2628 CN Delft, The Netherlands; m.chen-1@tudelft.nl (M.C.); m.b.tanis@tudelft.nl (B.T.); s.g.j.heijman@tudelft.nl (S.G.J.H.); j.b.vanlier@tudelft.nl (J.B.v.L.); r.e.f.lindeboom@tudelft.nl (R.E.F.L.)

2 Faculty of Ocean Engineering Technology, Universiti Malaysia Terengganu, Kuala Nerus 21030, Terengganu, Malaysia; shahrul.ismail@umt.edu.my

* Correspondence: s.a.a.al-muraisy@tudelft.nl

Abstract: Recovering oil and water from palm oil mill effluent reduces environmental pollution and promotes sustainable practices. An effective method to achieve this is ultrafiltration (UF), which uses semi-permeable membranes to separate oil, solids, and other contaminants from wastewater under pressure. To assess the most effective recovery method, an experimental comparison was conducted between PVDF and α-Al_2O_3 UF membranes at constant permeate of 20–50 LMH for PVDF and 20–70 LMH for α-Al_2O_3 membranes. Both membranes achieved 99.8% chemical oxygen demand (COD) rejection, with oil concentration factor (F_o) of 186.8% and 253.0%, and water recovery (R_w) of 46.6% and 60.5%, respectively. The permeate water quality was superior to the Malaysian discharge standards, and the fat, oil, and grease (FOG) content was suitable for phase separation processes. The optimal permeate fluxes, with stable transmembrane pressures (TMP), were observed at 40 LMH (PVDF) and 60 LMH (α-Al_2O_3). Total resistance (R_t) values were 1.30×10^{12} m^{-1} (PVDF) and 1.59×10^{12} m^{-1} (α-Al_2O_3). The ratio of irreversible to total resistances (R_{ir}/R_t) was 0.02 (PVDF) and 0.06 (α-Al_2O_3), indicating minimal irreversible fouling. Overall, the α-Al_2O_3 membrane demonstrated superior performance in oil and water recovery with more stable operation compared to the PVDF membrane. UF membrane technology emerges as an efficient technique for recovering oil and water compared to conventional methods.

Keywords: palm oil mill effluent; polymeric membranes; ceramic membranes; oil recovery; water recovery

1. Introduction

The global concern for the recovery of industrial effluents is increasing due to freshwater scarcity and environmental protection agreements [1–4]. Recently, more attention has been directed toward the discharge of oily wastewater, a major pollutant of the aquatic environment [4–8]. In this context, the removal of oil from palm oil industry waste streams is a major challenge, particularly in countries like Malaysia and Indonesia. Addressing this issue is crucial for developing more sustainable industrial practices [6,9]. Palm oil, identified as one of the rapidly growing industries due to its numerous product applications, continues to experience high demand [10,11]. Since 2016, global palm oil production has achieved continuous growth every year [10,12], simultaneously increasing its main brownish palm oil mill effluent (POME) [13,14].

 https://doi.org/10.3390/membranes15060176

POME is the primary waste stream generated during the processing of oil palm fresh fruit bunches (FFB) [15]. The latest reports show that global crude palm oil (CPO) production exceeded 79 million metric tons per year [16], which can be visualized by filling approximately 31,600 Olympic-sized swimming pools [17]. For every 1 ton of CPO produced, approximately 2.5 tons of POME are discharged [13]. POME typically contains 2.2 g/L–27.2 g/L of oil and grease [18], 11.5 g/L–79.0 g/L of total solids [18], and 894 g/L–986 g/L of water. The presence of residual oil and high organic matter in POME poses a serious environmental threat to receiving ecosystems [11]. The high organic matter content leads to elevated biological oxygen demand (BOD) and chemical oxygen demand (COD) levels in recipient surface waters. The discharge of water with high BOD and COD results in the rapid microbial consumption of dissolved oxygen, leading to the formation of anaerobic, or dead, zones in the surface water [11,19]. Further information on POME generation can be found in the literature [14,20,21].

The traditional ponding system, consisting of anaerobic or aerobic ponds, is insufficient in treating POME, as it requires a large open area and fails to meet discharge requirements or prevent environmental damage caused by oil adsorption and oxygen depletion [10,22]. For instance, the open ponding system can only remove approximately 95% of COD and BOD after a lengthy treatment period of 60 days or more [23], while the discharge requirements demand 99% and 99.5% removal of POME COD and BOD, respectively [14]. Moreover, the adsorption of oil residues on microbial surfaces leads to inhibition, biomass flotation, and washout of suspended solids [24]. Therefore, the presence of residual palm oil in rivers needs to be considered as a hazardous material for the environment that requires proper management [6,25,26]. At the same time, the oil content of POME holds potential value as it can serve as a raw material for products such as cosmetics, pharmaceuticals, and soap [27]. Recovered oil from POME can also be utilized as cooking oil or as concentrated feed for bioreactors, which reduces the reactor volume and surface area required for POME treatment.

The residual oil in POME can be categorized into four different particle size distributions (PSDs): free oil (>150 μm), dispersed oil mixture (20–150 μm), emulsified oil (5–20 μm), and soluble oil mixture (<5 μm) [11]. Free oil, which floats on the surface, can be easily removed using skimming or gravitational recovery technologies [11]. However, these conventional methods are insufficient for recovering oil from wastewater with low oil concentrations (i.e., <400 ppm) and smaller oil droplets (i.e., <20 μm) [6,28].

In addition to conventional open ponding systems [10,22], extensive research has been conducted on various technologies for treating POME, including biological wastewater treatment methods [10,22], Fenton-oxidation [29], coagulation and flocculation [30], and electrocoagulation [31,32]. Recent studies have also focused on resource recovery from POME, such as water reuse [14,33], residual oil recovery using an oil trap tank [34], and oil adsorption using polypropylene micro/nanofiber adsorbents [11]. However, the oil concentration in the effluent from these methods often exceeds the allowable discharge limits, making them unsuitable for disposal [6]. In addition, these techniques are not effective in separating low oil concentration wastewater (<400 ppm) and smaller oil droplets (<20 μm) [6,28]. Furthermore, these methods have limitations, including the requirement for a large operating surface area [6].

Membrane technology has proven to be a highly efficient method for recovering residual oil, achieving removal rates as high as 99.7%, significantly outperforming conventional approaches [4,6,7,14,35,36]. Ultrafiltration (UF) membranes, in particular, have been widely applied due to their ability to separate emulsified oil droplets and suspended solids effectively [6,37]. UF operates at low pressure, resulting in lower capital and operational costs compared to high-pressure membrane systems [6].

While several studies have demonstrated the effectiveness of UF in treating oily wastewater, challenges such as membrane fouling remain a concern [1,6,38]. The surface properties of the membrane play a crucial role in its antifouling performance [39]. In this line, Zhu et al. [40] showed that hydrophilic and oleophobic hollow fiber membranes significantly improved UF performance in separating oil and water from POME. Recent reviews by Aryanti et al. and Ho et al. provide an overview of membrane technologies applied in POME treatment, including progress in membrane materials, operating conditions, and fouling control through surface modifications and cleaning protocols [41,42].

There is growing evidence that UF membranes are particularly promising for POME treatment. For example, Ho et al. highlighted UF effectiveness in oil and water recovery when operated under optimized conditions [41]. Similarly, Samavati et al. investigated how variables such as pressure, temperature, and crossflow velocity influence membrane fouling and recovery efficiency during POME treatment, offering insights into process optimization [21]. Additionally, Aryanti et al. assessed advanced ceramic and polymeric UF membranes, identifying hybrid systems as a promising solution to the limitations of single-membrane processes [42].

While many studies have focused on reducing the fouling of UF membranes when treating oily wastewater, fewer studies have explored the oil and water recovery efficiency across different types of UF membranes. Polymeric membranes are commonly used in oily wastewater treatment due to their low manufacturing cost, ease of processing, and low energy requirements [1,7,43]. PVDF membranes offer lower upfront costs but may require more frequent replacement and cleaning, especially under harsh effluent conditions like those found in POME treatment, potentially increasing long-term operational expenditures [44]. However, they are more prone to fouling compared to ceramic membranes [1,7,43] and typically have limited thermal stability, with maximum operating temperatures generally below 60–70 °C, beyond which membrane deformation and degradation may occur [45]. Prolonged exposure to harsh cleaning agents, such as strong oxidizing solutions, can lead to degradation of PVDF membranes. For instance, a study demonstrated that PVDF membranes experienced a decline in performance after repeated cleaning cycles with sodium hypochlorite solutions, indicating potential chemical vulnerability under certain conditions [46].

Ceramic membranes, in contrast, exhibit greater fouling resistance, easier cleaning, lower maintenance, and higher mechanical strength [37,47–50]. They also offer excellent thermal and chemical resistance, allowing for stable operation at temperatures exceeding 90 °C without significant performance loss [37,47–50]. α-Al_2O_3 ceramic membranes exhibit superior chemical and thermal stability, making them more resilient in aggressive wastewater environments [37,47–50]. Recent advancements in ceramic membrane technology have highlighted their robustness and suitability for oil-water separation processes, including POME treatment [51,52]. These distinctions in chemical stability are critical when considering long-term operational efficiency and maintenance requirements. Moreover, ceramic membranes have a longer lifespan of around 20 years, whereas polymeric membranes typically last 7 to 10 years, resulting in lower life-cycle costs [7,43]. Therefore, ceramic membranes are used in various industrial applications, not only when polymeric membranes cannot perform properly, but also where high system integrity is needed [48]. However, the higher initial capital costs of ceramic membranes have limited their adoption in certain industrial sectors, such as food and beverage and oil and gas produced waters [43]. While ceramic membranes like α-Al_2O_3 typically have higher initial capital costs, ranging from 5 to 10 times that of PVDF, they are well known for their extended operational lifespan, superior chemical and thermal stability, and reduced cleaning frequency which are factors that can lead to lower overall lifecycle costs [53]. Therefore, the water industry is increasingly recognizing ceramic membranes as a viable and cost-competitive option in the long run, leading to growing industrial interest in their applications [43,48].

Most industrial UF systems operate at a constant permeate flux to maintain a consistent production rate of permeate water [54]. This approach allows for the adjustment and control of permeate flux to maximize water production, oil concentration, and recovery while minimizing fouling [54]. Moreover, UF membranes are more stable and require less frequent cleaning when operated in the constant permeate flux mode compared to constant transmembrane pressure (TMP) mode [54,55]. Therefore, this study adopts the constant permeate flux mode of filtration.

While some research has been conducted on ceramic UF membranes for oily wastewater treatment [6,51,52,56], there remains a critical gap in quantitative data on the recovery of oil and water and the operational performance of POME UF using ceramic membranes compared to their polymeric counterparts. In this regard, the objectives of the present study are as follows: (i) Analyze the performances of PVDF and α-Al_2O_3 membranes in terms of permeate flux, rejection capacity, fouling tendency, and the efficiency of membrane cleaning to recover hydraulic permeability. (ii) Determine the optimal operating conditions for both membranes within the specified range of variables, and (iii) Evaluate the quality of the recovered oil and permeate water for potential reuse applications, comparing them to relevant standards.

Ceramic membranes are typically capable of operating at higher temperatures [47], presenting an opportunity for thermal energy recovery, considering that POME is typically discharged at temperatures between 80 and 90 °C [57]. With the collective efforts towards enhancing the sustainable practices of palm oil production and maximizing the utilization of resources in waste streams, this study serves as a foundation for future POME treatment and resource recovery.

2. Materials and Methods

2.1. POME Emulsion Preparation

In the laboratory, POME emulsion was synthesized to replicate the main characteristics, including the particle size distribution (PSD) of oil droplets and the oil content. The PSD of oil droplets in oily wastewater has been reported to have an average range of 0.8 μm–1.4 μm [58], while the oil content in POME was measured on-site and found to be in the range of 2.2 g/L–27.2 g/L [18]. To achieve similar properties, unrefined palm oil (UPO) (KTC Edibles, Wednesbury, UK) was mixed with demineralized water at an initial concentration of 6 g UPO/L. The mixture was heated at 55 °C and simultaneously shaken at 150 RPM using an incubator shaker (New Brunswick Innova 43, Eppendorf, Hamburg, Germany) for 24 h. To enhance the emulsion process and minimize the hydrophobic characteristics of the unrefined palm oil, the mixture was further sonicated using an energy-intensive sonifier (Branson Digital Sonifier 450, Branson Ultrasonics, Brookfield, CT, USA) for 30 min at 40% intensity. The mixture was left to cool down to room temperature, then sieved using a 0.103 mm sieve (INTERLAB-BV, Putte, The Netherlands) to remove large oil aggregates. The oil content was measured via COD kits (Hach Lange, Loveland, CO, USA), and the PSD of oil droplets was analyzed with a particle size analyzer (Bluewave, Microtrac, York, PA, USA). POME emulsion was placed in a 10 L feed container prior to the experiments. It was used fresh for the UF experiments, after being cooled to room temperature for 2 h–4 h, to prevent any degradation or alteration of its physicochemical properties.

In this study, the prepared POME emulsion represents a solids-free, pretreated version of POME to prevent rapid membrane fouling caused by the high solids content present in raw POME. Pretreatment is essential to remove solids, protect the longevity and integrity of the membrane system [14,59–61], and focus on the recovery of oil and water during UF experiments.

2.2. Experimental Setup

Two types of tubular UF membranes were used in the present study: polymeric PVDF (Pentair, Golden Valley, MN, USA) and ceramic α-Al_2O_3 (Inopor, Veilsdorf, Germany). Table 1 illustrates the properties of the membranes. PVDF and α-Al_2O_3 membranes were installed in thermoplastic PVC-C and stainless-steel housings, respectively. Both membranes were operated in an inside-out configuration.

Table 1. The properties of PVDF and α-Al_2O_3 membranes.

	PVDF Membrane	α-Al_2O_3 Membrane
Inner diameter (mm)	5.2	7.0
Pore size (nm) [a]	30	70
Length (mm)	640.0	600.0
Cross-sectional area (m^2)	21.2×10^{-6}	38.5×10^{-6}
Membrane surface area (m^2)	1.05×10^{-2}	1.3×10^{-2}
Maximum TMP [b] (kPa)	500.0	800.0
MWCO [c] (kDa)	300.0	500.0
Operating pH range	2–10	2–12
Iso Electric Point	3–4 [62]	8.6–9.8 [63]
Initial Permeability [a,d] (LMH·bar)	≥750	≥800 [e]

[a] Manufacturer's design parameters, [b] trans-membrane pressure, [c] molecular weight cut-off, [d] clean water at 25 °C, [e] cited from literature [64].

A laboratory-scale constant permeate flux filtration system, compatible with both PVDF and α-Al_2O_3 membrane modules, was employed for the experiments (Figure 1). The system comprised two distinct pumps serving separate functions. The feed pump, a compact positive displacement diaphragm dosing pump (DDA12-10, Grundfos, Bjerringbro, Denmark), supplied the POME emulsion from the main feed container into the filtration circuit and controlled the permeate flux. Permeate flux was maintained at a constant setpoint via an integrated flow controller with a closed-loop feedback mechanism that continuously adjusted the pump speed based on real-time permeate flow rate measurements, thereby preventing the need for manual adjustments during operation.

Crossflow velocity (CFV) was regulated by a rotary gear circulation pump (VerderGear R, Verder Liquids, Utrecht, The Netherlands), which recirculated the feed solution from a separate circulation feed vessel through the membrane module. The flow rate within the circulation loop was continuously monitored using a flowmeter (YF-S402, Zhongjiang Energy-Efficient Electronics, Foshan, Guangdong, China) to ensure stable hydrodynamic conditions. The oil concentrate stream exiting the membrane module was collected separately after the end of each experiment and was not returned to the feed vessel, thereby maintaining a constant volume within the circulation loop.

To mitigate membrane fouling and maintain performance, a backwashing step was performed after each filtration cycle. Backwashing was carried out for 30 s at a fixed pressure of 3 bar using demineralized water supplied from a dedicated vessel pressurized by compressed air.

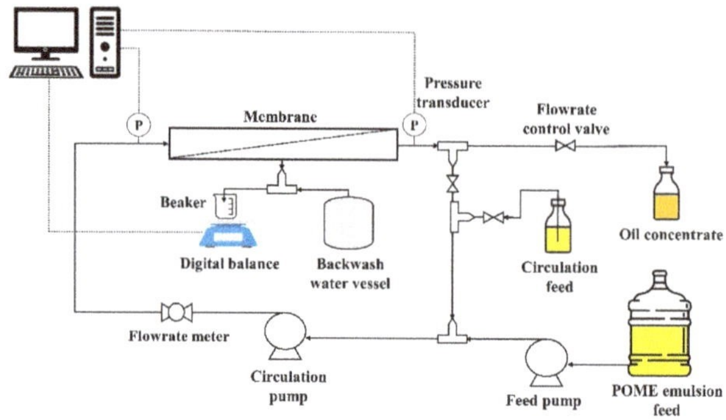

Figure 1. Schematic diagram of the ultrafiltration (UF) setup of POME emulsion using PVDF and α-Al_2O_3 membranes at a crossflow velocity (CFV) of 0.8 m/s.

Two pressure transducers (GS4200-USB, ESI Technology Ltd., Wales, UK) were connected to the two sides of the membrane module to monitor the filtration pressure continuously with a time interval of 30 s. The permeate water was weighed using an online digital balance (EWJ 600, Kern & Sohn GmbH, Balingen, Germany). Pressure transducers and the flow rate sensor were logged during the experiments, and data were monitored and recorded in real-time using DASYLab software (version 13.0). Since the permeate pressure was under atmospheric pressure, the values displayed by the transducers were taken as the transmembrane pressure (TMP), as described in the literature [65]. Further operational details of the setup can be found in [65].

2.3. Experimental Design

2.3.1. Constant Permeate Flux Crossflow Experiments

Filtration experiments were conducted at room temperature (~22 °C) with a constant permeate flux for each test. The steady-state TMP was typically reached within 1.5–2 min after the start of the filtration cycle. For the PVDF membrane, the permeate flux was incrementally increased starting from 20 LMH to 50 LMH. Due to its larger pore size of 70 nm, it was expected that the α-Al_2O_3 membrane would exhibit higher hydraulic permeability in comparison to the PVDF membrane. Consequently, the α-Al_2O_3 membrane was also expected to achieve a higher optimal permeate flux. Therefore, the α-Al_2O_3 membrane was tested at permeate fluxes ranging from 20 LMH to 70 LMH. The maximum flux for both membranes was determined based on the stability of TMP and the maximum pressure endured by the feed pump. Table 2 provides a summary of the operational parameters for each membrane.

To achieve and maintain constant flux operation, the system utilized a feed pump with an integrated flow controller, which actively measured and regulated the volumetric flow rate to meet the set flux targets. The flow rate was calibrated according to the membrane surface area to achieve the desired LMH. During filtration, any change in membrane resistance due to fouling was compensated for by the pump adjusting its power output to sustain the target permeate flow [65,66].

To determine the optimal permeate flux within the test range, the following criteria were considered: (1) maintaining the maximum TMP between 1 bar and 1.5 bar, (2) achieving the highest stable permeate flux that yields the maximum oil concentration factor (F_o) and water recovery (R_w), (3) minimizing TMP variation and irreversible fouling to reduce the frequency of chemical cleaning and extend membrane lifespan, and (4) ensuring a minimum irreversible resistance over filtration time, expressed as the ratio of irreversible to total resistances (R_{ir}/R_t) of $\leq$0.1.

Each UF experiment consisted of three filtration cycles and two hydraulic permeability tests using demineralized water, with the first test conducted before the first filtration cycle and the second test performed after the third filtration cycle. Further details on the filtration phases and hydraulic permeability tests are available in the Supplementary Materials (Section S1). Permeate water samples of the optimum UF condition were characterized. These characteristics of permeate water collected during the filtration cycles were measured in triplicates, and standard deviation was reported (Table 4).

Table 2. Operational parameters of PVDF and α-Al_2O_3 membranes.

Parameter/Membrane	CFV (m/s)	Flux 1 (LMH)	Flux 2 (LMH)	Flux 3 (LMH)	Flux 4 (LMH)
PVDF	0.8	20	40	50	-
α-Al_2O_3	0.8	20	40	60	70

2.3.2. Membrane Conditioning and Cleaning

Prior to the experiments, both PVDF and α-Al_2O_3 membranes were soaked in a 25% v/v ethanol solution (Sigma-Aldrich, St. Louis, MO, USA) for two hours to remove any residuals in the membrane wall and pores [67]. They were then washed with demineralized water [67].

After each UF experiment, the membranes underwent both physical and chemical cleaning. Physical cleaning involved flushing the filtration loop with demineralized water at the same crossflow velocity (CFV) used during the UF experiment until the water discharged from the outlet was clean.

Chemical cleaning was performed outside the system, using the membrane module after it was removed from the filtration setup. The PVDF membrane module was soaked in a freshly prepared 0.1 M NaOH solution (Sigma-Aldrich, St. Louis, MO, USA) for 12–15 h at room temperature [67]. The α-Al_2O_3 membrane module was similarly soaked in a 0.75 M NaOH solution (Sigma-Aldrich, Missouri, USA), which was maintained at 40 °C throughout the 12–15 h soaking period [68]. No transmembrane pressure, crossflow, or mixing was applied during chemical cleaning; the process relied solely on static soaking. The volume of the used cleaning solution matched the internal volume of the membrane module to ensure full submersion of the membrane. To ensure consistent cleaning conditions and reproducibility of the results, fresh NaOH solution was used for each cleaning cycle and was not reused.

Following chemical cleaning, the membrane modules were rinsed thoroughly with demineralized water before being reinstalled in the system. The effectiveness of both physical and chemical cleaning steps was evaluated by monitoring the recovery of the membranes hydraulic permeability.

2.3.3. Membrane Performance Evaluation Methods

Rejection Capacity (R_j)

The treatment efficiency of the membranes can be directly evaluated using the rejection rate, R_j, as shown in Equation (1).

$$R_j\ (\%) = \left(1 - \frac{C_p}{C_f}\right) \times 100 \quad (1)$$

where C_p, C_f are the concentrations of parameters of interest in permeate water and feed, respectively. In this study, the rejection capacity is calculated for chemical oxygen demand (COD), total solids (TS), turbidity, and electrical conductivity (EC) of feed and permeate water.

Oil Concentration Factor (F_o)

The oil concentration factor (F_o) measures the oil concentration and indirectly indicates the potential of oil recovery from the POME concentrate. Analyzing the quality of the recovered oil through F_o helps to determine any necessary applications or post-treatments. F_o is calculated by dividing the COD ratio of the oil in the POME concentrate by that in the POME emulsion feed (g concentrate COD/g POME emulsion COD feed) (Equation (2)). The COD value of the POME concentrate ($COD_{conc.}$) was determined by the equation of COD balance (Equation (3)). The equation represents the total COD involved before and after the filtration experiment with $COD_{conc.}$ calculated based on the assumption of a homogeneous POME concentrate.

$$F_o = \frac{COD_{conc.}}{COD_f} \times 100\% \quad (2)$$

$$COD_f \cdot \left(\dot{V}_f \times t_{cycle} + V_{loop}\right) = COD_{conc.} \times V_{loop} + COD_P \times \dot{V}_p \times t_{cycle} \quad (3)$$

where COD_f is the COD of POME emulsion feed (g POME COD/L), $COD_{conc.}$ is the COD of the POME concentrate discharged from the filtration system (g concentrate COD/L), and COD_p is the COD of the permeate water (g permeate water COD/L). $\dot{V}_f$ is the volumetric flowrate of POME emulsion fed into the filtration system (L/min). $\dot{V}_p$ is the volumetric flowrate of the permeate water (L/min). V_{loop} is the total volume (L) of the filtration system, i.e., the total internal volume of the tubes, connections, and membrane. V_{loop} using PVDF and Al_2O_3 membranes was calculated at 0.12 L and 0.13 L, respectively. t_{cycle} is the time for each filtration cycle (15 min).

Water Recovery (R_w)

R_w represents the ratio of permeate water volume to the total POME emulsion feed volume (L permeate water/L POME emulsion feed) (Equation (4)).

$$R_w\ \% = \frac{\dot{V}_p \times t_{cycle}}{\dot{V}_f \times t_{cycle} + V_{loop}} \times 100\% \quad (4)$$

Normalized Transmembrane Pressure (TMP_n)

TMP_n was calculated using Equation (5). This method is adopted from the literature [65,69].

$$TMP_n = \frac{TMP}{TMP_0} \quad (5)$$

where TMP is the POME emulsion UF transmembrane pressure (Pa) at a specified filtration time, and TMP_0 is the membrane initial transmembrane pressure (Pa) using demineralized water.

2.3.4. Membrane Fouling Evaluation

Resistance-in-series model [67,70], using Darcy's law (Equation (6)), was adopted to evaluate membrane fouling via membrane resistances. Further details on this method and the equations used are available in the Supplementary Materials (Section S3).

$$R_t = \frac{TMP}{\mu J} = R_m + R_r + R_{ir} \quad (6)$$

where R_t (m^{-1}) is the total resistance comprising intrinsic membrane resistance (R_m, m^{-1}), hydraulic reversible resistance (R_r, m^{-1}), and irreversible resistance (R_{ir}, m^{-1}). TMP represents the transmembrane pressure of POME emulsion filtration, and J is the permeate flux (m/s). The dynamic viscosity of the permeate water (μ) is assumed to be the same as that of pure water, 9.544×10^{-4} Pa·s, at the filtration temperature of 22 °C [54]. R_{ir}/R_t ratio was used in this study to analyze and compare the degree of irreversible fouling that occurred under each UF condition.

Hydraulic Permeability

The hydraulic permeability (L_h) was calculated using Darcy's Law (Equation (7)).

$$L_h = \frac{J}{TMP_0} \quad (7)$$

where J is the permeate flux of demineralized water (LMH), and TMP_o is the transmembrane pressure (Pa) using demineralized water.

2.4. Analytical Methods

POME emulsion feed and permeate water samples were characterized in triplicates. Total and soluble chemical oxygen demand (COD_t, and COD_s) were determined using Hach-Lange kits (Hach Lange, USA). Fat, oil, and grease (FOG) content was directly calculated based on the measured COD of POME emulsion using the theoretical ratio of 2.71 g COD/g FOG. More details on the ratio calculation can be found in the Supplementary Materials (Section S2).

pH was measured using a multi-meter (Multi 3430, WTW inoLab_IDS, Xylem Analytics, Weilheim, Germany), whereas electrical conductivity (EC) was measured using a digital meter (Multi 9420, WTW inoLab_IDS, Xylem Analytics, Germany). Turbidity was analyzed using a digital turbidimeter (2100 N, Hach Lange, USA).

The particle size distribution of the oil droplets was determined using a particle size analyzer (S3500 Bluewave, Microtrac MRB, Haan/Duesseldorf, Germany). Total nitrogen and ammoniacal nitrogen were characterized using Hach-Lange kits (Hach Lange, USA). The measurements of total solids (TS) and total suspended solids (TSS) were performed according to the standard methods [71], whereas total dissolved solids (TDS) were calculated from the difference between TS and TSS. The analysis of these parameters is essential to assess the efficiency of the UF experiments using PVDF and α-Al_2O_3 membranes. Additionally, these analyses allowed for a comparison of the permeate water quality with industrial water discharge standards [72,73].

3. Results and Discussion

3.1. POME Emulsion Characteristics

The physicochemical characteristics of POME emulsion compared to raw POME are presented in Table 3. The POME emulsion has an acidic pH of 5.4, which is consistent with the reported literature on raw POME pH ranging from 3.4 to 5.5 [18]. The high amount of organic matter in the POME emulsion can be observed by COD_t of 15.0 ± 0.2 gO_2/L, which is solely attributed to unrefined palm oil. Similarly, the concentration of fat, oil, and grease (FOG) in POME emulsion was found to be 5.5 ± 0.1 g/L, falling within the range of raw POME FOG values of 2.2 g/L–27.2 g/L [18]. COD of raw POME typically exhibits a higher COD range of 15 g/L–100 g/L [18], compared to the POME emulsion prepared in this study. Raw POME also contains other constituents such as protein (0.5 g/L–1.6 g/L), cellulose (4.0 g/L–14.3 g/L), lignin (9.0 g/L–15.2 g/L), extractives (4.4 g/L–14.5 g/L), and ash (1.2 g/L–5.9 g/L) [18].

The turbidity of POME emulsion was measured at 580.7 ± 9.0 NTU, which is lower than the turbidity of raw POME, in the range of (65.6–69.4) × 10^3 NTU [18]. Other physicochemical characteristics of POME emulsion were also quantitatively lower than those of raw POME. For instance, EC of POME emulsion was measured at 18.3 ± 0.1 μS/cm, which is less than 137 μS/cm of raw POME [74]. Similarly, total nitrogen and ammoniacal nitrogen of POME emulsion of (5.9 ± 0.1) × 10^{-3} g/L, and 0.2 × 10^{-3} g/L, respectively, were lower than the respective values in raw POME, in the range of 0.2–1.7 and (17–254) × 10^{-3} g/L, respectively [18].

The presence of TDS at 1.8 ± 0.1 g/L in the POME emulsion, along with EC, was associated with monovalent and divalent ions, which could be from the dissociation of FOG [70]. Moreover, the TSS of 1.9 ± 0.1 g/L in the POME emulsion could be attributed to the colloids and suspended organic matter [70]. Furthermore, approximately two-thirds of the droplet size of the POME emulsion was 1.54 μm, which is consistent with the average particle size of oily wastewater [58]. The synthesis of POME emulsion with such a small particle size distribution of oil droplets is crucial for investigating the efficiency of oil and water recovery using PVDF and α-Al_2O_3 membranes, especially considering the challenges associated with recovering emulsified and soluble oil using conventional methods [6].

Table 3. Physicochemical characteristics of POME emulsion.

Parameters	POME Emulsion	Raw POME [18,74]
pH	5.4	3.4–5.5
Electrical Conductivity (μS/cm)	18.3 ± 0.1	137
Turbidity (NTU)	580.7 ± 9.0	(65.6–69.4) × 10^3
COD_t (g/L)	15.0 ± 0.2	15–100
COD_s (g/L)	2.1	-
TS (g/L)	3.7 ± 0.1	11.5–79.0
TDS [a] (g/L)	1.8 ± 0.1	20.6–41.1
TSS (g/L)	1.9 ± 0.1	5–71.3
FOG [b] (g/L)	5.5 ± 0.1	2.2–27.2
Total nitrogen (g TN/L)	(5.9 ± 0.1) × 10^{-3}	0.2–1.7
Ammoniacal nitrogen (g NH_4-N/L)	0.2 × 10^{-3}	(17–254) × 10^{-3}
Particle size distribution (PSD) (μm, % v/v)	1.54 (61.5%), 0.04 (38.5%)	-

[a] Value of POME emulsion was calculated from the difference between TS and TSS. [b] Value of POME emulsion was calculated using the ratio of 2.71 g COD_t/g FOG.

A comparison between PVDF and α-Al_2O_3 membranes was made based on various performance variables, including the rejection capacity (R_j), oil concentration factor (F_o), water recovery (R_w), the effect of permeate flux on normalized TMP (TMP_n) and fouling resistances (R), and the efficiency of membranes cleaning. The key results of this comparison are summarized in Table 5.

3.2. *Comparison on the Quality of Oil and Water Recovery*

3.2.1. Rejection Capacity (Rj)

Rejection capacity for COD, turbidity, TS, and EC was analyzed at the optimum UF conditions for each membrane, as shown in Figure 2. Both membranes achieved high rejection rates for the selected parameters. Specifically, COD rejection of 99.8% was achieved with both membranes. Similarly, PVDF and α-Al_2O_3 membranes removed 99.6% and 99.9%

of turbidity, respectively. The COD rejection obtained in this study was higher than that reported in the literature [70,75–77]. Hernández et al. [70] reported COD rejection of 80–85% and turbidity rejection of 90–95% using nanofiltration (NF) membranes. Moreover, Huang et al. [78] achieved a turbidity rejection of 98.3% for raw surface water using UF membranes. The high COD and turbidity rejections observed in this study suggest that suspended particles and colloids larger than the membrane's pore sizes may be the primary foulants, which aligns with the findings of Huang et al. [78].

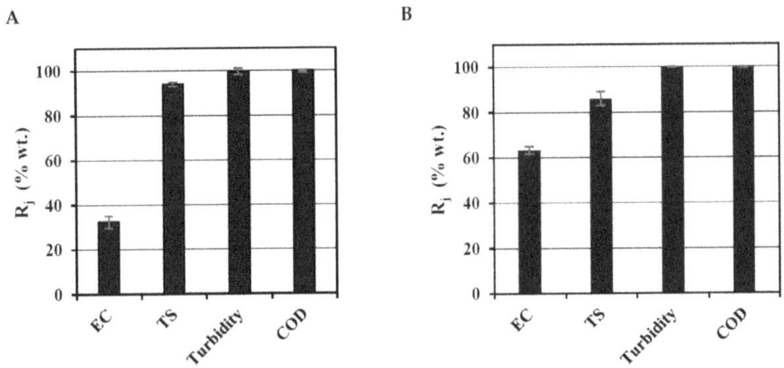

Figure 2. Steady-state rejection capacity (R_j, % wt.) of EC, TS, turbidity, and COD at the optimum UF conditions, (**A**) PVDF membrane at 40 LMH, and (**B**) α-Al_2O_3 membrane at 60 LMH.

Furthermore, PVDF and α-Al_2O_3 membranes removed 94.0% and 86.1% of TS, respectively. It can be inferred that some oil droplets were smaller (0.04 μm) than the α-Al_2O_3 membrane pore size (0.07 μm), allowing them to pass through the membrane pores. Additionally, oil droplets may have been distorted during filtration, breaking into smaller particles that passed through the membrane pores [69,79,80]. However, rejections higher than 80% are considered adequate [70]. Furthermore, it was anticipated that the formation of an oil layer on the membrane surface would provide additional resistance to the passage of solutes [1]. This phenomenon was observed by the increasing values of rejection capacities from the first to the third filtration cycles using both membranes.

Rejection of EC was achieved at 32.2% and 63.4% using PVDF and α-Al_2O_3 membranes, respectively. The observed low EC rejection might be attributable to the fact that UF membranes have the ability to retain only insoluble suspended solids, colloids, and soluble macromolecular substances that are larger than the membrane pore size [69,70], but not mono acids such as volatile fatty acids (VFAs) [69]. Therefore, POME emulsion was analyzed for VFAs and found that it contained 2.2 g/L–4.6 g/L acetic acid, 0.4 g/L–1.2 g/L propionic acid, and 0.5 g/L–1.2 g/L butyric acid. VFAs in POME emulsion could be formed due to the breakdown of lipids during the sonication of unrefined palm oil (UPO). Nonetheless, EC rejection achieved in this study was higher than some reported values of 20% using NF membranes [70]. This enhanced rejection may be explained by the formation of an oil layer on the membrane surface driven by electrostatic and drag forces, creating steric hindrance that improves retention beyond what the membrane's nominal pore size would suggest. Further details are available in the Supplementary Material (Section S4).

3.2.2. Oil Concentration Factor (F_o) and Water Recovery (R_w)

F_o and R_w were analyzed at the operating conditions chosen for PVDF and α-Al_2O_3 membranes. As illustrated in Figure 3, F_o is proportionally related to R_w for both membranes.

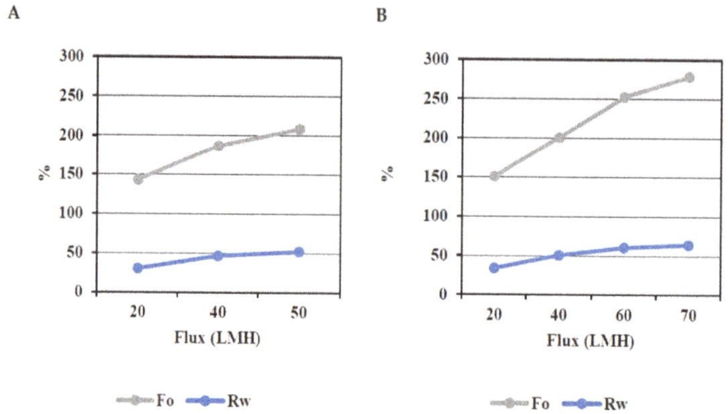

Figure 3. Oil concentration factor (F_o), and water recovery (R_w) at CFV = 0.8 m/s. (**A**) PVDF membrane, and (**B**) α-Al_2O_3 membrane.

Figure 3 also illustrates that F_o and R_w are both influenced by the rejection of oil droplets and were found to be proportional to the permeate flux. PVDF membrane achieved F_o ranging from 143.4% at 20 LMH to 208.6% at 50 LMH. Consequently, R_w ranged from 30.4% at 20 LMH to 52.2% at 50 LMH. At the optimum condition of 40 LMH, the PVDF membrane achieved R_w of 46.6%, resulting in F_o of 186.8%.

Comparatively, α-Al_2O_3 membrane achieved F_o ranging from 150.7% at 20 LMH to 278.6% at 70 LMH, which resulted in R_w ranging from 33.8% at 20 LMH to 64.1% at 70 LMH. It can be observed that the α-Al_2O_3 membrane obtained higher F_o and R_w at similar UF conditions compared to the PVDF membrane. For instance, at 40 LMH, F_o and R_w of 186.8% and 46.6%, respectively, were achieved for the PVDF membrane and 200.9% and 50.5% for the α-Al_2O_3 membrane. At the optimum condition of 60 LMH, the α-Al_2O_3 membrane achieved R_w of 60.5%, resulting in an F_o of 253.0%. According to Ren, et al. [69], F_o of 500% could be achieved using the α-Al_2O_3 membrane at a higher permeate flux, but this is on account of process stability, i.e., constant TMP_n and permeate flux over filtration time [81]. Furthermore, F_o is dependent on the feed and the membrane characteristics.

Under the optimum UF conditions, a $COD_{conc.}$ of 17.1 gO_2/L and 20.4 gO_2/L was calculated, corresponding to a FOG content of 6.3 g/L and 7.5 g/L for PVDF and α-Al_2O_3 membranes, respectively. This observation indicates that the POME concentrate could be suitable for traditional phase separation processes (gravity separation, centrifugation, hydrocyclones, and gas flotation, among others) since the oil concentration exceeds 400 ppm [6,28,37]. Another potential application for the recovered POME concentrate is biodegradation via anaerobic digestion to produce methane-rich biogas. Reducing the water content in the POME emulsion by 46.6% and 60.5% for PVDF and α-Al_2O_3 membranes, respectively, can decrease the bioreactor volume and surface area, or the membrane filtration area in the case of membrane bioreactor application, and subsequently reduce capital and operational costs.

3.2.3. Permeate Water Characteristics

Permeate water quality characteristics at the optimum conditions of both membranes were analyzed and reported in Table 4. Moreover, industrial water discharge requirements were aimed at comparing the recovered water with the global standardized requirements for industrial water discharge. Overall, the treatment of POME emulsion exhibited high COD rejection capacities ranging from 94.0% to 99.8% using the PVDF membrane and 86.1% to 99.8% using the α-Al_2O_3 membrane, as illustrated in Figure 2.

The quality of permeate water using PVDF and α-Al_2O_3 membranes showed low values of turbidity and COD of 0.5–2.5 NTU and 33.1 mg/L–35.7 mg/L, respectively, demonstrating the retention of colloidal particles. Similar trends were observed for TS, TSS, FOG, ammoniacal nitrogen, and total nitrogen, where high retention was achieved using both membranes. A similar observation was reported in the literature for UF-RO treatment of food industry wastewater [70]. Although EC rejection was the lowest parameter, ranging from 32.2% to 63.4%, the permeate water still exhibited low EC values. The remaining EC in the permeate water could be attributed to the presence of volatile fatty acids and/or dissociated acids that the membranes' MWCOs are not able to retain [70]. This was further supported by the TDS concentration in the permeate water, which reached 186.0 ± 44.2 mg/L and 488.5 ± 42.4 mg/L using PVDF and α-Al_2O_3 membranes, respectively. TDS could be associated with monovalent ions originating from the dissociation of FOG [70]. The slightly acidic permeate water, with a pH of 5.4 when using PVDF membranes and 6.0 with α-Al_2O_3 membranes, also contributed modestly to the observed EC value.

It is worth noting that all quality parameters are below the discharge limits of industrial water and specifically POME discharge standards, whereas pH was within the accepted range, as shown in Table 4. Therefore, the produced permeate water can be considered to have high-quality characteristics. The permeate water could be reused as boiler feed water in the palm oil mill, as service water within the mill, or potentially as a source for drinking water production [73]. It can also be discharged into the river system [73]. These findings indicate that the UF process to treat POME emulsion is a viable option for recovering clean water and oil.

Table 4. Comparison of permeate water characteristics of POME emulsion UF using PVDF and α-Al_2O_3 membranes.

Parameters	Permeate Water Characteristics		Industrial Water Discharge Requirements	POME Discharge Standards [14,82]
	PVDF Membrane	α-Al_2O_3 Membrane		
pH	5.4 ± 0.20	6.0 ± 0.1	6.5–8.5 [70,83]	5–9
EC (μS/cm)	12.4 ± 0.1	6.7 ± 0.1	260 [70]	-
Turbidity (NTU)	2.5 ± 0.04	0.5 ± 0.1	<5 [70]	-
COD (mg/L)	35.7 ± 1.5	33.1 ± 1.8	<50 [70]	<1000
TS (mg/L)	219.3 ± 50.0	509.0 ± 43	<1000 [70]	<1500
TSS (mg/L)	33.3 ± 5.8	20.3 ± 0.6	<400 [83]	<400
TDS [a] (mg/L)	186.0 ± 44.2	488.5 ± 42.4	-	-
FOG [b] (mg/L)	13.2 ± 0.6	12.2 ± 0.7	<50 [83]	<50
Ammoniacal nitrogen (mg NH_4-N/L)	<0.001	<0.001	-	<100
Total nitrogen (mg TN/L)	<0.001	<0.001	<150 [83]	-

[a] Value calculated from the difference between TS and TSS. [b] Value calculated using the ratio of 2.71 g COD/g FOG.

3.3. *Comparison of the Rate of Oil and Water Recovery*

3.3.1. Effect of Permeate Flux on Normalized Transmembrane Pressure (TMP_n)

UF experiments at a constant permeate flux of POME emulsion were conducted using PVDF and α-Al_2O_3 membranes. PVDF membrane was tested at permeate fluxes of 20, 40, and 50 LMH, whereas the α-Al_2O_3 membrane was tested at permeate fluxes of 20 LMH, 40 LMH, 60 LMH, and 70 LMH.

Figure 4A shows that the PVDF membrane had a stable TMP_n of 1.2 at both 20 LMH and 40 LMH. This observation indicated that either no fouling occurred or there was no filtration resistance from the accumulation of oil droplets. Therefore, backwash was not necessary at 20 LMH and 40 LMH, although it was performed as part of the experimental procedure. However, at 50 LMH, TMP_n gradually increased in each filtration cycle, reaching about 1.5 in the second filtration cycle and exponentially increasing in the third filtration cycle, indicating serious fouling. Despite the backwash being performed after the third filtration cycle, it was not effective in stabilizing TMP_n, suggesting that the fouling was irreversible. This was further supported by the high R_{ir}/R_t ratio of 0.58 at 50 LMH, compared to 0.07 and 0.02 at 20 LMH and 40 LMH, respectively (Figure 5A).

A

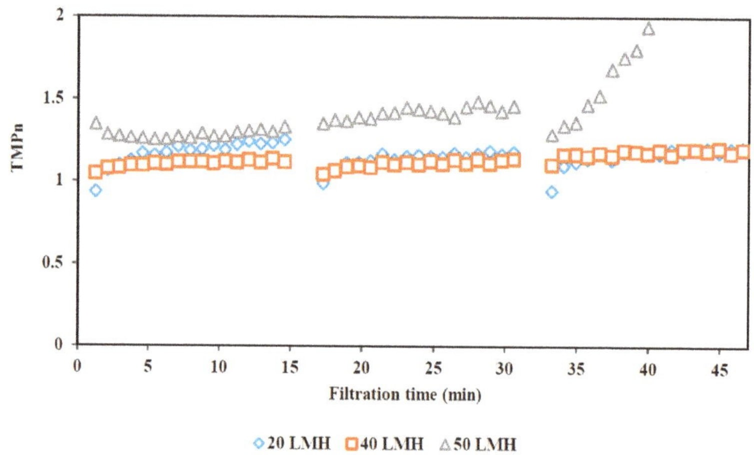

B

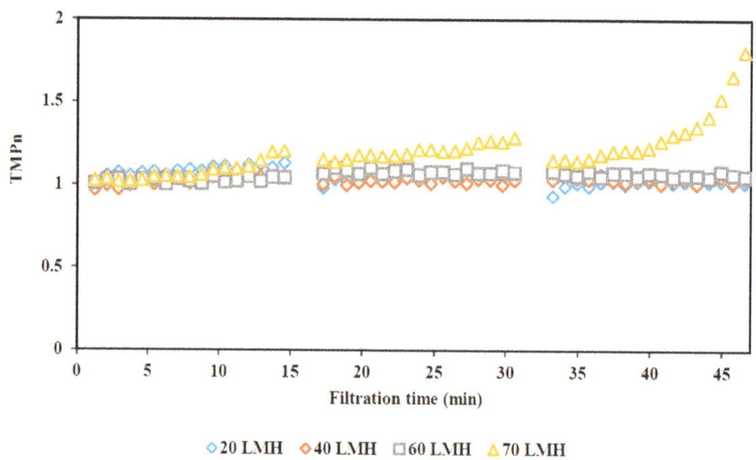

Figure 4. TMP_n evolution during constant flux of POME UF at CFV = 0.8 m/s. (**A**) PVDF membrane, and (**B**) α-Al_2O_3 membrane.

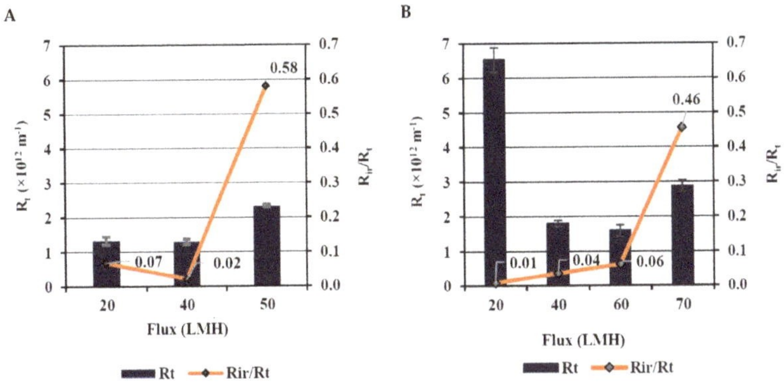

Figure 5. R_t and R_{ir}/R_t variation during constant flux of POME emulsion UF at CFV = 0.8 m/s. (**A**) PVDF membrane, and (**B**) α-Al_2O_3 membrane.

He, et al. [84] found that for 1500 ppm soybean oil emulsions, the threshold flux for PVDF microfiltration (MF) and PS UF membranes is between 55 and 62 LMH [84]. According to He et al. [84] Stoller et al. [85], and Yang et al. [86], threshold flux is defined as the flux that separates a low fouling region, characterized by a nearly constant rate of fouling (i.e., constant TMP_n), from a rapid fouling region (i.e., unstable TMP_n). Therefore, it can be assumed that the threshold flux of PVDF UF membrane treating POME emulsion is approximately 50 LMH. As shown in Figures 4A and 5A, 40 LMH resulted in the least variation in TMP_n, and correspondingly, the lowest irreversible fouling. Therefore, in this study, 40 LMH was chosen as the optimum condition for the PVDF membrane treating the POME emulsion. It is worth noting that the PVDF membrane showed satisfactory performance at 20 LMH.

On the other hand, the α-Al_2O_3 membrane showed stable TMP_n values of 1.00 at both 20 and 40 LMH, and 1.08 at 60 LMH, respectively (Figure 4B). This was further supported by the low R_{ir}/R_t ratio of 0.01, 0.04, and 0.06 at 20 LMH, 40 LMH, and 60 LMH, respectively (Figure 5B). Oil droplets were gradually brought to the membrane surface, but crossflow shear forces enhanced the back diffusion of solutes and prevented buildup, reducing concentration polarization and minimizing the oil concentration on the membrane surface [37,54]. However, at 70 LMH, there was a gradual increase in TMP_n in the first and second filtration cycles, reaching 1.3, and an exponential increase in TMP_n in the third filtration cycle. This sudden rise in TMP_n and the ineffective backwash to stabilize TMP_n indicated that serious and rapid fouling occurred due to thickening and compression of the oil layer on the membrane surface wall and a potential distortion of the deposited oil particles [69,79,80]. Oil droplets were brought to the membrane wall surface, forming a layer at a faster rate than they could be removed by the crossflow shear forces [54]. Similar TMP_n evolution profiles have been reported in the literature [54,87]. The rapid increase in TMP_n likely corresponded to the growth of cake formation [54,88], which increased the resistance of permeate water transport through the membrane pores [54,88]. Further details of the proposed filtration mechanism are available in the Supplementary Materials (Section S4, Figure S1, Table S4). It is also observed in Figure 5B that R_{ir}/R_t drastically increased to 0.46 at 70 LMH. Therefore, it can be claimed that the threshold flux of the α-Al_2O_3 membrane treating POME emulsion is approximately 70 LMH. Due to the low and stable TMP_n, high fouling resistance, and low R_{ir}/R_t ratio, the optimum condition of the α-Al_2O_3 membrane was selected at 60 LMH. However, satisfactory performance was also observed at 20 and 40 LMH.

It is worth noting that the α-Al_2O_3 membrane generally had lower and more stable TMP_n than the PVDF membrane at the various permeate fluxes applied. To illustrate, at the optimum condition of 40 LMH applied using PVDF membrane, TMP_n reached around 1.2, compared to 1.05 using α-Al_2O_3 membrane under at a similar conditions. Furthermore, the optimum condition for the α-Al_2O_3 membrane was at 60 LMH, compared to 40 LMH for the PVDF membrane. This difference may be attributed to the surface properties, such as pore size, hydrophilicity, and surface charge, of PVDF and α-Al_2O_3 membranes [89]. It was also reported that α-Al_2O_3 membranes are more hydrophilic than PVDF membranes [89], indicating that α-Al_2O_3 membranes are less prone to fouling. These findings align with the reported literature [5,89]. These observations demonstrate the potential of each membrane and some of the key advantages of using the α-Al_2O_3 membrane in the treatment of the POME emulsion.

3.3.2. Effect of Permeate Flux on Fouling Resistances (R)

The fouling behavior of PVDF and α-Al_2O_3 membranes was evaluated based on the total filtration resistance (R_t) and the ratio of irreversible to total resistances (R_{ir}/R_t). R_{ir} and R_t were calculated using the measured TMP of each filtration condition. Figure 5 illustrates the results in terms of R_t and R_{ir}/R_t. In addition, the filtration resistances (R_m, R_{ir}, and R_r) for each UF condition were also calculated and reported in the Supplementary Material (Tables S2 and S3).

Generally, increasing the permeate flux to values below the threshold flux resulted in a decrease in the total resistance (R_t) the over filtration time [69]. To illustrate, for the PVDF membrane, at 20 LMH, and 40 LMH, R_t reached 1.34×10^{12}, and 1.30×10^{12} m^{-1}, whereas the R_{ir}/R_t ratio was 0.07 and 0.02, respectively. However, at 50 LMH, the R_t and R_{ir}/R_t ratio increased to 2.32×10^{12} m^{-1} and 0.58, respectively, due to the exponential increase in TMP_n. The increment of filtration resistances can be attributed to pore blocking, causing a severe loss of hydraulic permeability [69].

For the α-Al_2O_3 membrane, R_t of 6.53×10^{12} m^{-1} was observed at 20 LMH, which is higher than those of higher fluxes. This phenomenon could be attributed to the reported trend that increasing the permeate flux below the threshold flux results in decreasing the total resistance (R_t) over the filtration time [69]. Interestingly, increasing the permeate flux between 40 LMH and 60 LMH resulted in a gradual decrease in R_t from 1.80×10^{12} to 1.59×10^{12} m^{-1}, respectively. On the contrary, the R_{ir}/R_t ratio slightly increased with the permeate flux from 0.01 to 0.06 for permeate fluxes of 20–60 LMH, respectively. However, at 70 LMH, R_t increased to 2.86×10^{12} m^{-1}, with a rise in the R_{ir}/R_t ratio reaching 0.46. The sudden increase in the R_{ir}/R_t ratio was supported by the exponential increase in TMP_n which indicated serious fouling and a potential pore-blocking [69,90].

The rapid increase in resistance at 50 LMH for PVDF and 70 LMH for α-Al_2O_3 membrane could be due to the formation of an oil layer on the membrane surface [54]. During constant flux filtration, fouling appears to be a self-accelerating phenomenon [54]. Miller et al. [54] and Ognier et al. [91] reported that as the membrane pores gradually become blocked by the oil droplets, the local permeate flux in the surrounding pores must compensate to maintain a similar flux over the membrane filtration area. This increase in local permeate flux directs the oil droplets more rapidly into the other open pores, resulting in faster fouling. As more pores are blocked with oil droplets and the local permeate flux is high, cake formation occurs, causing an increase in filtration resistance.

One of the reasons that the optimum permeate flux of the PVDF membrane (40 LMH) is lower than that of the α-Al_2O_3 membrane (60 LMH) is the pore size. The PVDF membrane has a pore size of 30 nm, which allows it to capture more oil particles, leading to the accumulation of oil particles on the membrane surface and a thicker oil layer. This phenomenon could cause more severe concentration polarisation on the membrane surface [37,69].

Since the selected optimum conditions for both membranes were below their selected threshold flux, the accumulation of oil layers may not contribute to the membrane resistance [54]. Similar findings were observed by Miller et al. (2014), where increasing the permeate flux did not affect the resistance of the membranes during constant flux filtration below the threshold flux of emulsified oil wastewater and whey protein using PVDF and polysulfone (PS) UF membranes [54].

3.4. Efficiency of Membrane Cleaning

PVDF and α-Al_2O_3 membranes were cleaned after each UF test using the cleaning methods described in Section 2.4. To recover the hydraulic permeability of the α-Al_2O_3 membrane, a stronger chemical cleaning agent (0.75 M NaOH solution) along with heating at 40 °C was required, as the cleaning method used for the PVDF membrane (0.1 M NaOH solution) was not sufficient. The reason for this could be due to the higher likelihood of oil particles being distorted into the larger pores (0.07 μm) of the α-Al_2O_3 membrane [69,79,80], compared to the pores of the PVDF membrane (0.03 μm). This can be further supported by the measured PSD of the POME emulsion, which indicated that around 38.5% v/v have a PSD of 0.04 μm, which is smaller than the pores of the α-Al_2O_3 membrane.

Figure 6 demonstrates the effectiveness of each cleaning method in recovering the hydraulic permeability of PVDF and α-Al_2O_3 membranes. A comparison between the initial and fouled membrane hydraulic permeability conditions was also conducted. Physical cleaning through backwashing and flushing the membranes with demineralized water was able to recover 17.4% and 9.8% of the hydraulic permeability of the PVDF and α-Al_2O_3 membranes, respectively. The low efficiency of physical cleaning can be attributed to the hydrophobic nature of the POME emulsion feed, which mainly consists of oil particles that water cannot efficiently remove. Although physical cleaning was not sufficient to fully recover the hydraulic permeability, it indicated that the fouling layer was looser in the PVDF membrane than in the α-Al_2O_3 membrane [1]. Chemical cleaning recovered the hydraulic permeability of both membranes, reaching over 97%, confirming that fouling was mostly organic [92]. Additionally, a hydraulic permeability recovery of over 95% is considered indicative of a clean membrane [1].

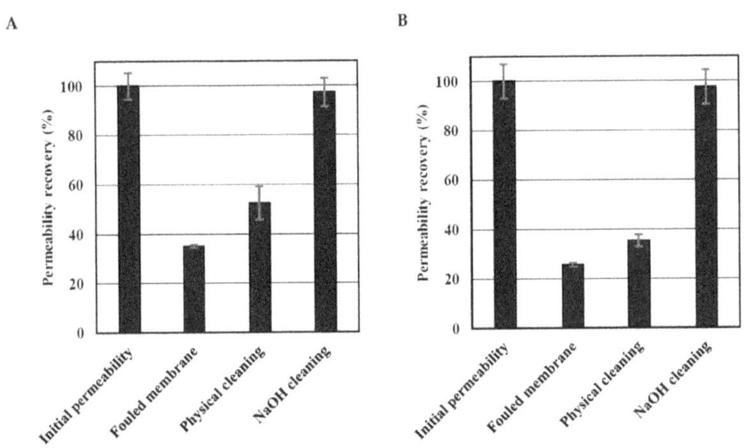

Figure 6. Efficiency of membrane cleaning methods on the recovery of hydraulic permeability (L_h); (**A**) PVDF membrane and (**B**) α-Al_2O_3 membrane.

3.5. Comparison Summary of PVDF and α-Al_2O_3 Membranes Performance

Table 5 highlights the key findings of the performance comparison between PVDF and α-Al_2O_3 membranes in recovering oil and water from the POME emulsion.

Table 5. Summary of the performance comparison between PVDF and α-Al_2O_3 membranes to treat POME emulsion at 40 LMH and 60 LMH, respectively.

Performance Variables	Experimental Results		Key Findings
	PVDF Membrane	α-Al_2O_3 Membrane	
Normalized transmembrane pressure (TMP_n)	1.2	1.08	α-Al_2O_3 membrane showed more stability and slightly lower in magnitude of TMP_n at a higher flux than PVDF membrane. At similar fluxes, α-Al_2O_3 membrane had TMP_n that is lower and more stable than those of PVDF membrane.
Total resistance (R_t)	1.3×10^{12} m^{-1}	1.59×10^{12} m^{-1}	Both membranes showed relatively low total resistance at their respective optimum conditions.
Irreversible to total resistances ratio (R_{ir}/R_t)	0.02	0.06	Both membranes demonstrated minimum irreversible fouling at their respective optimum conditions.
COD rejection (R_j)	99.8%	99.8%	Both membranes illustrated efficient removal of COD at their respective optimum conditions.
Oil concentration factor (F_o)	186.8%	253.0%	α-Al_2O_3 membrane achieved higher concentration of oil per filtration cycle than PVDF membrane.
Water recovery (R_w)	46.6%	60.5%	α-Al_2O_3 membrane recovered more water per filtration cycle than PVDF membrane.
Membrane cleaning efficiency	97.3%	97.4%	Cleaning methods used in this study achieved an efficient recovery of hydraulic permeability.

4. Conclusions

This study provides a detailed comparative assessment of PVDF and α-Al_2O_3 ultrafiltration (UF) membranes for the treatment of palm oil mill effluent (POME) emulsion, with a focus on recovering both oil and water. The synthesis of the POME emulsion successfully replicated the characteristics of raw POME, including a fat, oil, and grease (FOG) content of 5.5 ± 0.1 g/L and a particle size distribution (PSD) of 1.54 μm. The optimum UF conditions were determined at permeate fluxes of 40 LMH and 60 LMH using PVDF and α-Al_2O_3 membrane, respectively.

Under these conditions, the total resistance (R_t) was found to be 1.30×10^{12} and 1.59×10^{12} m^{-1} for PVDF and α-Al_2O_3 membranes, respectively. Moreover, the irreversible to total resistances (R_{ir}/R_t) ratio was calculated to be 0.02 and 0.06 for PVDF and α-Al_2O_3 membranes, respectively, indicating that most of the developed resistances were reversible and that no severe fouling occurred.

Both membranes demonstrated an effective rejection capacity of COD, reaching 99.8% at their respective optimum UF conditions. The FOG content in POME concentrates, reaching 6.3 g/L and 7.5 g/L for PVDF and α-Al_2O_3 membranes, respectively, indicated that it could be suitable for traditional phase separation processes. The produced permeate water from both UF membranes achieved high-quality standards that were superior to the global discharge requirements of POME. Therefore, the permeate water can be discharged into the river system.

The α-Al_2O_3 membrane showed more advantages than the PVDF membrane in treating the POME emulsion at their respective optimum UF conditions. For instance, the α-Al_2O_3 membrane achieved an oil concentration factor (F_o) of 253% and a water recovery (R_w) of 60.5% at 60 LMH, whereas the PVDF membrane achieved F_o of 186.8% and R_w of 46.6% at 40 LMH. Moreover, α-Al_2O_3 membranes are more hydrophilic than PVDF membranes, indicating that α-Al_2O_3 membranes are less prone to fouling. Therefore, α-Al_2O_3 membranes exhibited lower irreversible fouling at higher flux. For these reasons, α-Al_2O_3 membranes are more promising regarding oil and water recovery from POME emulsion, with a more stable performance and a higher rejection capacity.

Hydraulic permeability was efficiently recovered by more than 97% by combining the physical and chemical cleaning of PVDF and α-Al_2O_3 membranes at 40 LMH and 60 LMH, respectively. These findings indicate that UF is a viable option for the treatment of POME emulsion, allowing for the recovery of clean water and retention of oil. The UF process for POME could represent a promising method for more sustainable wastewater management in the palm oil industry.

Supplementary Materials: The following supporting information can be downloaded at: https://www.mdpi.com/article/10.3390/membranes15060176/s1, Figure S1: Proposed POME emulsion UF fouling mechanism. (A) Forces on oil droplets near membrane surface, (B) Surface interactions with PVDF membrane, (C) Surface interactions with α-Al_2O_3 membrane. F_x: shear force, F_{y1}: surface interactions, and F_{y2}: drag force. Table S1: COD calculations of the dominant LCFAs in POME emulsion. Table S2: Filtration resistances variation during constant flux of POME emulsion UF at CFV = 0.8 m/s using PVDF membrane. Table S3: Filtration resistances variation during constant flux of POME emulsion UF at CFV = 0.8 m/s using α-Al_2O_3 membrane. Table S4. Summary of the proposed filtration mechanism of PVDF and α-Al_2O_3 membranes.

Author Contributions: Conceptualization, S.A.A.A.-M., J.W. and R.E.F.L.; methodology, S.A.A.A.-M., J.W., M.C. and R.E.F.L.; software, S.A.A.A.-M. and M.C.; validation, S.A.A.A.-M. and J.W.; formal analysis, S.A.A.A.-M. and J.W.; investigation, S.A.A.A.-M. and J.W.; resources, S.A.A.A.-M., M.C., S.G.J.H., J.B.v.L. and R.E.F.L.; data curation, S.A.A.A.-M. and J.W.; writing—original draft preparation, S.A.A.A.-M. and J.W.; writing—review and editing, S.A.A.A.-M., M.C., B.T., S.G.J.H., J.B.v.L. and R.E.F.L.; visualization, S.A.A.A.-M. and J.W.; supervision, S.A.A.A.-M., S.b.I., J.B.v.L. and R.E.F.L.; project administration, S.A.A.A.-M. and R.E.F.L.; funding acquisition, S.b.I. and R.E.F.L. All authors have read and agreed to the published version of the manuscript.

Funding: This work was supported by the TU Delft | Global Initiative (project No. 15DGF220), a program of the Delft University of Technology to boost science and technology for global development.

Data Availability Statement: All relevant data have been reported in the manuscript and the Supplementary Materials. Raw data are not published, but they are available upon request.

Conflicts of Interest: The authors declare no conflict of interest. The funders had no role in the design of the study; in the collection, analyses, or interpretation of data; in the writing of the manuscript, or in the decision to publish the results.

References

1. Cifuentes-Cabezas, M.; Carbonell-Alcaina, C.; Vincent-Vela, M.C.; Mendoza-Roca, J.A.; Álvarez-Blanco, S. Comparison of different ultrafiltration membranes as first step for the recovery of phenolic compounds from olive-oil washing wastewater. *Process Saf. Environ. Prot.* **2021**, *149*, 724–734. [CrossRef]
2. Al-Muraisy, S.A.; Soares, L.A.; Chuayboon, S.; Ismail, S.B.; Abanades, S.; van Lier, J.B.; Lindeboom, R.E. Solar-driven steam gasification of oil palm empty fruit bunch to produce syngas: Parametric optimization via central composite design. *Fuel Process. Technol.* **2022**, *227*, 107118. [CrossRef]
3. Avornyo, A.; Chrysikopoulos, C.V. Applications of graphene oxide (GO) in oily wastewater treatment: Recent developments, challenges, and opportunities. *J. Environ. Manag.* **2024**, *353*, 120178. [CrossRef]
4. Ihsanullah, I.; Bilal, M.; Sajid, M.; Mohammad, A.W.; Atieh, M.A.; Ghaffour, N. Emerging MXenes: Revolutionizing oily wastewater treatment—A comprehensive and critical review. *Sep. Purif. Technol.* **2024**, *329*, 125181. [CrossRef]

5. Murić, A.; Petrinić, I.; Christensen, M.L. Comparison of ceramic and polymeric ultrafiltration membranes for treating wastewater from metalworking industry. *Chem. Eng. J.* **2014**, *255*, 403–410. [CrossRef]
6. Ahmad, T.; Guria, C.; Mandal, A. A review of oily wastewater treatment using ultrafiltration membrane: A parametric study to enhance the membrane performance. *J. Water Process Eng.* **2020**, *36*, 101289. [CrossRef]
7. Romero, A.S.; Innocentini, M.D.; Oliveira, J.V.; Lider, A.; Fey, T.; Travitzky, N.; Hotza, D. Unveiling the potential of silicon carbide as a support material and membranes for oily wastewater remediation. *Sep. Purif. Technol.* **2024**, *354*, 129044. [CrossRef]
8. Tian, Q.; Jiang, Y.; Li, Z.; Zhao, B.; Qiu, F.; Zhang, T. A dual wastes-based aerogel with inverse beetles-like structure for enhanced oily wastewater treatment. *Sep. Purif. Technol.* **2024**, *333*, 125896. [CrossRef]
9. Sidabutar, R.; Trisakti, B.; Michael, M.; Vanness, V.; Alexander, V.; Natasya, Y.; Alamsyah, V.; Zaiyat, M.Z.Z.; Syafriandy, S.; Al Fath, M.T. Synergistic integration of zeolite engineering and fixed-bed column design for enhanced biogas upgrading: Adsorbent synthesis, CO_2/CH_4 separation kinetics, and regeneration assessment. *Sep. Purif. Technol.* **2025**, *355*, 129772. [CrossRef]
10. Saad, M.S.; Wirzal, M.D.H.; Putra, Z.A. Review on current approach for treatment of palm oil mill effluent: Integrated system. *J. Environ. Manag.* **2021**, *286*, 112209. [CrossRef]
11. Semilin, V.; Janaun, J.; Chung, C.H.; Touhami, D.; Haywood, S.K.; Chong, K.P.; Yaser, A.Z.; Zein, S.H. Recovery of oil from palm oil mill effluent using polypropylene micro/nanofiber. *J. Hazard. Mater.* **2021**, *404*, 124144. [CrossRef] [PubMed]
12. Hassan, M.A.; Farid, M.A.A.; Zakaria, M.R.; Ariffin, H.; Andou, Y.; Shirai, Y. Palm oil expansion in Malaysia and its countermeasures through policy window and biorefinery approach. *Environ. Sci. Policy* **2024**, *153*, 103671. [CrossRef]
13. Hosseini, S.E.; Bagheri, G.; Wahid, M.A.; Saat, A. Clean fuel, clean energy conversion technology: Experimental and numerical investigation of palm oil mill effluent biogas flameless combustion. *BioResources* **2015**, *10*, 6597–6609. [CrossRef]
14. Mahmod, S.S.; Takriff, M.S.; AL-Rajabi, M.M.; Abdul, P.M.; Gunny, A.A.N.; Silvamany, H.; Jahim, J.M. Water reclamation from palm oil mill effluent (POME): Recent technologies, by-product recovery, and challenges. *J. Water Process Eng.* **2023**, *52*, 103488. [CrossRef]
15. Sadhukhan, J.; Martinez-Hernandez, E.; Murphy, R.J.; Ng, D.K.; Hassim, M.H.; Ng, K.S.; Kin, W.Y.; Jaye, I.F.M.; Hang, M.Y.L.P.; Andiappan, V. Role of bioenergy, biorefinery and bioeconomy in sustainable development: Strategic pathways for Malaysia. *Renew. Sustain. Energy Rev.* **2018**, *81*, 1966–1987. [CrossRef]
16. Soo, P.L.; Bashir, M.J.; Wong, L.-P. Recent advancements in the treatment of palm oil mill effluent (POME) using anaerobic biofilm reactors: Challenges and future perspectives. *J. Environ. Manag.* **2022**, *320*, 115750. [CrossRef]
17. Gouveia, P.; Felgueiras, F.; Mourão, Z.; Fernandes, E.D.O.; Moreira, A.; Gabriel, M.F. Predicting health risk from exposure to trihalomethanes in an Olympic-size indoor swimming pool among elite swimmers and coaches. *J. Toxicol. Environ. Health Part A* **2019**, *82*, 577–590. [CrossRef] [PubMed]
18. Cheng, Y.W.; Chong, C.C.; Lam, M.K.; Leong, W.H.; Chuah, L.F.; Yusup, S.; Setiabudi, H.D.; Tang, Y.; Lim, J.W. Identification of microbial inhibitions and mitigation strategies towards cleaner bioconversions of palm oil mill effluent (POME): A review. *J. Clean. Prod.* **2021**, *280*, 124346. [CrossRef]
19. Devlin, M.; Brodie, J. Nutrients eutrophication. In *Marine Pollution–Monitoring, Management and Mitigation*; Springer: Berlin/Heidelberg, Germany, 2023; pp. 75–100.
20. Yusof, M.A.B.M.; Chan, Y.J.; Chong, C.H.; Chew, C.L. Effects of operational processes and equipment in palm oil mills on characteristics of raw Palm Oil Mill Effluent (POME): A comparative study of four mills. *Clean. Waste Syst.* **2023**, *5*, 100101. [CrossRef]
21. Samavati, Z.; Goh, P.S.; Ismail, A.F.; Lau, W.J.; Samavati, A.; Ng, B.C.; Abdullah, M.S. Advancements in membrane technology for efficient POME treatment: A comprehensive review and future perspectives. *J. Environ. Sci.* **2024**, *155*, 730–761. [CrossRef]
22. Iskandar, M.J.; Baharum, A.; Anuar, F.H.; Othaman, R. Palm oil industry in South East Asia and the effluent treatment technology—A review. *Environ. Technol. Innov.* **2018**, *9*, 169–185. [CrossRef]
23. Cheng, Y.W.; Chong, C.C.; Lam, M.K.; Ayoub, M.; Cheng, C.K.; Lim, J.W.; Yusup, S.; Tang, Y.; Bai, J. Holistic process evaluation of non-conventional palm oil mill effluent (POME) treatment technologies: A conceptual and comparative review. *J. Hazard. Mater.* **2021**, *409*, 124964. [CrossRef]
24. Szabo-Corbacho, M.A.; Pacheco-Ruiz, S.; Míguez, D.; Hooijmans, C.M.; García, H.A.; Brdjanovic, D.; van Lier, J.B. Impact of solids retention time on the biological performance of an AnMBR treating lipid-rich synthetic dairy wastewater. *Environ. Technol.* **2019**, *42*, 597–608. [CrossRef]
25. Ohimain, E.I.; Izah, S.C. A review of biogas production from palm oil mill effluents using different configurations of bioreactors. *Renew. Sustain. Energy Rev.* **2017**, *70*, 242–253. [CrossRef]
26. Choong, Y.Y.; Chou, K.W.; Norli, I. Strategies for improving biogas production of palm oil mill effluent (POME) anaerobic digestion: A critical review. *Renew. Sustain. Energy Rev.* **2018**, *82*, 2993–3006. [CrossRef]
27. Izah, S.; Oduah, A.; Ohimain, E. Effects of temperature and fermentation period on the recovery of second grade palm oil from palm press fiber. *Int. J. Eng. Sci. Innov. Technol.* **2014**, *3*, 131–138.

28. Padaki, M.; Murali, R.S.; Abdullah, M.S.; Misdan, N.; Moslehyani, A.; Kassim, M.; Hilal, N.; Ismail, A. Membrane technology enhancement in oil–water separation. A review. *Desalination* **2015**, *357*, 197–207. [CrossRef]
29. Saeed, M.O.; Azizli, K.A.M.; Isa, M.H.; Ezechi, E.H. Treatment of POME using Fenton oxidation process: Removal efficiency, optimization, and acidity condition. *Desalination Water Treat.* **2016**, *57*, 23750–23759. [CrossRef]
30. Zahrim, A.; Dexter, Z.; Joseph, C.; Hilal, N. Effective coagulation-flocculation treatment of highly polluted palm oil mill biogas plant wastewater using dual coagulants: Decolourisation, kinetics and phytotoxicity studies. *J. Water Process Eng.* **2017**, *16*, 258–269. [CrossRef]
31. Bashir, M.J.; Han, T.M.; Wei, L.J.; Aun, N.C.; Amr, S.S.A. Polishing of treated palm oil mill effluent (POME) from ponding system by electrocoagulation process. *Water Sci. Technol.* **2016**, *73*, 2704–2712. [CrossRef]
32. Das, P.P.; Sharma, M.; Purkait, M.K. Recent progress on electrocoagulation process for wastewater treatment: A review. *Sep. Purif. Technol.* **2022**, *292*, 121058. [CrossRef]
33. Azmi, N.S.; Yunos, K.F.M. Wastewater treatment of palm oil mill effluent (POME) by ultrafiltration membrane separation technique coupled with adsorption treatment as pre-treatment. *Agric. Agric. Sci. Procedia* **2014**, *2*, 257–264. [CrossRef]
34. Hosseini, S.E.; Wahid, M.A. Pollutant in palm oil production process. *J. Air Waste Manag. Assoc.* **2015**, *65*, 773–781. [CrossRef] [PubMed]
35. Yu, L.; Han, M.; He, F. A review of treating oily wastewater. *Arab. J. Chem.* **2017**, *10*, S1913–S1922. [CrossRef]
36. Yalcinkaya, F.; Boyraz, E.; Maryska, J.; Kucerova, K. A review on membrane technology and chemical surface modification for the oily wastewater treatment. *Materials* **2020**, *13*, 493. [CrossRef]
37. Amaral, M.C.S.; Lebron, Y.; Moreira, V. Oily wastewater treatment by membrane-assisted technologies. In *Advanced Technologies in Wastewater Treatment*; Elsevier: Amsterdam, The Netherlands, 2023; pp. 213–255.
38. Salahi, A.; Abbasi, M.; Mohammadi, T. Permeate flux decline during UF of oily wastewater: Experimental and modeling. *Desalination* **2010**, *251*, 153–160. [CrossRef]
39. Asatekin, A.; Mayes, A.M. Oil industry wastewater treatment with fouling resistant membranes containing amphiphilic comb copolymers. *Environ. Sci. Technol.* **2009**, *43*, 4487–4492. [CrossRef] [PubMed]
40. Zhu, X.; Tu, W.; Wee, K.-H.; Bai, R. Effective and low fouling oil/water separation by a novel hollow fiber membrane with both hydrophilic and oleophobic surface properties. *J. Membr. Sci.* **2014**, *466*, 36–44. [CrossRef]
41. Ho, Q.N.; Lau, W.J.; Jaafar, J.; Othman, M.H.D.; Yoshida, N. Membrane Technology for Valuable Resource Recovery from Palm Oil Mill Effluent (POME): A Review. *Membranes* **2025**, *15*, 138. [CrossRef]
42. Aryanti, P.T.P.; Harsono, B.; Biantoro, M.F.W.; Romariyo, R.; Putri, T.A.; Hakim, A.N.; Setia, G.A.; Saputra, D.I.; Khoiruddin, K. The role of membrane technology in palm oil mill effluent (POME) decontamination: Current trends and future prospects. *J. Environ. Manag.* **2025**, *374*, 124094. [CrossRef]
43. Alresheedi, M.T.; Barbeau, B.; Basu, O.D. Comparisons of NOM fouling and cleaning of ceramic and polymeric membranes during water treatment. *Sep. Purif. Technol.* **2019**, *209*, 452–460. [CrossRef]
44. Ilyas, A.; Vankelecom, I.F. Designing sustainable membrane-based water treatment via fouling control through membrane interface engineering and process developments. *Adv. Colloid Interface Sci.* **2023**, *312*, 102834. [CrossRef]
45. Turk, O.K.; Zoungrana, A.; Cakmakci, M. Performances of PTFE and PVDF membranes in achieving the discharge limit of mixed anodic oxidation coating wastewaters treated by membrane distillation. *Environ. Sci. Pollut. Res.* **2024**, *31*, 39663–39677. [CrossRef]
46. Muhamad, N.A.S.; Mokhtar, N.M.; Naim, R.; Lau, W.J.; Ismail, N.H. Treatment of wastewater from oil palm industry in Malaysia using polyvinylidene fluoride-bentonite hollow fiber membranes via membrane distillation system. *Environ. Pollut.* **2024**, *361*, 124739. [CrossRef] [PubMed]
47. Wadekar, S.S.; Vidic, R.D. Comparison of ceramic and polymeric nanofiltration membranes for treatment of abandoned coal mine drainage. *Desalination* **2018**, *440*, 135–145. [CrossRef]
48. Lee, S.-J.; Dilaver, M.; Park, P.-K.; Kim, J.-H. Comparative analysis of fouling characteristics of ceramic and polymeric microfiltration membranes using filtration models. *J. Membr. Sci.* **2013**, *432*, 97–105. [CrossRef]
49. Subramaniam, M.; Goh, P.; Lau, W.; Tan, Y.; Ng, B.; Ismail, A. Hydrophilic hollow fiber PVDF ultrafiltration membrane incorporated with titanate nanotubes for decolourization of aerobically-treated palm oil mill effluent. *Chem. Eng. J.* **2017**, *316*, 101–110. [CrossRef]
50. Jiang, Y.; Xu, R.; Jiang, R.; Yang, F.; Liu, H.; Sun, X. High-temperature mechanical properties and thermal shock resistance of an alumina-fiber-reinforced alumina ceramic matrix composite. *Ceram. Int.* **2025**, *51*, 5459–5469. [CrossRef]
51. Akash, F.A.; Shovon, S.M.; Rahman, W.; Rahman, M.A.; Chakraborty, P.; Prasetya, T.A.E.; Monir, M.U. Advancements in ceramic membrane technology for water and wastewater treatment: A comprehensive exploration of current utilizations and prospective horizons. *Desalination Water Treat.* **2024**, *319*, 100569. [CrossRef]
52. Sutrisna, P.D.; Khoiruddin, K.; Mustika, P.C.W.; Ismadji, S.; Wenten, I.G. Advancements in ceramic membranes for robust oil-water separation. *J. Environ. Chem. Eng.* **2024**, *12*, 113658. [CrossRef]

53. Mountoumnjou, O.; Szymczyk, A.; Mbambyah, E.E.L.; Njoya, D.; Elimbi, A. New low-cost ceramic microfiltration membranes for bacteria removal. *Membranes* **2022**, *12*, 490. [CrossRef] [PubMed]
54. Miller, D.J.; Kasemset, S.; Paul, D.R.; Freeman, B.D. Comparison of membrane fouling at constant flux and constant transmembrane pressure conditions. *J. Membr. Sci.* **2014**, *454*, 505–515. [CrossRef]
55. Lv, M.; Feng, H.; Ding, Y.; Pan, S.; Qiao, H. Comparison of the formation, filtration performance, and structural characteristic of self-forming dynamic membranes under constant transmembrane pressure and constant filtration flux. *J. Environ. Chem. Eng.* **2022**, *10*, 108691. [CrossRef]
56. Goh, P.; Ismail, A. A review on inorganic membranes for desalination and wastewater treatment. *Desalination* **2018**, *434*, 60–80. [CrossRef]
57. Khadaroo, S.N.; Grassia, P.; Gouwanda, D.; Poh, P.E. Is the dewatering of Palm Oil Mill Effluent (POME) feasible? Effect of temperature on POME's rheological properties and compressive behavior. *Chem. Eng. Sci.* **2019**, *202*, 519–528. [CrossRef]
58. Zhong, Z.; Xing, W.; Zhang, B. Fabrication of ceramic membranes with controllable surface roughness and their applications in oil/water separation. *Ceram. Int.* **2013**, *39*, 4355–4361. [CrossRef]
59. Xu, Y.; Li, Y.; Hou, Y. Reducing ultrafiltration membrane fouling during recycled paper mill wastewater treatment using pretreatment technologies: A comparison between coagulation and Fenton. *J. Chem. Technol. Biotechnol.* **2019**, *94*, 804–811. [CrossRef]
60. Moser, P.B.; Ricci, B.C.; Reis, B.G.; Neta, L.S.; Cerqueira, A.C.; Amaral, M.C. Effect of MBR-H_2O_2/UV Hybrid pre-treatment on nanofiltration performance for the treatment of petroleum refinery wastewater. *Sep. Purif. Technol.* **2018**, *192*, 176–184. [CrossRef]
61. Elhady, S.; Bassyouni, M.; Mansour, R.A.; Elzahar, M.H.; Abdel-Hamid, S.; Elhenawy, Y.; Saleh, M.Y. Oily wastewater treatment using polyamide thin film composite membrane technology. *Membranes* **2020**, *10*, 84. [CrossRef]
62. Nguyen, M.; Loulergue, P.; Karpel, N.; Teychene, B. Electron beam irradiation of polyvinylidene fluoride/polyvinylpyrrolidone ultrafiltration membrane in presence of zwitterions molecules evaluation of filtration performances. *Radiat. Phys. Chem.* **2019**, *159*, 101–110. [CrossRef]
63. Kobayashi, Y.; Yasuda, Y.; Morita, T. Low-temperature synthesis of α-alumina based on sol-gel processes. *Adv. Mater. Process Technol.* **2020**, *7*, 482–513. [CrossRef]
64. Liu, Y.; Zhu, W.; Guan, K.; Peng, C.; Wu, J. Preparation of high permeable alumina ceramic membrane with good separation performance via UV curing technique. *RSC Adv.* **2018**, *8*, 13567–13577. [CrossRef] [PubMed]
65. Chen, M.; Shang, R.; Sberna, P.M.; Luiten-Olieman, M.W.; Rietveld, L.C.; Heijman, S.G. Highly permeable silicon carbide-alumina ultrafiltration membranes for oil-in-water filtration produced with low-pressure chemical vapor deposition. *Sep. Purif. Technol.* **2020**, *253*, 117496. [CrossRef]
66. Chen, M.; Heijman, S.G.; Luiten-Olieman, M.W.; Rietveld, L.C. Oil-in-water emulsion separation: Fouling of alumina membranes with and without a silicon carbide deposition in constant flux filtration mode. *Water Res.* **2022**, *216*, 118267. [CrossRef]
67. Xing, J.; Liang, H.; Chuah, C.J.; Bao, Y.; Luo, X.; Wang, T.; Wang, J.; Li, G.; Snyder, S.A. Insight into Fe (II)/UV/chlorine pretreatment for reducing ultrafiltration (UF) membrane fouling: Effects of different natural organic fractions and comparison with coagulation. *Water Res.* **2019**, *167*, 115112. [CrossRef]
68. Elmaleh, S.; Abdelmoumni, L. Cross-flow filtration of an anaerobic methanogenic suspension. *J. Membr. Sci.* **1997**, *131*, 261–274. [CrossRef]
69. Ren, Q.; Chen, X.; Yumminaga, Y.; Wang, N.; Yan, W.; Li, Y.; Liu, L.; Shi, J. Effect of operating conditions on the performance of multichannel ceramic ultrafiltration membranes for cattle wastewater treatment. *J. Water Process Eng.* **2021**, *41*, 102102. [CrossRef]
70. Hernández, K.; Muro, C.; Ortega, R.E.; Velazquez, S.; Riera, F. Water recovery by treatment of food industry wastewater using membrane processes. *Environ. Technol.* **2021**, *42*, 775–788. [CrossRef] [PubMed]
71. APHA. *Standard Methods for the Examination of Water and Wastewater. APHA, American Water Works Association and Water Environment Federation*, 21st ed.; American Public Health Association: Washington, DC, USA, 2005.
72. Rupani, P.F.; Singh, R.P.; Ibrahim, M.H.; Esa, N. Review of current palm oil mill effluent (POME) treatment methods: Vermicomposting as a sustainable practice. *World Appl. Sci. J.* **2010**, *11*, 70–81.
73. Ahmad, A.L.; Ismail, S.; Bhatia, S. Water recycling from palm oil mill effluent (POME) using membrane technology. *Desalination* **2003**, *157*, 87–95. [CrossRef]
74. Okogbenin, O.; Anisiobi, G.; Okogbenin, E.; Okunwaye, T.; Ojieabu, A. Microbiological assessment and physiochemical parameters of palm oil mill effluent collected in a local mill in Ovia North East area of Edo State, Nigeria. *Her. J. Microbiol. Biotechnol.* **2014**, *1*, 001–009.
75. Kanani, D.M.; Sun, X.; Ghosh, R. Reversible and irreversible membrane fouling during in-line microfiltration of concentrated protein solutions. *J. Membr. Sci.* **2008**, *315*, 1–10. [CrossRef]
76. Bennani, C.F.; Ousji, B.; Ennigrou, D.J. Reclamation of dairy wastewater using ultrafiltration process. *Desalination Water Treat.* **2015**, *55*, 297–303. [CrossRef]

77. Bortoluzzi, A.C.; Faitão, J.A.; Di Luccio, M.; Dallago, R.M.; Steffens, J.; Zabot, G.L.; Tres, M.V. Dairy wastewater treatment using integrated membrane systems. *J. Environ. Chem. Eng.* **2017**, *5*, 4819–4827. [CrossRef]
78. Huang, B.; Gu, H.; Xiao, K.; Qu, F.; Yu, H.; Wei, C. Fouling mechanisms analysis via combined fouling models for surface water ultrafiltration process. *Membranes* **2020**, *10*, 149. [CrossRef] [PubMed]
79. Tonova, K.; Lazarova, M.; Dencheva-Zarkova, M.; Paniovska, S.; Tsibranska, I.; Stanoev, V.; Dzhonova, D.; Genova, J. Separation of glucose, other reducing sugars and phenolics from natural extract by nanofiltration: Effect of pressure and cross-flow velocity. *Chem. Eng. Res. Des.* **2020**, *162*, 107–116. [CrossRef]
80. Cifuentes-Cabezas, M.; Vincent-Vela, M.C.; Mendoza-Roca, J.A.; Álvarez-Blanco, S. Use of ultrafiltration ceramic membranes as a first step treatment for olive oil washing wastewater. *Food Bioprod. Process.* **2022**, *135*, 60–73. [CrossRef]
81. Guo, C.; Chang, H.; Liu, B.; He, Q.; Xiong, B.; Kumar, M.; Zydney, A.L. A combined ultrafiltration–reverse osmosis process for external reuse of Weiyuan shale gas flowback and produced water. *Environ. Sci. Water Res. Technol.* **2018**, *4*, 942–955. [CrossRef]
82. Kamyab, H.; Chelliapan, S.; Din, M.F.M.; Rezania, S.; Khademi, T.; Kumar, A. Palm oil mill effluent as an environmental pollutant. *Palm Oil* **2018**, *13*, 13–28.
83. Azmi, N.S.; Yunos, K.F.M.; Baharuddin, A.S.; Dom, Z.M. The effect of operating parameters on ultrafiltration and reverse osmosis of palm oil mill effluent for reclamation and reuse of water. *BioResources* **2013**, *8*, 76–87. [CrossRef]
84. He, Z.; Miller, D.J.; Kasemset, S.; Paul, D.R.; Freeman, B.D. The effect of permeate flux on membrane fouling during microfiltration of oily water. *J. Membr. Sci.* **2017**, *525*, 25–34. [CrossRef]
85. Stoller, M.; Ochando-Pulido, J.M. About merging threshold and critical flux concepts into a single one: The boundary flux. *Sci. World J.* **2014**, *2014*, 656101. [CrossRef] [PubMed]
86. Yang, Q.; Luo, J.; Guo, S.; Hang, X.; Chen, X.; Wan, Y. Threshold flux in concentration mode: Fouling control during clarification of molasses by ultrafiltration. *J. Membr. Sci.* **2019**, *586*, 130–139. [CrossRef]
87. Cai, Z.; Wee, C.; Benjamin, M.M. Fouling mechanisms in low-pressure membrane filtration in the presence of an adsorbent cake layer. *J. Membr. Sci.* **2013**, *433*, 32–38. [CrossRef]
88. Cifuentes-Cabezas, M.; Bohórquez-Zurita, J.L.; Gil-Herrero, S.; Vincent-Vela, M.C.; Mendoza-Roca, J.A.; Álvarez-Blanco, S. Deep study on fouling modelling of ultrafiltration membranes used for OMW treatment: Comparison between semi-empirical models, response surface, and artificial neural networks. *Food Bioprocess Technol.* **2023**, *16*, 2126–2146. [CrossRef]
89. Jeong, Y.; Kim, Y.; Jin, Y.; Hong, S.; Park, C. Comparison of filtration and treatment performance between polymeric and ceramic membranes in anaerobic membrane bioreactor treatment of domestic wastewater. *Sep. Purif. Technol.* **2018**, *199*, 182–188. [CrossRef]
90. Lv, J.; Zhang, G.; Zhang, H.; Zhao, C.; Yang, F. Improvement of antifouling performances for modified PVDF ultrafiltration membrane with hydrophilic cellulose nanocrystal. *Appl. Surf. Sci.* **2018**, *440*, 1091–1100. [CrossRef]
91. Ognier, S.; Wisniewski, C.; Grasmick, A. Membrane bioreactor fouling in sub-critical filtration conditions: A local critical flux concept. *J. Membr. Sci.* **2004**, *229*, 171–177. [CrossRef]
92. Dereli, R.K.; Heffernan, B.; Grelot, A.; van der Zee, F.P.; van Lier, J.B. Influence of high lipid containing wastewater on filtration performance and fouling in AnMBRs operated at different solids retention times. *Sep. Purif. Technol.* **2015**, *139*, 43–52. [CrossRef]

membranes

MDPI

Article

Development and Application of a Novel Ultrafiltration Membrane for Efficient Removal of Dibutyl Phthalate from Wastewater

Qiang Zhou [1,2,*], Meiling Chen [1], Yushan Jiang [1,3], Linnan Zhang [3] and Yanhong Wang [1,2,*]

[1] Institute of Applied Ecology, Chinese Academy of Sciences, Shenyang 110162, China; zhangdz@iae.ac.cn (M.C.); stcs@iae.ac.cn (Y.J.)
[2] Shenyang Key Laboratory of Food Safety Testing and Control Technology of Shenyang, Shenyang 110016, China
[3] Faculty of Science, Shenyang University of Technology, Shenyang 110870, China; zhanglinnan@sut.edu.cn
* Correspondence: zhouqiang@iae.ac.cn (Q.Z.); wangyh@iae.ac.cn (Y.W.)

Abstract: This study successfully developed a novel molecularly imprinted ultrafiltration membrane (MIUM) for energy-efficient and selective removal of dibutyl phthalate (DBP) from wastewater. Guided by Gaussian simulations, methacrylic acid (MAA) was identified as the optimal functional monomer, achieving the strongest binding energy ($\Delta E = -0.0698$ a.u.) with DBP at a 1:6 molar ratio, providing a foundation for precise cavity construction. DBP-imprinted polymers (MIPs) synthesized via bulk polymerization were integrated into polysulfone membranes through phase inversion. The optimized MIUM (81.27% polymer content) exhibited exceptional performance under low-pressure operation (0.2 MPa), with a water flux of 111.49 $L \cdot m^2 \cdot h^{-1}$ and 92.87% DBP rejection, representing a 43% energy saving compared to conventional nanofiber membranes requiring 0.4 MPa. Structural characterization confirmed synergistic effects between imprinted cavities and membrane transport properties as the key mechanism for efficient separation. Notably, MIUM demonstrated remarkable selectivity, achieving 91.57% retention for DBP while showing limited affinity for structurally analogous phthalates (e.g., diethyl/diisononyl phthalates). The membrane maintained > 70% retention after 10 elution cycles, highlighting robust reusability. These findings establish a paradigm for molecular simulation-guided design of selective membranes, offering an innovative solution for low-energy removal of endocrine disruptors. The work advances wastewater treatment technologies by balancing high permeability, targeted pollutant removal, and operational sustainability, with direct implications for mitigating environmental risks and improving water quality management.

Keywords: molecularly imprinted ultrafiltration membrane; retention rate; sewage treatment

1. Introduction

Dibutyl phthalate (DBP) is an environmental endocrine disruptor commonly used as a plasticizer to enhance the flexibility, transparency, and durability of plastics [1,2]. It is extensively employed in the industrial production of plastic products. DBP primarily bonds with polymers through hydrogen bonds and van der Waals forces, making it susceptible to release from plastic materials. Once released, DBP can enter aquatic environments through various channels, including sewage discharge, urban and agricultural runoff, and landfill leakage [3,4]. DBP interferes with the synthesis, secretion, transportation, and metabolism of natural hormones. When it enters the food chain, its health impacts can be magnified, posing significant risks to both the environment and human health [5].

https://doi.org/10.3390/membranes15050142

The widespread use of DBP has resulted in its frequent detection in natural water bodies. In some rivers and lakes in China, such as the Songhua River, Taihu Lake, and Chaohu Lake [6–8], the DBP concentration exceeds the surface water quality standard of 3 μg/L. Consequently, sewage treatment plants have become the primary facilities for DBP treatment. Current methods for treating DBP include adsorption [9–11], microbial degradation [12], photocatalytic degradation [13–15], and oxidative degradation [16]. However, these approaches often involve complicated treatment processes. For example, adsorption can take between 2 and 12 h, and pressure differentials can reach up to 0.5 MPa. Additionally, substantial amounts of reagents, such as hydrogen peroxide or persulfate, are required. These reagents generate reactive radicals through activation, enhancing DBP decomposition via advanced oxidation pathways. Their use compensates for adsorption capacity limitations, but increases operational complexity and chemical consumption. This contrasts with MIUM's reagent-free, low-pressure operation. While these methods may be effective for various pollutants, their efficiency in removing DBP is not ideal. Effective removal of DBP from wastewater is crucial for ensuring water quality safety and is a necessary measure for reducing the ecological risks associated with water environments.

In recent years, research on the removal of DBP from wastewater has concentrated on the development, characterization, and evaluation of new adsorbents and membrane separation materials. For instance, Chen et al. [17] created a cellulosic biochar with a hollow structure, demonstrating that it achieved a retention rate of 67.32% for 40 mg/L of DBP. Wei et al. [18] developed a nanofiber membrane that retained 86.7% of 1000 μg/L DBP at a pressure of 0.4 MPa. While these technologies can effectively remove DBP from wastewater to some degree, improvements are still needed in material selectivity, adsorption time, pressure differentials, and utilization rates. Therefore, it is essential to focus on developing a simple, efficient, and highly selective material for DBP removal.

Molecular imprinting technology is based on the principle of antigen–antibody interactions synthesizing polymers that are tailored to match the shape, size, functional groups, and spatial arrangements of target molecules [19]. This technology allows for the creation of specific polymer materials that can selectively adsorb target molecules, facilitating their enrichment or purification. Molecularly imprinted polymers (MIPs) are durable, exhibiting resistance to high temperatures, high pressures, acids, alkalis, and organic solvents. They are not easily biodegradable, can be reused multiple times, and possess strong recognition capabilities [20]. As a result, MIPs have found wide applications in fields such as solid-phase extraction and membrane separation [21,22], demonstrating significant potential in the separation of biological macromolecules and chiral compounds. For instance, Sahar et al. [23] successfully prepared molecularly imprinted nanomembranes that specifically adsorb paclitaxel, achieving an extraction efficiency of 48% from taxol extracts. Similarly, Ali et al. [24] developed naphthalene molecularly imprinted membrane materials that exhibited a high affinity for naphthalene in wastewater, with a retention rate of up to 52.64%. However, molecularly imprinted membrane materials still face critical challenges in integrating recognition sites with membrane transport characteristics. Therefore, the development of ultrafiltration membranes that offer high directional adsorption efficiency and excellent membrane flux represents a crucial area for advancing wastewater treatment technologies.

This study aims to develop a molecularly imprinted ultrafiltration membrane (MIUM) for the efficient separation and removal of DBP from wastewater. DBP was used as the template molecule, and optimal types and proportions of functional monomers were selected through software simulations. Molecularly imprinted polymers (MIPs) with a high adsorption rate were prepared using the bulk polymerization method. The ideal ratio of MIPs to polysulfone casting liquid was investigated to create a MIUM that effectively adsorbs DBP under low pressure. Unlike conventional membranes relying solely on size exclusion, MIUM combines molecularly imprinted cavities with ultrafiltration properties, enabling selective adsorption under low-pressure conditions. The imprinting effect of the MIUM was validated by comparing it to non-molecularly imprinted membranes. This study offers a new technological reference for the effective removal of DBP from sewage.

2. Materials and Method

2.1. Reagents and Instruments

Polysulfone (PSF) was purchased from Shanghai Plastics Industry Union Company Shuguang Chemical Plant (Shanghai, China); N-methylpyrrolidone (NMP), azoisobutyronitrile (AIBN), trichloromethane, methanol and acetic acid were purchased from Tianjin Damao Chemical Reagent Factory (Tianjin, China); chlorpyrifos was purchased from Shanghai Maclin Co., Ltd (Shanghai, China). Methacrylic acid (MAA), acrylic acid (AA), acrylamide (AM), dibutyl phthalate (DBP), diethyl phthalate (DEP), diisononyl phthalate (DINP) and ethylene glycol dimethacrylate (EGDMA) were purchased from Alfa Aesar Chemical Co., LTD., (Ward Hill, MA, USA). Deionized water was used in all experiments.

A KTQ-I adjustable film applicator (Shanghai Jiuran Instrument & Equipment Co., Ltd., Shanghai, China); DZF vacuum drying oven (Shanghai Yuejin Medical Instrument Co., Ltd., Shanghai, China); TD4Z-WS centrifuge (Shanghai Luxiangyi Centrifuge Instrument Co., Ltd., Shanghai, China); DF-101S oil bath pan (Gongyi Yuhua Instrument Co., Ltd., Henan, China); SCQ-2000 ultrasonic cleaner (Shanghai Shengyan Ultrasonic Instrument Co., Ltd., Shanghai, China); FEI Quanta 250 scanning electron Microscope (Thermo Fisher Scientific Inc., Waltham, MA, USA); Thermo Scientific NicoLet 6700 Fourier Transform Infrared Spectrometer(Thermo Fisher Scientific Inc., Waltham, MA, USA); Thermo Scientific UltiMate 3000 Ultra-high performance liquid chromatography(Thermo Fisher Scientific Inc., Waltham, MA, USA); Gaussian 09(Gaussian Inc., Wallingford, CT, USA) and SF-SA film performance evaluation instrument (Wuxi Saibo LLC, Jiangsu, China) were used.

2.2. The Optimal Functional Monomer and Its Ratio Were Determined Using Calculation Software

Gaussian 09 software was utilized to select the optimal functional monomers imprinted by DBP. For this investigation, three functional monomers were chosen, as follows: the acidic monomers methacrylic acid and acrylic acid, along with the neutral monomer acrylamide. Molecular simulation techniques were employed to aid in this selection process. The simulation conditions were deliberately simplified to represent the interactions between atoms as analogous to the movement of a particle system. This allows for the simulation of molecular movement and the microscopic behavior associated with molecular binding during chemical reactions. By simulating the binding modes of the functional monomers with the template molecules, we were able to calculate binding energies and identify the optimal functional monomers.

2.3. MIUM Preparation

2.3.1. Preparation of DBP MIPs

DBP MIPs were prepared through bulk polymerization [25]. Initially, a specific amount of DBP and MAA was dissolved in 8 mL of trichloromethane. This mixture was then placed in a refrigerator at 4 °C for 12 h to allow for pre-polymerization of DBP and MAA. Following this, 4.92 mL of glycol dimethacrylate and 0.02 g of azodiisobutyronitrile were added. The mixture was subjected to ultrasonic oscillation for 10 min, and nitrogen gas was introduced for 15 min to remove any air. The hot polymerization reaction was then carried out in an oil bath set at 60 °C for 24 h. After polymerization, the resulting polymer was centrifuged and extracted using a Soxhlet apparatus with a mixture of methanol and acetic acid (9:1, *v:v*) for

24 h to eliminate the template molecules. The polymer was subsequently washed with a methanol solution until neutral and then vacuum-dried at 60 °C to yield DBP MIPs. The preparation of the non-molecularly imprinted polymer (NIP) followed the same procedure as that for MIPs, but without the addition of DBP.

2.3.2. Preparation of DBP MIUM

MIUM is prepared using phase conversion technology [26–28]. To begin, dissolve 14 g of dried PSF in 83.28 mL of NMP (N-methyl-2-pyrrolidone). Next, add the required amount of MIPs (molecularly imprinted polymers) and stir vigorously with a magnetic stirrer until the mixture is uniform. Once well-mixed, cast the resulting solution evenly onto a glass plate to achieve a predetermined thickness, using an applicator. After allowing the mixture to evaporate for 10 s, immediately place the plate in a water bath to form a film. The formed film should then be submerged in pure water for 12 h to remove excess NMP. Finally, the film is vacuum-dried at room temperature to obtain MIUM.

2.4. Membrane Water Flux

Membrane water flux is defined as the volume flow rate per unit area of membrane through a solution under a specified pressure over a certain period of time. To evaluate the performance of the prepared MIUM, it is placed in a membrane performance analyzer at room temperature. The membrane is subjected to a pre-pressure of 0.2 MPa with distilled water for 30 min. After this period, the volume of distilled water that passes through the membrane within a specified unit of time is measured. The membrane water flux can then be calculated using the following formula:

$$J = \frac{V}{AT} \tag{1}$$

where J represents the flux ($L/m^2/h$), V denotes the volume of solution that permeates (L), A indicates the effective area of the membrane (m^2), and T signifies the duration of permeation (h).

2.5. Membrane Retention Rate

The membrane retention rate, similar to water flux, is a key indicator of membrane performance. To evaluate this, prepare a DBP solution and test the DBP retention rate of the blended membrane at room temperature and a pressure of 0.2 MPa. Collect the filtrate that passes through the membrane and measure its absorbance. Use the standard curve to determine the concentration of the DBP in the filtrate, and then calculate the retention rate using the following formula:

$$R = \frac{C_f - C_p}{C_f} \tag{2}$$

where R (%) represents the retention percentage, C_p indicates the DBP concentration in the permeating solution (mg/L), and C_f denotes the DBP concentration in the feed solution (mg/L).

2.6. Membrane Porosity

Cut a diaphragm to a specific size and soak it in water. Once the surface moisture of the membrane is absorbed using filter paper, measure the wet film mass (W_w). Next, place the film in a clean dish and dry it in an oven at 60 °C until it reaches a constant weight.

$$\varepsilon = \frac{W_W - W_d}{\rho_W A d_0} \tag{3}$$

where ε is the sample porosity (%), W_w is the wet sample mass (g), W_d is the dry sample mass (g), ρ_W is the pure water density (g/m^3), A is the film area (m^2), and d_0 is the film thickness (m).

3. Results and Discussion

3.1. Screening of Optimal Functional Monomers and Simulation of Polymerization Scale

In this paper, we optimized the molecular configurations of DBP, methacrylic acid, acrylamide, and acrylic acid using density functional theory with Gaussian 09 and GaussView 5 software, as illustrated in Figure 1. We also calculated the electrostatic potential for these four molecular configurations. From the electrostatic scale, we can observe that colors closer to red indicate stronger electronegativity of the molecule, while colors approaching blue signify lower electronegativity [29,30]. For DBP, the red regions are primarily concentrated around the four oxygen atoms, which indicates strong electronegativity and a negative charge, making it prone to losing electrons when reacting with nucleophiles. In contrast, the four hydrogen atoms bonded to the benzene ring exhibit a light blue color, suggesting they carry little positive charge and are less likely to react with electrophilic reagents or engage in unstable reactions. Therefore, the active sites of DBP are located on the four oxygen atoms. Due to the molecule's symmetrical structure, there are two types of oxygen atoms present. Similarly, methacrylic acid and acrylamide have their active sites located on the carboxyl oxygen atom and the hydroxyl hydrogen atom. In the case of acrylic acid, the active sites can be found on the amino hydrogen atom and the carbonyl oxygen atom.

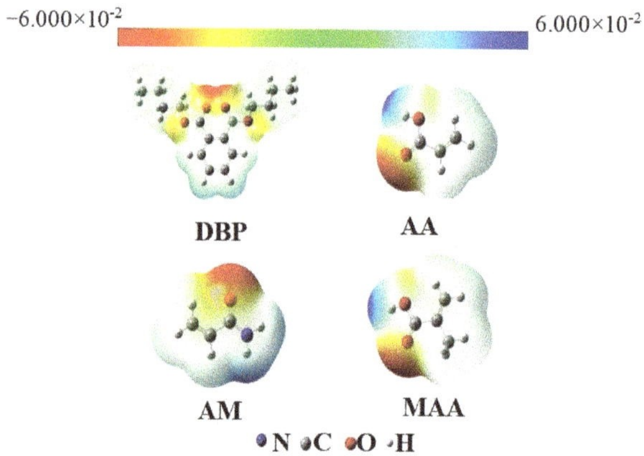

Figure 1. The surface electrostatic potentials of the template molecule DBP and three functional monomers.

After identifying suitable binding sites, we constructed simulated geometric configurations for DBP and each functional monomer. As illustrated in Figure 2, DBP was combined with methacrylic acid, acrylamide, and acrylic acid in a 1:1 ratio. The calculated binding energies are presented in Table 1. Specifically, the binding energies are as follows: ΔE(DBP-AA) = 0.054 a.u., ΔE(DBP-AM) = 0.04 a.u., and ΔE(DBP-MAA) = 0.07 a.u. This indicates that ΔE(DBP-MAA) > ΔE(DBP-AA) > ΔE(DBP-AM). A higher binding energy signifies a more stable bond between the template molecule and the functional monomer, leading to a better re-imprinting effect after elution. This results in the formation of pores with moderate rigidity and good structure. Consequently, methacrylic acid was selected as the optimal functional monomer for this study.

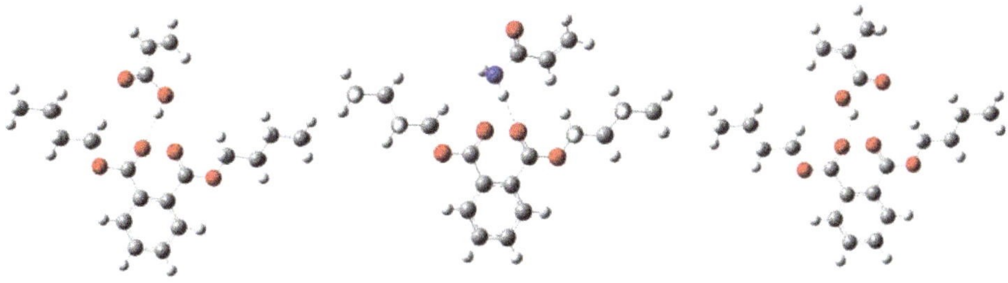

Figure 2. The geometric arrangement of template molecules and functional monomer complexes.

Table 1. The binding energy of the template molecule to the functional monomer.

Molecules	E(a.u.)	ΔE(a.u.)
DBP	−923.47	—
AA	−267.16	—
AM	−247.29	—
MAA	−306.48	—
DBP-AA	−1190.68	−0.054
DBP-AM	−1170.80	−0.04
DBP-MAA	−1230.02	−0.070

To investigate the optimal binding ratio of DBP and methacrylic acid, molecular simulations were conducted to calculate the binding energy at various ratios: 1:2, 1:4, 1:6, and 1:8. Given DBP's symmetrical structure, the results in Table 2 indicate that the binding energy decreases in the following order: ΔE(1:6) > ΔE(1:8) > ΔE(1:4) > ΔE(1:2). Therefore, the optimal binding ratio of 1:6 was selected in this study for the preparation of DBP molecularly imprinted polymers (MIPs). To our knowledge, this is the first study integrating Gaussian simulation-guided monomer selection with phase inversion to fabricate MIUMs for phthalate removal, addressing critical gaps in energy efficiency and specificity.

Table 2. Binding energy of DBP-MAA in different proportions.

Proportion	E(a.u.)	ΔE(a.u.)
1:2	−1536.77	−0.34
1:4	−1843.28	−0.37
1:6	−2763.08	−0.73
1:8	−1894.56	−0.41

3.2. Analysis of the Principles and Structure of MIUM Retained DBP

In this study, DBP was used as the template, methacrylic acid served as the functional monomer, EGDMA acted as the crosslinking agent, and AIBN was employed as the initiator. DBP interacted with the functional monomer through hydrogen bonding. During the aggregation process, DBP was encapsulated within the polymer, and the hydrogen bonds were disrupted by washing the mixture with a methanol-acetic acid solution. This procedure yielded a polymer material that exhibited specific adsorption of DBP. The preparation process for molecularly imprinted polymers (MIPs) is illustrated in Figure 3. After MIPs were mixed with the polysulfone casting membrane solution, a membrane was formed, featuring interconnected pore structures that facilitate the transport and diffusion of molecules. When an aqueous solution containing DBP flows through the membrane, the mass transfer mechanism follows the Piletsky gate model [31]. Molecules primarily enter the membrane through molecular diffusion and pressure. In this process, the MIPs within the membrane selectively adsorb DBP molecules from wastewater. As a result, a clean aqueous solution is collected from the other side of the membrane [32].

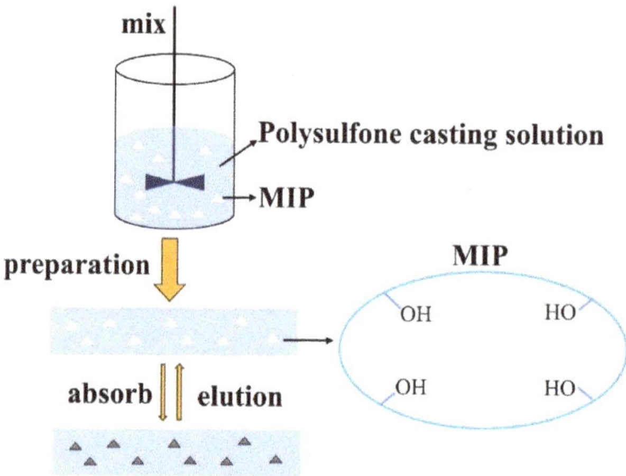

Figure 3. DBP MIUM preparation diagram.

The microstructure of the polymers and MIUM was characterized using scanning electron microscopy, as illustrated in Figure 4. In Figure 4a, the MIPs prepared in the previous stage are shown to be in an amorphous state. Figure 4b displays the micromorphology of the MIUM micro-surface, which has developed a fine membrane pore structure that facilitates liquid flow. Lastly, Figure 4c depicts the microstructure of the MIUM cross-section, where the polymer can be seen embedded within the MIUM structure.

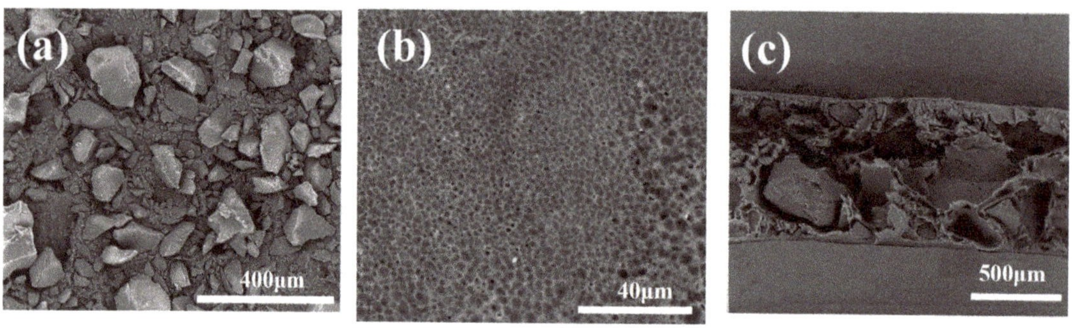

Figure 4. (**a**) polymer microstructure (**b**) MIUM surface morphology (**c**) MIUM cross-section morphology.

The structure of DBP MIPs was examined using infrared spectroscopy. The infrared spectra of MAA, EGDMA, and the MIPs before and after elution are presented in Figure 5. Curve 5a shows the infrared spectrum of MAA, which reveals a stretching vibration peak of the C=C double bond at 1623 cm^{-1}, a stretching vibration peak of the C=O in the carboxyl group at 1701 cm^{-1}, and a stretching vibration peak of the -OH in the hydroxyl group at 2999 cm^{-1}. Curve 5b represents the infrared spectrum of EGDMA, displaying the C=C double bond stretching vibration peak at 1642 cm^{-1}, and the C=O stretching vibration peak for the crosslinker at 1715 cm^{-1} [23]. From curves a and b, it is evident that there are notable differences in the infrared spectra of the functional monomers and the crosslinker.

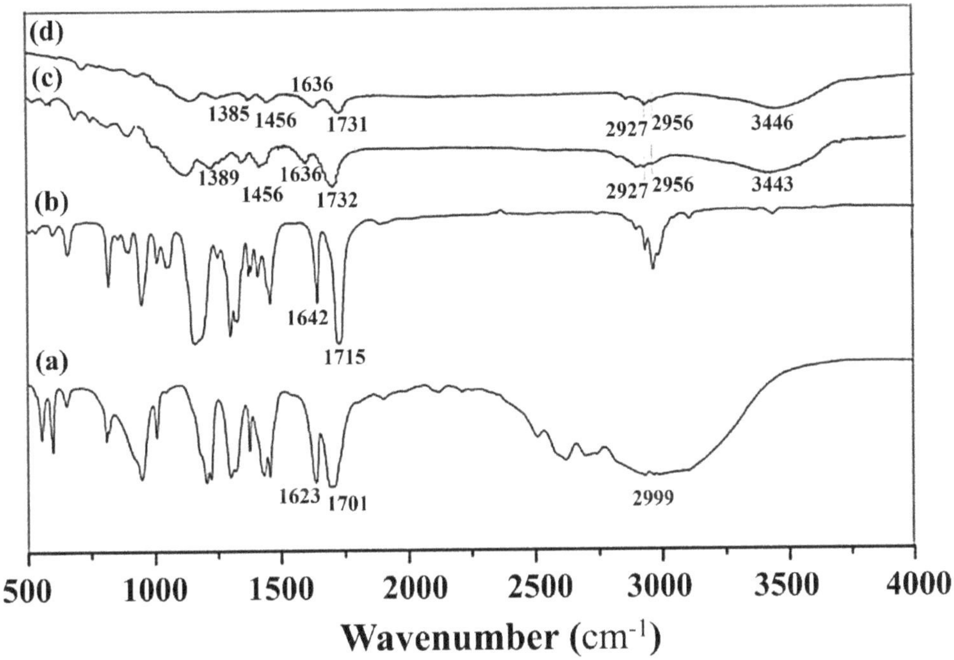

Figure 5. Infrared spectra of (**a**) MAA (**b**) EGDMA (**c**) DBP MIPs before elution (**d**) DBP MIPs after elution.

From curve 5d, we can observe that the main bands of the molecularly imprinted polymer are aligned in similar locations, indicating a resemblance between curves c and d. The figure shows a strong peak at 3443 cm^{-1} and another at 3446 cm^{-1}, which correspond to the tensile vibrations of the -OH group. Additionally, the prominent peaks at 1732 cm^{-1} and 1731 cm^{-1} are attributed to the tensile vibrations of C=O. This confirms that the crosslinking agent EGDMA has been successfully incorporated into the imprinted polymer. Furthermore, the weak peak at 1636 cm^{-1} indicates C=C tensile vibrations, suggesting that most of the functional monomer methacrylic acid and the crosslinking agent EGDMA have reacted during the crosslinking process. These results demonstrate that MIPs have been successfully synthesized and that their composition remains unchanged after elution.

The exceptional performance of the MIUM arises from the synergistic interplay of two distinct mechanisms: chemical affinity and spatial matching. First, FTIR spectroscopy confirmed the critical role of hydrogen bonding between the carboxyl groups of MAA and the ester oxygen atoms of DBP, as evidenced by the characteristic peak shift from 1732 to 1728 cm^{-1} after adsorption. Second, Gaussian simulations demonstrated that the optimized 1:6 DBP/MAA molar ratio created geometrically complementary cavities, enhancing host–guest recognition. This ratio maximized binding energy ($\Delta E = -0.07$ a.u.), ensuring stable spatial alignment between the imprinted sites and the target molecule. Collectively, these mechanisms—directional chemical interactions and cavity geometry—enable MIUM to achieve high selectivity and efficiency in DBP removal, surpassing conventional ultrafiltration membranes reliant solely on pore-size exclusion.

3.3. *Effect of Polymer Addition on MIUM Properties*

The mass percentages of MIPs in the MIUM dry film are 0%, 12.84%, 23.45%, 38.09%, 47.06%, 62.44%, 75.01%, 81.27%, and 85.7% when mixed with DBP MIPs and casting liquid in varying proportions. As the proportion of polymers in the MIUM increases, more DBP can be retained at once, resulting in a higher utilization of MIUM. However, when the mass percentage of MIPs exceeds 85.7%, the mechanical properties of MIUM are compromised, making it prone to breakage, which does not meet the requirements for sewage treatment. At a room temperature of 25 °C and under a pressure of 0.2 MPa, we investigated the impact of different proportions on the adsorption properties of the film. Three films of the same size were prepared according to the same proportions, and their properties were evaluated three times. The average values were then used to create a graph, as illustrated in Figure 6.

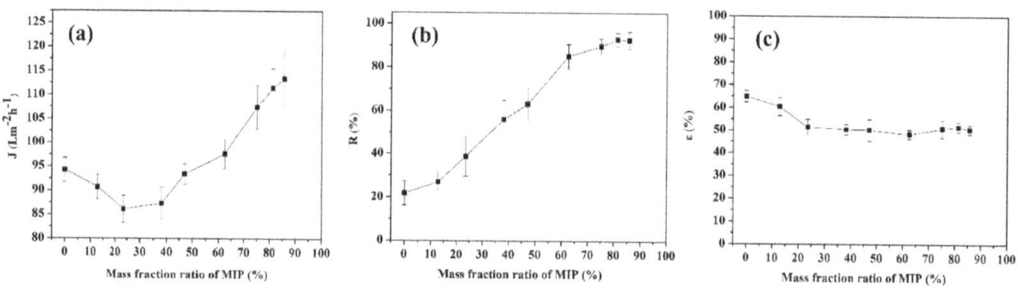

Figure 6. Comparison of MIUM properties (**a**) membrane water flux (**b**) membrane retention (**c**) membrane porosity.

Figure 6a illustrates the trend in membrane water flux. As the amount of polymer added increases, the water flux displays a noticeable decreasing trend. Initially, when no polymer was included, the water flux measured at 94.27 L·m^{-2}h^{-1}. However, with the addition of a small amount of molecularly imprinted polymers (MIPs), the overall water flux began to decline. The MIPs have a significant number of small pores at the molecular scale, which exhibit strong adsorption properties. These MIPs interact with the pore structure of the membrane formed by the polysulfone casting solution during film formation, causing the condensation between the membrane pores to become denser. As a result, the water flux decreases with the addition of a small amount of MIPs. In this study, the water flux reaches its lowest value of 86.15 L·m^{-2}h^{-1} when the mass fraction of MIPs is 23.45%. After this point, as the proportion of MIPs continues to increase, the water flux begins to rise overall until it stabilizes. This increase is due to the large amount of MIPs disrupting the density of the cavernous pore structure within the membrane layers. The impact of this disruption outweighs the constraining effect of the small pores, transforming the originally dense structure into a more interconnected structure between the membrane and MIPs.

Figure 6b illustrates the retention rate of molecularly imprinted polymers (MIPs) on DBP. A simulated wastewater sample with a concentration of 0.1 g/L DBP was prepared, and methanol was used as a co-solvent to pass through the membrane. As the content of MIPs increased, the retention rate of DBP also rose gradually, reaching a maximum of 92.87% when the polymer content was at 81.27%. However, when the MIPs content was higher than this level, the retention rate did not change significantly. The dense structure of MIPs causes the blended film to become increasingly compact as more MIPs are added, resulting in a gradual decrease in porosity (as shown in Figure 6c). This increased density leads to a decline in the membrane's water flux. When the MIPs content reached 81.27%, there was not a significant change in membrane porosity, and the water flux stabilized at 111.49 $L \cdot m^{-2} h^{-1}$. The increased water flux (111.49 $L \cdot m^{-2} \cdot h^{-1}$) compared to pure PSF membranes (94.27 $L \cdot ^{m-2} \cdot h^{-1}$) suggests enhanced hydrophilicity from MIPs incorporation, as hydrophilic surfaces typically exhibit higher permeability.

In summary, with a polymer addition of 81.27%, the membrane's retention rate, water flux, and porosity remained relatively stable. Considering the mechanical properties and the intended service life of the membrane in practical applications, 81.27% is selected as the optimal polymer addition ratio in this study. The synergistic interplay of steric exclusion, directional chemical interactions, and cavity geometry complementarity collectively mirrors the Piletsky model's multi-gate recognition system, which integrates physical and chemical gates for energy-efficient selective transport. This mechanism enables the molecularly imprinted membrane to achieve 92.87% rejection at 0.2 MPa, surpassing conventional ultrafiltration membranes reliant solely on pore-size exclusion, as shown in Table 3.

Table 3. Comparison of the performance of sewage treatment materials.

Material Type	Removal/Adsorption Efficiency	Operating Pressure	Reusability	Primary Mechanism	Selectivity	Reference
Mesoporous cellulose biochar	84.15% removal rate	Not applicable	Not explicitly addressed	π-π electron donor-acceptor (EDA) interaction and pore-filling	Linear-chain PAEs > branched-chain PAEs	[17]
Nanofiltration hollow fiber membrane	91.5% rejection rate	0.4 MPa	Stable performance over 30 days	Steric hindrance	Higher rejection for larger PAEs (e.g., DnOP/DEHP)	[18]
Molecularly imprinted ultrafiltration membrane	92.87% rejection rate	0.2 MPa	>70% retention after 10 elution cycles	Synergy of imprinted cavities (π-π EDA and hydrophobic interactions)	High selectivity for DBP (91.57%), limited affinity for structural analogs	This paper

3.4. A Selective Investigation of MIUM

The unique feature of MIUM is its ability to selectively remove DBP from sewage under low-pressure conditions. In this study, we prepared a sewage simulation solution using 81.27% polymer addition, selecting diethyl phthalate (DEP), diisononyl phthalate (DINP), and chlorpyrifos, which have different molecular structures. We investigated MIUM's selectivity towards four types of molecules, as illustrated in Figure 7. The retention rate of MIUM for DBP is significantly higher than for the other three substances, reaching up to 91.57%, demonstrating strong directional selectivity for DBP [33]. MIUM also shows some capability to intercept DEP and DINP, as these two compounds share structural similarities with DBP, allowing the polymer's adsorption sites to effectively interact with them.

In contrast, the interception of chlorpyrifos molecules by MIUM is similar to that of a filter membrane made with non-imprinted polymers. NIPM refers to non-imprinted polymer membrane, synthesized as a control material without the DBP template. This is because MIUM's interaction with chlorpyrifos primarily relies on the physical adsorption properties of polysulfone, lacking a specific interception mechanism.

The imprinted cavities act as molecular gates, preferentially adsorbing DBP via shape complementarity and hydrogen bonding, while non-target molecules (e.g., DEP, DINP) are excluded due to mismatched functional group alignment. This synergistic mechanism explains MIUM's 2.3× higher DBP rejection compared to non-imprinted membranes.

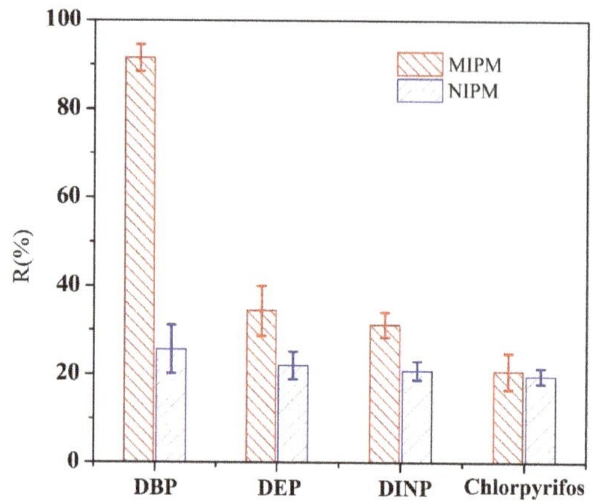

Figure 7. MIUM selectivity.

In summary, MIUM exhibits a strong removal capability for phthalate molecules in sewage, particularly showing high selectivity for the target molecule DBP, making it suitable for the specific removal of DBP.

3.5. *Effects of Sewage pH on MIUM Properties*

This paper investigates the effects of different pH values on the MIUM properties of sewage. The results, illustrated in Figure 8, reveal a linear relationship between the MIUM water flux and the sewage pH; specifically, the water flux increases as pH rises. This occurs because, in an acidic environment, the charge state on the surface of the polysulfone filter membrane changes, leading to a reduction in pore size, which subsequently affects the water transmission rate [34]. In contrast, pH has minimal impact on MIUM retention, as the polymer material is largely unaffected by changes in pH. To enhance treatment efficiency in sewage treatment, it is advisable to adjust the sewage pH to a neutral or weakly alkaline level.

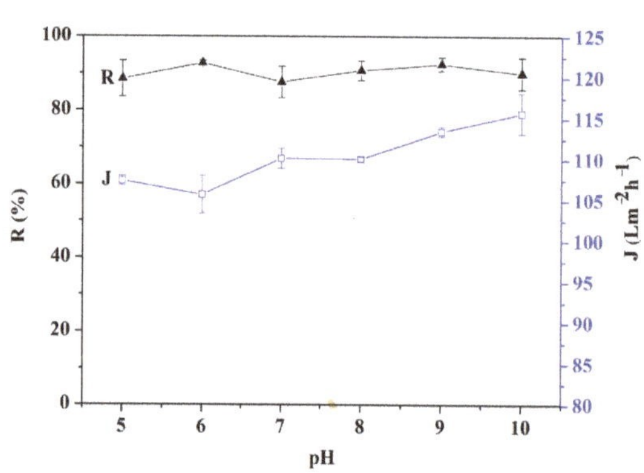

Figure 8. Influence of sewage pH value on MIUM retention rate and water flux.

3.6. *MIUM Reusability Study*

MIPs material exhibits good physical and chemical stability, allowing for reuse through a simple elution process using a mixture of methanol and acetic acid. This method demonstrates excellent reusability. MIUM, prepared from the polymer, can also be reused multiple times, and the repeated elution process helps reduce contamination on the film's surface.

In this study, we investigated the performance of MIUM with repeated elution for ten cycles, as illustrated in Figure 9. Figure 9a shows the change in water flux for MIUM across different elution cycles. The water flux decreases as the number of elution cycles increases. This decrease can be attributed to the compaction of the MIUM film structure under prolonged pressure during multiple elutions, which leads to only minor changes in water flux.

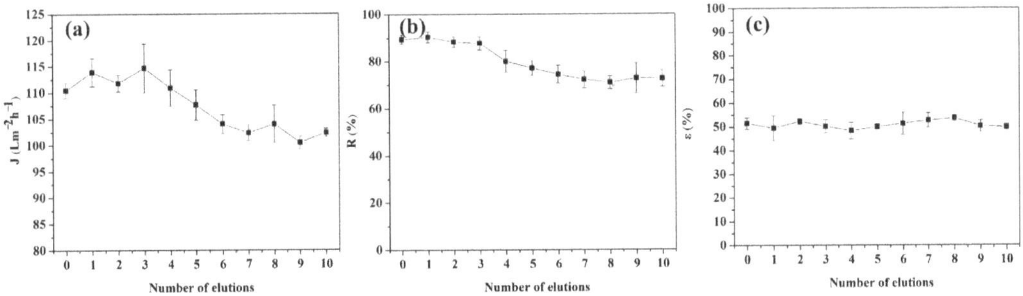

Figure 9. Effect of elution times on MIUM properties; (**a**) membrane water flux, (**b**) membrane retention, (**c**) membrane porosity.

Figure 9b illustrates the effect of the number of elution cycles on the DBP retention rate. The data indicates that the DBP retention rate decreases with an increasing number of elution cycles, leveling off after seven cycles. This stabilization occurs due to the elution process damaging some binding sites or causing an inclusion phenomenon, which results in a minor loss of polymer material. Nevertheless, the retention rate remains above 70%.

Overall, the elution process impacts the MIUM water flux and DBP retention rate. However, as the number of elution times increases, the film performance parameters gradually stabilize and show good performance. The hydrophilic MIPs and smooth surface reduce foulant adhesion. Additionally, the imprinted cavities minimize non-specific interactions, as evidenced by <5% flux decline after 10 cycles.

During repeated elution, the MIUM structure becomes more tightly compressed. Figure 10 compares the cross-sectional morphology of the MIUM before and after elution has occurred 10 times. It is evident that the MIUM structure is partially compressed after elution compared to its previous state. This observation further supports the explanation for the decrease in water flux, as it indicates changes to the overall structure of the MIUM, which can lead to ruptures. However, the structure can still be reused through the elution process.

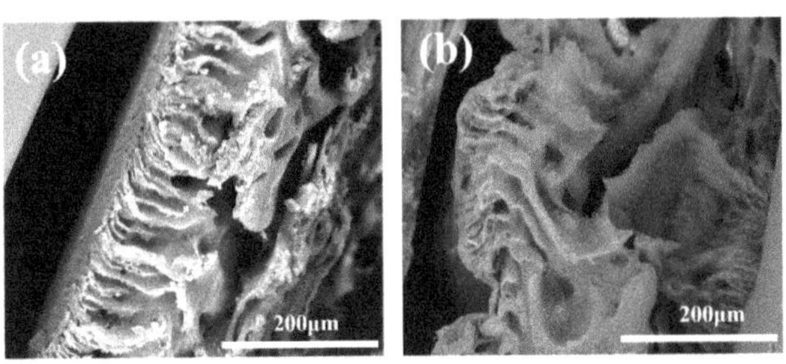

Figure 10. Comparison of cross-section morphology of MIUM before and after elution: (**a**) before elution; (**b**) after 10 elution times.

4. Conclusions

This study establishes a groundbreaking paradigm for molecularly imprinted membrane technology by integrating Gaussian simulation-guided design with scalable phase inversion fabrication. The MIUM achieved 92.87% DBP rejection at 0.2 MPa-43% lower pressure than conventional nanofiber membranes—while maintaining ultrahigh permeability (111.49 $L \cdot m^{-2} \cdot h^{-1}$) and selectivity (91.57% retention for DBP vs. <30% for structural analogs). Crucially, the 1:6 DBP/MAA ratio, optimized via binding energy calculations ($\Delta E = -0.07$ a.u.), created geometrically tailored cavities that synergize size exclusion (20–50 nm pores) and hydrogen-bond-driven adsorption, enabling multi-mechanistic contaminant capture under energy-efficient conditions.

The MIUM's engineering viability is underscored by its >70% retention after 10 cycles and <3.2% flux decline during reuse, attributed to hydrophilic MAA-functionalized surfaces and anti-fouling morphology. With 81.27% MIP loading, this design not only addresses endocrine disruptors in complex wastewater but also reduces chemical/energy consumption by ~50% compared to adsorption-based methods. These innovations position MIUM as a transformative solution for sustainable water treatment, with immediate applicability in industrial settings and scalability for global deployment. Future work will expand this platform to target emerging contaminants (e.g., PFAS, microplastics), leveraging simulation-driven precision for next-generation environmental remediation. Future work will prioritize comprehensive surface characterization (e.g., contact angle, AFM, zeta potential) to further elucidate structure-performance relationships, with collaborations already planned to access the required instrumentation.

Author Contributions: Conceptualization, Q.Z. and Y.W.; software, Y.J.; formal analysis, M.C.; data curation, M.C. and L.Z.; writing—original draft preparation, Q.Z. and Y.J.; writing—review and editing, Q.Z., Y.W. and L.Z.; supervision, Y.W.; project administration, Q.Z. and Y.W.; funding acquisition, Q.Z. All authors have read and agreed to the published version of the manuscript.

Funding: This research was funded by Liaoning Province Science and Technology Project grant number 2022JH2/101300096 and Shenyang Science and Technology Project grant number 20-204-3-30.

Data Availability Statement: The raw data supporting the conclusions of this article will be made available by the authors on request.

Acknowledgments: We sincerely acknowledge Lei Ye (Lund University, Sweden) for his invaluable technical mentorship throughout this project. We gratefully acknowledge the financial support received from the Biomedicine Division of Liaoning Provincial Department of Science and Technology and the Rural Affairs Office of Shenyang Municipal Bureau of Science and Technology for this research project. Their contributions were essential to the completion of this work.

Conflicts of Interest: The authors declare that they have no known competing financial interests or personal relationships that could have appeared to influence the work reported in this paper.

References

1. Liu, J.; Wu, J.; Sun, L.; Liu, L.; Li, B.; Liu, S.; Chen, M.; Wang, T.; Fang, J.; Zhu, N.; et al. A newly-constructed technology to remove and recover diethyl phthalate from wastewater by using the instant plasticization assembly ability of PVC. *J. Environ. Manag.* **2024**, *370*, 122599. [CrossRef] [PubMed]
2. Jiang, J.; Han, D.; Xiao, Y.; Song, X. Occurrence; migration, and assessment of human health and ecological risks of PFASs and EDCs in groundwater of Northeast China. *Water Res.* **2025**, *269*, 122810. [CrossRef] [PubMed]
3. Lourenço, A.C.S.; Galbiati, V.; Corti, D.; Papale, A.; Martino-Andrade, A.J.; Corsini, E. The plasticizer dibutyl phthalate (DBP) potentiates chemical allergen-induced THP-1 activation. *Toxicol. In Vitro* **2015**, *29*, 2001–2008. [CrossRef]
4. Wei, C.; Ding, S.; You, H.; Zhang, Y.; Wang, Y.; Yang, X.; Yuan, J. An immunoassay for dibutyl phthalate based on direct hapten linkage to the polystyrene surface of microtiter plates. *PLoS ONE* **2011**, *6*, e29196. [CrossRef]

5. Xu, Y.; Gye, M.C. Developmental toxicity of dibutyl phthalate and citrate ester plasticizers in *Xenopus laevis* embryos. *Chemosphere* **2018**, *204*, 523–534. [CrossRef]
6. Lokhande, S.; Kalbar, P.P. Economic and environmental benefits of natural treatment systems for sewage treatment: A life cycle perspective. *Water Res.* **2024**, *257*, 121710. [CrossRef]
7. Paluselli, A.; Aminot, Y.; Galgani, F.; Net, S.; Sempere, R. Occurrence of phthalate acid esters (PAEs) in the northwestern Mediterranean Sea and the Rhone River. *Prog. Oceanogr.* **2018**, *163*, 221–231. [CrossRef]
8. He, W.; Qin, N.; Kong, X.; Liu, W.; He, Q.; Ouyang, H.; Yang, C.; Jiang, Y.; Wang, Q.; Yang, B.; et al. Spatio-temporal distributions and the ecological and health risks of phthalate esters (PAEs) in the surface water of a large, shallow Chinese lake. *Sci. Total Environ.* **2013**, *4612*, 672–680. [CrossRef]
9. Jiang, Z.; Li, X.; Wang, R.; Zhang, X.; Luo, Y.; Feng, J.; Sun, M. Extraction performance of SiO_2@COF before and after carbonization and online SPE-HPLC-DAD methods for tetrabromobisphenol A derivatives and phthalates. *Talanta* **2025**, *285*, 127285. [CrossRef]
10. Zhang, C.; Zeng, G.; Huang, D.; Cui, L.; Huang, C.; Li, N.; Xu, P.; Cheng, M.; Zhou, Y.; Tang, W.; et al. Combined removal of di(2-ethylhexyl)phthalate (DEHP) and Pb(ii) by using a cutinase loaded nanoporous gold-polyethyleneimine adsorbent. *RSC Adv.* **2014**, *4*, 55511–55518. [CrossRef]
11. Yang, G.C.C.; Chen, Y.-C.; Yang, H.-X.; Yen, C.-H. Performance and mechanisms for the removal of phthalates and pharmaceuticals from aqueous solution by graphene-containing ceramic composite tubular membrane coupled with the simultaneous electrocoagulation and electrofiltration process. *Chemosphere* **2016**, *155*, 274–282. [CrossRef] [PubMed]
12. Zheng, D.; Chang, Q.; Li, Z.; Gao, M.; She, Z.; Wang, X.; Guo, L.; Zhao, Y.; Jin, C.; Gao, F. Performance and microbial community of a sequencing batch biofilm reactor treating synthetic mariculture wastewater under long-term exposure to norfloxacin. *Bioresour. Technol.* **2016**, *222*, 139–147. [CrossRef] [PubMed]
13. Wang, C.; Zeng, T.; Gu, C.; Zhu, S.; Zhang, Q.; Luo, X. Photodegradation Pathways of Typical Phthalic Acid Esters Under UV, UV/TiO_2, and UV-Vis/Bi2WO_6 Systems. *Front. Chem.* **2019**, *7*, 852. [CrossRef] [PubMed]
14. Chung, Y.C.; Chen, C.Y. Degradation of Di-(2-ethylhexyl) Phthalate (DEHP) by TiO_2 Photocatalysis. *Water Air Soil Pollut* **2009**, *200*, 191–198. [CrossRef]
15. Huang, J.; Li, X.; Ma, M.; Li, D. Removal of di-(2-ethylhexyl) phthalate from aqueous solution by UV/peroxymonosulfate: Influencing factors and reaction pathways. *Chem. Eng. J.* **2017**, *314*, 182–191. [CrossRef]
16. Khan, M.H.; Jung, J.Y. Ozonation catalyzed by homogeneous and heterogeneous catalysts for degradation of DEHP in aqueous phase. *Chemosphere* **2008**, *72*, 690–696. [CrossRef]
17. Chen, Q.; Zheng, J.; Yang, Q.; Dang, Z.; Zhang, L. Effect of carbon chain structure on the phthalic acid esters (PAEs) adsorption mechanism by mesoporous cellulose biochar. *Chem. Eng. J.* **2019**, *362*, 383–391. [CrossRef]
18. Wei, X.; Shi, Y.; Fei, Y.; Chen, J.; Lv, B.; Chen, Y.; Zheng, H.; Shen, J.; Zhu, L. Removal of trace phthalate esters from water by thin-film composite nanofiltration hollow fiber membranes. *Chem. Eng. J.* **2016**, *292*, 382–388. [CrossRef]
19. Ton, X.-A.; Acha, V.; Haupt, K. Bernadette Tse Sum Bui, Direct fluorimetric sensing of UV-excited analytes in biological and environmental samples using molecularly imprinted polymer nanoparticles and fluorescence polarization. *Biosens. Bioelectron.* **2012**, *36*, 22–28. [CrossRef]
20. Shen, X.; Xu, C.; Ye, L. Molecularly Imprinted Polymers for Clean Water: Analysis and Purification. *Ind. Eng. Chem. Res.* **2013**, *52*, 13890–13899. [CrossRef]
21. Akbari-adergani, B.; Sadeghian, G.-H.; Alimohammadi, A.; Esfandiari, Z. Integrated photografted molecularly imprinted polymers with a cellulose acetate membrane for the extraction of melamine from dry milk before HPLC analysis. *J. Sep. Sci.* **2017**, *40*, 1361–1368. [CrossRef] [PubMed]
22. Di Bello, M.P.; Mergola, L.; Scorrano, S.; Del Sole, R. Towards a new strategy of a chitosan-based molecularly imprinted membrane for removal of 4-nitrophenol in real water samples. *Polym. Int.* **2017**, *66*, 1055–1063. [CrossRef]
23. Ghasemi, S.; Nematollahzadeh, A. Molecularly imprinted ultrafiltration polysulfone membrane with specific nano-cavities for selective separation and enrichment of paclitaxel from plant extract. *React. Funct. Polym.* **2018**, *126*, 9–19. [CrossRef]
24. Ghasemi, S.; Nematollahzadeh, A. Molecularly imprinted polymer membrane for the removal of naphthalene from petrochemical wastewater streams. *Adv. Polym. Technol.* **2018**, *37*, 2288–2293. [CrossRef]
25. Cirillo, G.; Puoci, F.; Curcio, M.; Parisi, O.I.; Iemma, F.; Spizzirri, U.G.; Picci, N. Molecular imprinting polymerization by Fenton reaction. *Colloid Polym. Sci.* **2010**, *288*, 689–693. [CrossRef]
26. Yang, Z.; Sun, Z.; Cui, D.; Li, P.; Wang, Z. TIPS behavior for IPP/nano-SiO2 blend membrane formation and its contribution to membrane morphology and performance. *Chin. J. Chem. Eng.* **2019**, *27*, 467–475. [CrossRef]
27. Sun, L. The removal of typical pollutants in secondary effluent by the combined process of powdered activated carbon–ultrafiltration. *Water Sci. Technol.* **2017**, *75*, 1485–1493. [CrossRef]
28. Lin, D.; Xu, D.; Fang, K.; Wang, Z.; Du, X.; Ke, Q.; Liang, H.; Ulbricht, M. How in situ fabricated polyzwitterionic hydrogel coating on nanofiltration and ultrafiltration membranes influences mass transfer processes and fouling development. *J. Membr. Sci.* **2024**, *711*, 123174. [CrossRef]

29. Hamerton, I.; Howlin, B.J.; Mitchell, A.L.; Hall, S.A.; McNamara, L. Using Molecular Simulation to Predict the Physical and Mechanical Properties of Polybenzoxazines. In *Handbook of Benzoxazine Resins*; Elsevier: Amsterdam, The Netherlands, 2017; pp. 127–142. [CrossRef]
30. Yankelov, R.; Yungerman, I.; Srebnik, S. The selectivity of protein-imprinted gels and its relation to protein properties: A computer simulation study. *J. Mol. Recognit.* **2017**, *30*, 2607–2617. [CrossRef]
31. Sergeyeva, T.; Slinchenko, O.; Gorbach, L.; Matyushov, V.; Brovko, O.; Piletsky, S.; Sergeeva, L.; Elska, G. Catalytic molecularly imprinted polymer membranes: Development of the biomimetic sensor for phenols detection. *Anal. Chim. Acta* **2010**, *659*, 274–279. [CrossRef]
32. Jorgensen, T.C.; Weatherley, L.R. Ammonia removal from wastewater by ion exchange in the presence of organic contaminants. *Water Res.* **2003**, *37*, 1723–1728. [CrossRef] [PubMed]
33. Wu, Y.; Lu, J.; Lin, X.; Gao, J.; Chen, L.; Lv, P.; Liu, X.; Meng, M.; Lia, C.; Yan, Y. Bioinspired synthesis of SiO_2/pDA-based nano composite-imprinted membranes with sol-gel imprinted layers for selective adsorption and separation applications. *Phys. Chem. Chem. Phys.* **2018**, *20*, 15775–15783. [CrossRef] [PubMed]
34. Nabe, A.; Staude, E.; Belfort, G. Surface modification of polysulfone ultrafiltration membranes and fouling by BSA solutions. *J. Membr. Sci.* **1997**, *133*, 57–72. [CrossRef]

membranes

MDPI

Article

Influence of Silane Treatment on CNM/PAC/PVDF Properties and Performance for Water Desalination by VMD

Samraa R. Khaleel [1], Salah S. Ibrahim [1], Alessandra Criscuoli [2,*], Alberto Figoli [2], Dahiru U. Lawal [3,4] and Qusay F. Alsalhy [1,*]

1 Membrane Technology Research Unit, Chemical Engineering Department, University of Technology-Iraq, Alsinaa Street 52, Baghdad 10066, Iraq; che.21.02@grad.uotechnology.edu.iq (S.R.K.); salah.s.ibrahim@uotechnology.edu.iq (S.S.I.)

2 Institute on Membrane Technology (CNR-ITM), Via P. Bucci 17/C, 87036 Rende, CS, Italy; a.figoli@itm.cnr.it

3 Interdisciplinary Research Center for Membrane and Water Security, King Fahd University of Petroleum and Minerals, Dhahran 31261, Saudi Arabia; dahiru.lawal@kfupm.edu.sa

4 Mechanical Engineering Department, King Fahd University of Petroleum and Minerals, Dhahran 31261, Saudi Arabia

* Correspondence: a.criscuoli@itm.cnr.it (A.C.); qusay.f.abdulhameed@uotechnology.edu.iq or qusay_alsalhy@yahoo.com (Q.F.A.)

Abstract: Vacuum membrane distillation (VMD) is a promising process for water desalination. However, it suffers some obstacles, such as fouling and wetting, due to the inadequate hydrophobicity of the membrane and high vacuum pressure on the permeate side. Therefore, improving surface hydrophobicity and roughness is important. In this study, the effect of 1H,1H,2H,2H-Perfluorodecyltriethoxysilane (PFTES) on the morphology and performance of CNM/PAC/PVDF membranes at various concentrations was investigated for the first time. Membrane characteristics such as FTIR, XRD, FE-SEM, EDX, contact angle, and hydrophobicity before and after modification were analyzed and tested using VMD for water desalination. The results showed that the membrane coated with 1 wt.% PFTES had a higher permeate flux and lower rejection than the membranes coated with the 2 wt.% PFTES. The 2 wt.% PFTES enhanced the contact angle to 117° and increased the salt rejection above 99.9%, with the permeate flux set to 23.2 L/m^2·h and at a 35 g/L NaCl feed solution, 65 °C feed temperature, a 0.6 L/min feed flow rate, and 21 kPa (abs) vacuum pressure. This means that 2 wt.% PFTES-coated PVDF membranes exhibited slightly lower permeate flux with higher hydrophobicity, salt rejection, and stability over long-term operation. These outstanding results indicate the potential of the novel CNM/PAC/PVDF/PFTES membranes for saline water desalination. Moreover, this study presents useful guidance for the enhancement of membrane structures and physical properties in the field of saline water desalination using porous CNM/PAC/PVDF/PFTES membranes.

Keywords: PFTES; vacuum membrane distillation; membrane; coating layer; desalination; carbon nanomaterials

1. Introduction

Over the past few decades, freshwater availability has been under high stress due to climate change, growing industrialization, and population [1,2]. Globally, more than two billion people do not have access to drinking water, and by 2050, it is anticipated that it will be scarce for fifty percent of the world's population. Desalination technologies can help increase the supply of fresh water by recovering water from contaminated or salty water, such as brackish water, seawater, and industrial wastewater [3]. Membrane distillation is a promising technique for treating high-salt water and wastewater [4–6]. MD is a non-isothermal process that operates based on the vapor pressure difference between the feed and permeate sides of the microporous hydrophobic membrane [7–9].

The literature mostly concerns four main configurations of MD: direct contact membrane distillation (DCMD), air gap membrane distillation (AGMD), sweep gas membrane distillation (SGMD), and vacuum membrane distillation (VMD), depending on the way water accumulates on the permeate side. Compared with other desalination techniques, VMD has attracted more attention due to the low heat loss by conduction and higher permeate flux [10–13]. Despite these advantages, it has a low resistance to pore wetting and fouling [14,15]. For ideal rejection and permeability in MD, a hydrophobic microporous membrane is the most important property of MD membrane, which is fabricated from hydrophobic polymers, such as polytetrafluoroethylene (PTFE), polypropylene (PP), and polyvinylidene fluoride (PVDF). It has low surface-free energy and high water repellency [16–18]. More research has been developed to improve the wetting resistance by increasing the surface hydrophobicity, and scientific researchers have noticed that decreasing surface energy and increasing roughness are the main ways to enhance the surface hydrophobicity by surface fluorination [17].

Silane is an effective agent for low surface-free energy materials with hydrophobic ends, which makes it a more suitable monomer for hydrophobic surfaces [19]. It is used for modifying different types of material surfaces by incorporating fluorinated groups on the membrane surface, resulting in reduced pore wetting and enhanced hydrophobicity, salt rejection, stability flux, and fouling resistance [20]. Fluoro-silane has a small atomic radius that attempts to lower surface energy on the outer surface by forming a covalent bond that is stable with carbon. Furthermore, non-polar bonds of carbon–silicon cause reduced surface energy and enhanced hydrophobicity [21]. The long reaction time and low concentration of silane solution make it efficient for the surface modification method [22,23].

In this study, high hydrophobic carbon nanomaterial powder-activated carbon (CNM/PAC) was embedded using different amounts of PVDF polymer. The optimum membrane was modified chemically using different concentrations of Perfluorodecyltriethoxysilane (PFTES) for the first time to enhance the hydrophobicity of the membrane and the salt rejection. The membranes were characterized using several techniques and tested in vacuum membrane distillation (VMD) for the desalination of highly saline water (35 g/L). The effect of operating parameters on the permeate flux of the VMD system was also evaluated.

2. Materials and Methods

2.1. Chemicals Materials

Polyvinylidene fluoride (PVDF) with a molecular weight of 500–600 g/mol was acquired from Hebei luoxing Tech Co., Ltd., Shijiazhuang, China. N-methyl-2-pyrrolidone (NMP) was purchased from the Chinese company Shandong Zhishang Chemical Co., Ltd., Jinan, China, CNM/PAC high-hydrophobic nanomaterials were synthesized via chemical vapor deposition [24]. 1H,1H,2H,2H-Perfluorodecyltriethoxysilane (PFTES) ($C_{13}H_{13}F_{17}O_3Si$ at 98% purity and 568.3 g/mol molecular weight was purchased from Jinan Future Chemical Co., Ltd., Jinan, China. N-hexane (C_6H_{14}) with a molecular weight of 86.18 g/mol), sodium chloride (NaCl), and isopropanol with a density of $\rho = 0.786\ g{\cdot}cm^{-3}$ were purchased from Loba company (Bronx, NY, USA).

2.2. Membrane Preparation

PVDF flat sheet membranes were prepared using the phase inversion method with varying concentrations of CNM/PAC. Firstly, PVDF powder was dried at 50 °C for 2–3 h to remove any moisture content before fabricating the casting solution. Various amounts of CNM/PAC (0.1, 0.2, 0.3, 0.4, and 0.5%) were separately loaded in NMP solvent. To facilitate dispersion, the mixtures (CNM/PAC and NMP) were sonicated ultrasonically for 30 min in a water bath (24 kHz, Fyang, Fuyang, China). After sonication, PVDF powder at 14% was added gradually to the mixture (CNM/PAC/NMP solution), then stirred magnetically (hot plate, Joan lab, Huzhou, China) at 100–150 rpm and 60 °C for 24 h to achieve a homogenous mixture. Then, it was stored at 50 °C for 2–3 h to remove any trapped air bubbles before casting.

Following that, the dope solution was cast onto a clean glass plate using a film applicator (AFA-IV, Yantai, China) with a 0.25 mm gap of clearance. The membrane casting was immediately immersed in a water bath as a coagulation at room temperature. After some time, the membrane was separated from the glass plate and transferred to a distilled water bath, where it remained for at least 24 h. It was then replaced with other distilled water to ensure the removal of any residual solvent completely. Finally, the prepared membranes were stored in a container with distilled water for further analysis. Details of the dope solution composition are provided in Table 1.

Table 1. The composition of CNM/PAC/PVDF membrane.

Sample	PVDF (wt.%)	NMP (wt.%)	CNM/PAC (wt.%)
P0	14	86	--
P1	14	86	0.1
P2	14	86	0.2
P3	14	86	0.3
P4	14	86	0.4
P5	14	86	0.5

The prepared CNM/PAC/PVDF membranes are chemically modified using a fluorosilane solution, specifically, 1H,1H,2H,2H-Perfluorodecyltrimethoxysilane (PFTES, $C_{13}H_{13}F_{17}O_3Si$) by the dip-coating method. To obtain the coating solution, PFTES was dissolved in hexane at two different concentrations (1 and 2 vol./vol.%). Membranes were dip-coated in the PFTES solution for 24 h to obtain sufficient coverage of the surface and enhance the hydrophobicity. After this treatment, the membranes were removed and rinsed several times with distilled water to remove any unreacted material. They were then dried at ambient temperature for 2 h before further analysis

3. Membrane Characterization

The chemical composition of the membrane was characterized before and after a long time of operation in VMD. The Attenuated total reflectance Fourier-Transform Infrared Spectroscopy (ATR-FTIR) instrument (IRT-5200 FT-IR, JASCO, Tokyo, Japan) was used to analyze the functional group of the prepared membrane before and after the VMD experiments and to investigate the possible interaction between the membrane material and the components of the feed solution. Also, X-ray diffraction (XRD, Shimadzu, Kyoto, Japan) was used to display the composition of the membranes and the salt deposition after VMD experiments.

The thickness of the membrane was measured by a micrometer (HDT, Beijing, China) with an accuracy of 0.001 mm and estimated by SEM for more accuracy. The value of the average thickness was calculated for five measurements taken at various locations on the sample of the membrane [25]. The porosity of the membrane is defined as the ratio of pore volume to total membrane volume, which is determined by a gravimetric method dependent on the density of the prepared membrane to polymer density according to Equation (1) [26,27].

$$\varepsilon(\%) = \frac{(W_w - W_d)/\rho_{ISO}}{(W_w - W_d)/\rho_{ISO} + (W_d/\rho_m)} \tag{1}$$

where ρ_m and ρ_{pISO} are the membrane density and isopropanol density (0.786 g/cm^3), and W_w and W_d are the weight and dry membrane

$$\rho_m = \frac{W_m}{Vm} \tag{2}$$

where W_m = weight of the membrane (g), Vm = membrane volume (Vm = A × l) A = membrane area (cm^2), and l is the thickness of the membrane (cm).

The mean pore radius is calculated based on the porosity and permeate flux according to the Guerout–Elford–Ferry Equation (3):

$$r_m = \sqrt{\frac{(2.9 - 1.75\varepsilon)8\eta lJ}{\varepsilon A \Delta p}} \tag{3}$$

where r_m is the mean pore size; η is the viscosity of the water 1×10^{-3} (at 65 °C); l is the membrane thickness (m); J is the water permeate flux (L/m^2·h); ε is the porosity of the membrane; A is the membrane effective area (m^2); and Δp is the transmembrane pressure (Pa) [23].

The wettability of all the membranes prior and after the VMD experiment is evaluated by the measurement water contact angle (WCA) using Theta Lite-TL100 and TL101- Finland. The WCA will instantaneously indicate the hydrophobic nature of the prepared membrane. A 3 μL sessile droplet of distilled water was applied carefully to the surface of the flat membrane using a synergy, then the WCA between the membrane surface and the drop of water was evaluated. Five readings were measured at various locations for each sample of the membrane to eliminate errors, then the average contact angle was determined at room temperature.

The liquid entry pressure (LEP), or pore entry pressure, is the maximum pressure at which water does not pass through the pores of the membrane. LEP is affected by pore geometry (size, shape) and hydrophobicity [28,29]. The LEP is calculated to evaluate membrane stability over long-term MD applications, using the Laplace (Cantor) Equation (4) [30–32]:

$$LEP = P_f - P_p = \frac{-2\,\beta\,\gamma\,\cos\theta}{r_{max}} \tag{4}$$

where P_f and P_p represent the feed and permeate sides' hydraulic pressures of the system, β is the pore coefficient geometric (equal to 1 for cylindrical pores and $0 < \beta < 1$ for non-cylindrical pores), γ is the surface tension of the liquid surface tension, θ is the contact angle on the surface of the membrane, and r_{max} is the maximum pore radius.

The mechanical properties of the prepared membrane were evaluated using a tensile strength testing machine (Tinius Olsen H50KT, England) equipped with computer software for data analysis. The samples of the membrane, measuring 2 cm × 10 cm, were clamped at both ends of the machine and subjected to tension at an elongation speed of 5 mm/min. The load cell had a force of 5 N, and the tests were conducted at room temperature. The tensile strength of each sample was calculated by dividing the force at the point of fracture (N) by the cross-sectional area of the membrane (m^2) [33].

The morphology and the elemental composition of the membrane surface were tested before and after the VMD experiments by Field Emission Scanning Electron Microscopy (FE-SEM, Zeiss sigma- Germany) combined with Energy Dispersive X-ray spectroscopy analysis (EDS). For cross-section imaging, the samples of the membrane were dried at room temperature for 48 h before being fractured by exposure to the liquid nitrogen, and then the surface and cross-section of the membrane were subsequently coated with a uniform thin layer of gold.

4. Membrane Performance

The performance of all the prepared membranes was evaluated using a vacuum membrane distillation (VMD) system shown in Figure 1. The permeate flux in VMD was determined using Equation (5) [33,34]:

$$J = \frac{Q}{A \times t} \tag{5}$$

where J represents the permeate flux ($L/m^2 \cdot h$); Q is the volume of vapor collected (L); A is the effective membrane area (m^2) calculated as width (W) times length (L); t is the time over which the vapor is collected (h); and the salt rejection (%R) was determined using Equation (6) [29,31]:

$$\%R = (1 - \frac{C_p}{C_f}) \times 100 \tag{6}$$

where C_f and C_p represent the feed and permeate concentrations, respectively. To ensure the stability of nanomaterials over a long time of VMD operation, a leaching ratio experiment was performed. Membrane samples (3 × 5 cm) were first dried at 55–60 °C for two hours, weighed, and then immersed in distilled water for one month, with regular water changes. Following the soaking time, the membranes were weighed and dried once more in an oven. The following Equation (7) was used for calculating the leaching ratio [35]:

$$\text{Leaching ratio}(\%) = \frac{M_1 - M_2}{M_1} \times 100\% \tag{7}$$

where M_1 and M_2 represent the membranes' relative weights prior to and following soaking.

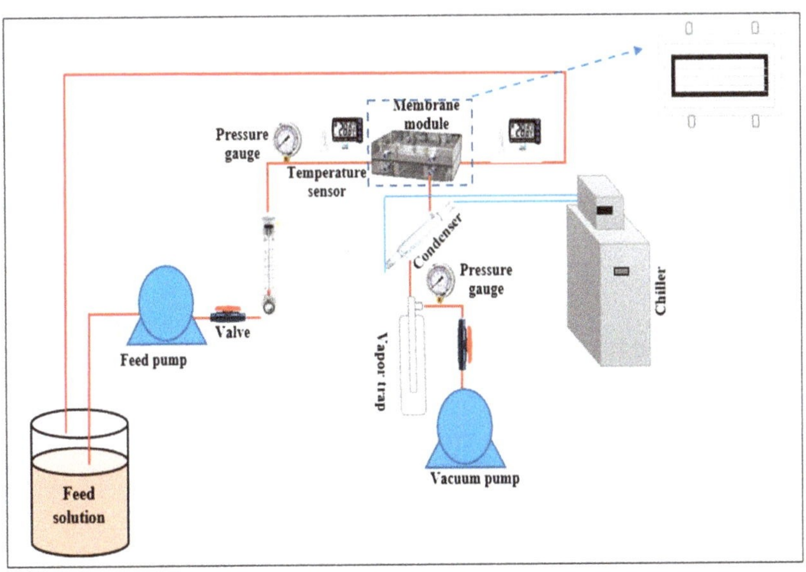

Figure 1. Schematic diagram of VMD system.

5. Results and Discussion

5.1. FTIR Analysis

The FT-IR spectra were used to analyze changes in the structure of PVDF membranes after modification with nanomaterials and chemical modifications, as displayed in Figure 2. The peak at 1068 cm^{-1} was related to CH_2 groups, while the peaks at 1180 cm^{-1} and 1274 cm^{-1} were attributed to carbon–fluorine (C-F) vibration bonds. In addition, the α-phase characteristic is found at 782 cm^{-1}, which means the phase is crystalline. The β-phase characteristic peaks were seen at 875 cm^{-1}, which means the CF_2 is stretching in a symmetric way [36]. Also, the 1660 peaks cm^{-1} were linked to the carbonyl (C-O) group [37], and the bands at 2980 cm^{-1} and 3020 cm^{-1} were linked to the –CH and CH_2 stretching vibrations [38]. When we compared the pure PVDF with the modified matrix carbon nanomaterial powder-activated carbon (CNM/PAC), we observed weaker peaks at 1274 cm^{-1}, 1180 cm^{-1}, and 1040 cm^{-1} for the symmetrical and asymmetrical C-F stretching bands, as well as the C-H vibrations peak Figure 2a. These peaks indicate a change in the crystalline structure of polymers after the addition of CNM/PAC.

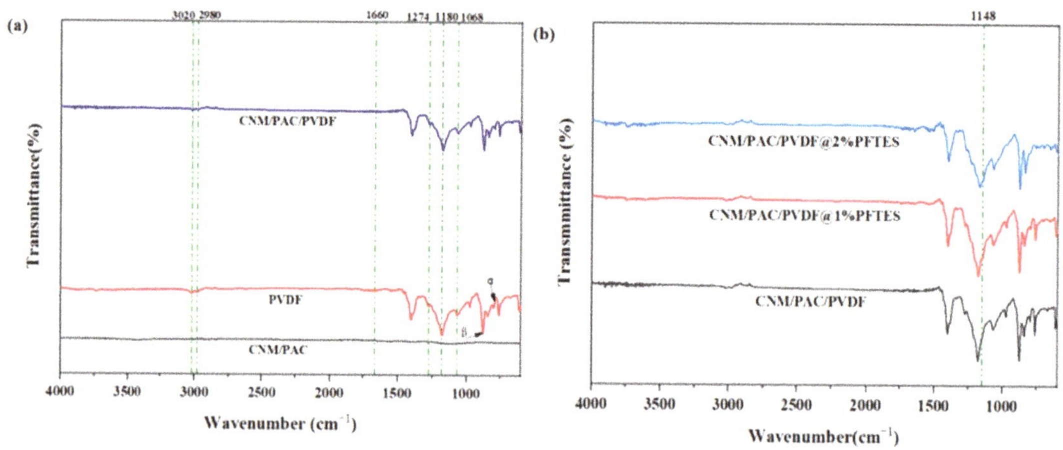

Figure 2. FTIR spectra for (**a**) CNM/PAC, P0, and P4 membranes; (**b**) spectra of control P4 and P4 membranes treated with 1 and 2 wt.% PFTES.

Figure 2b displays the FT-IR spectra of the CNM/PAC/PVDF membranes before and after chemical modification by the dip-coating method with 1–2 wt.% PFTES for 24 h. As can be seen from that figure, the chemically modified membrane showed a peak at 1148 cm^{-1}, which indicated a successfully coated membrane surface with silane solution and Si-O stretching vibrations. Moreover, the membranes coated with 2 wt.% PFTES had a strong Si-O absorption peak and improved the spectral features of fluorine compounds, which confirms efficient modification with PFTES. These results highlight the transformation of the PVDF structure by being embedded with CNM/PAC and coated with PFTES, displaying improved chemical features for enhanced hydrophobicity and membrane performance.

5.2. X-Ray Diffraction (XRD)

The XRD analysis was utilized to determine the crystal phase for the pristine PVDF membrane before and after modification with various concentrations of CNM/PAC membranes, as shown in Figure 3. It can be seen that peaks at 18.2°, 20.32°, and 26.04° are related to α and β phases, respectively [39]. After the addition of CNM/PAC, the intensity of peaks 18.2° (002), 20.23° (110), and 26.04° (021) reduced slightly compared with the crystal structure of the pristine PVDF, as shown in Figure 3a. The intensity of the peak was enhanced after chemically modifying the membrane with 1–2 wt.% PFTES, as seen in Figure 3b.

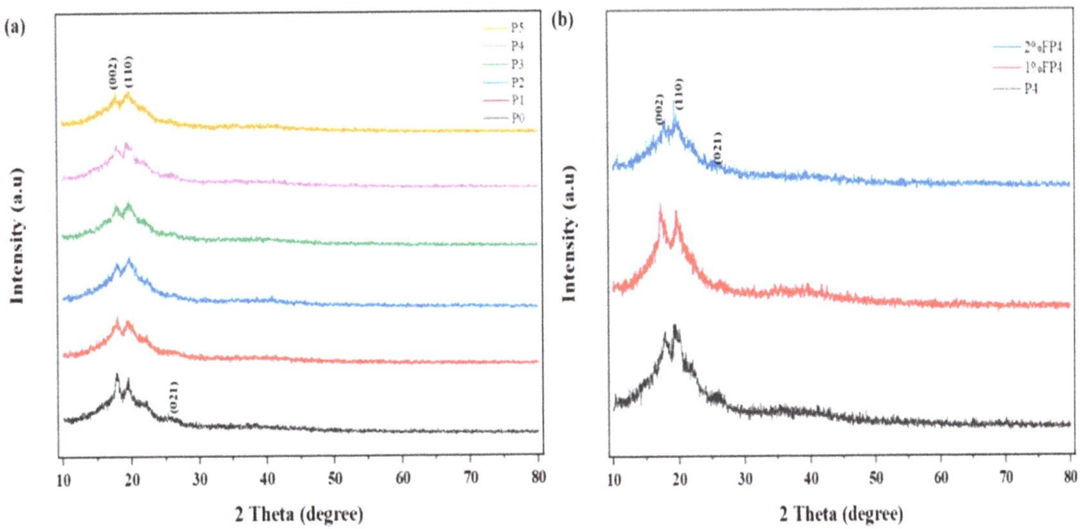

Figure 3. XRD spectra of (**a**) P0, P1, P2, P3, P4, and P5 membranes; (**b**) P4 before and after being chemically modified with 1–2% PFTES (1% FP4) and (2 wt.% FP4) membranes.

5.3. Field Emission Scanning Electron Microscopy and Energy Dispersive X-Ray Spectrometer (FE-SEM and EDX)

The morphology changes of the PVDF membrane before and after modification with CNM/PAC nanomaterials and chemical modification with PFTES were analyzed by FE-SEM. In this study, 0.1 wt.% (P1), 0.2 wt.% (P2), 0.3 wt.% (P3), 0.4 wt.% (P4), and 0.5 wt.% (P5) of CNM/PAC were embedded with PVDF by phase inversion. The FE-SEM images of the membrane surface of all the prepared membranes are shown in Figure 4a–f. The pore size increased as the CNM/PAC increased from 0.1 wt.% to 0.4 wt.%. Upon loading 0.5 wt.%, the pore size on the membrane surface decreased, as seen in Figure 4b–f. The change in the structure of the membrane surface after being embedded with CNM/PAC is related to a delay in the rate of exchange between the solvent (NMP) and the non-solvent (water) through the membrane [35,40]. Also, Figure 4g,h shows the FE-SEM images for the top surface of the 0.4 wt.% CNM/PAC/PVDF membrane when treated with 1–2 wt.% PFTES. When the membrane surface was coated with 1% PFTES, the pore size of the membrane surface increased slightly, and no effect after coating with 1% PFTES was observed due to the small molecule structure and concentration of PFTES, whereas the coating with 2 wt.% PFTES resulted in a slight decrease in pore size. This decrease in pore size may be attributed to the narrowed pore size on the surface of the chemically modified membrane, which would lead to a reduction in the permeate flux [17,41].

It was observed that the P3 membrane and 1FP4 membrane surfaces show large amounts of big holes, while other membranes show much fewer holes. This observation was attributed to the increment in the content of CNM/PAC in the casting solution. 1FP4 shows a large pore surface due to the coated P4, with a low concentration of PFTES caused by the evaporation of a hexane-silane solution, which leads to an increase in the surface pore size and porosity.

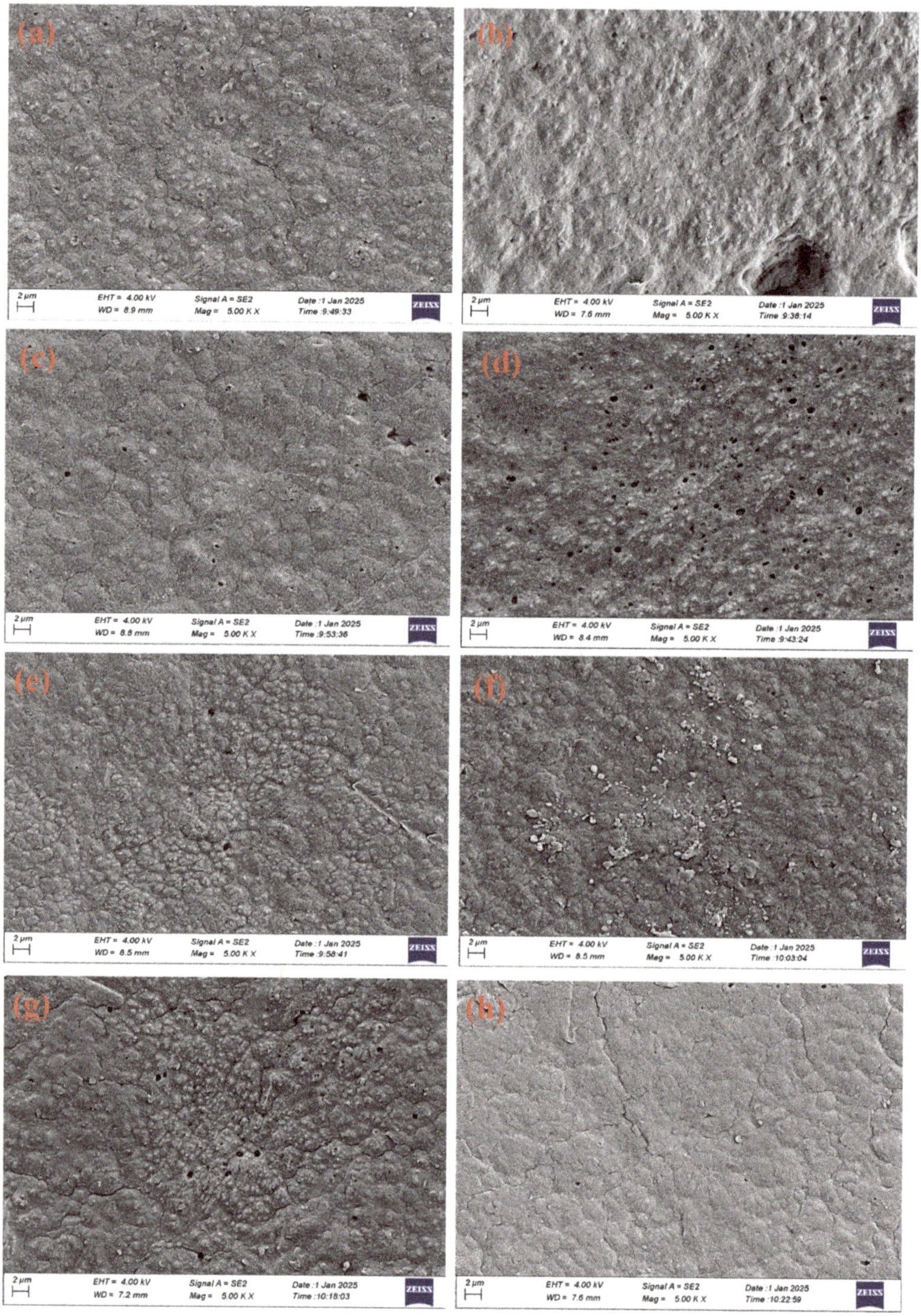

Figure 4. FE-SEM surface image of various concentrations of CNM/PAC/PV.DF membranes (**a**) P0 (0%), (**b**) P1 (0.1 wt.%), (**c**) P2 (0.2 wt.%), (**d**) P3 (0.3 wt.%), (**e**) P4 (0.4 wt.%), (**f**) P5 (0.5 wt.%), (**g**) 1 wt.% FP4, (**h**) 2% FP4 membranes.

The cross-sectional images of the PVDF membranes before and after being modified with different concentrations of CNM/PAC and compared with the chemically modified PVDF are shown in Figure 5. The images show a typical construction with two layers: finger-like macro-voids near the upper surface and sponge-like structures towards the bottom. After loading the nanomaterials, both the size and the number of these macro-voids increase. The macrovoid structure increases after loading 0.1–0.4 wt.% of CNM/PAC (Figure 5a–e). But when the CNM/PAC goes up to 0.5 wt.%, the membrane's appearance has fewer and smaller macrovoids (Figure 5f). These observations suggest that the addition of nanomaterials significantly influences membrane morphology by increasing the viscosity of the casting solution [42,43]. Additionally, Figure 5g,h illustrates the impact of chemical modification using PFTES at 1–2 wt% concentrations by the dip-coating method. The cross-sectional images of P4 membranes show that the macrovoid structure reduced slightly after coating with 1 wt.% PFTES Figure 5g. When coated with 2 wt.% PFTES, a dense skin layer appears near the top surface, as shown in Figure 5h [44,45].

To evaluate the elemental composition of pristine PVDF membranes before and after the addition of CNM/PAC, Energy Dispersive X-ray Spectroscopy (EDX) was used to analyze the chemical structure of the membranes before and after modification with CNM/PAC and subsequent modification with fluoro-silane (PFTES) solution. EDX spectra were recorded in the binding energy range of 0 to 10 keV, as shown in Figure 6. The elemental analysis of the pristine PVDF membrane (P0) revealed clear peaks corresponding to fluorine (F) and carbon (C), confirming their presence at 59.2 wt.% and 39.8 wt.%, respectively (Figure 6a). For the P4 membrane, which embedded 0.4 wt.% CNM/PAC, an increase in the carbon content to 57.7 wt.% and a slight decrease in fluorine content to 41.4 wt.% was observed, as well as the detection of oxygen (O) at 0.9 wt.% (Figure 6b). This confirms the incorporation of CNM/PAC into the membrane matrix. The presence of PFTES on the surface of the P4 membranes is illustrated by the appearance of silicon (Si) peaks, as shown in Figure 6c,d. At higher PFTES concentrations (2 wt.%), the EDX spectra displayed a slightly higher Si content, confirming increased surface modification (Figure 6c,d). Overall, EDX analysis confirms the successful modification of the PVDF membranes, as indicated by the increase in fluorine content and the detection of Si and O, validating the effective chemical modification with PFTES and the incorporation of CNM/PAC.

5.4. Membrane Thickness, Porosity, Pore Size, and LEP

The properties of the membranes in terms of thickness, porosity, mean pore size, and LEP are presented in Figure 7. Figure 7a-left shows the thickness (82.85 μm) of the pristine PVDF (P0) membranes. Firstly, the thickness rose to 93.79 μm at 0.1% of CNM/PAC (P1) loading. As the concentration of CNM/PAC grew, the thickness reduced. The thickness reduced to 65.41 μm for P4 and then increased to 73.1 μm for P5. Following chemical modification, the thickness of the P4 membrane increased from 65.41 μm to 67.75 μm and 68.73 μm after being coated with 1–2% PFTES. A hydrophobic layer deposited on the membrane surface in proportion to the silane solution concentration (Figure 7b-right) probably causes this rise in thickness.

The porosity of all the prepared membranes was calculated by the dry and wet method (gravimetric analysis). As shown in Figure 7a-right, the porosity of the pristine PVDF (P0) was the lowest, 83.3%, which progressively increased with the incorporation of CNM/PAC, reaching 87.2% for the P4 and slightly reducing to 86.1% for P5. This increase in porosity is probably due to the CNM/PAC dispersion within the polymer matrix, which enhances the rate of exchange between solvent and non-solvent during phase inversion, thus promoting the formation of a more porous structure. Following the chemical modification, the P4 membrane porosity decreased slightly after modification with 1 wt.% and 2 wt.% PFTES.

The chemically modified membrane decreased from 87.2% to 86.95% and 85.4% (Figure 7b-right). As the concentration of PFTES solution increased, the modified membrane's porosity slightly decreased because some small or large pores of the membranes were clogging, resulting in a slight reduction in the water permeate flux.

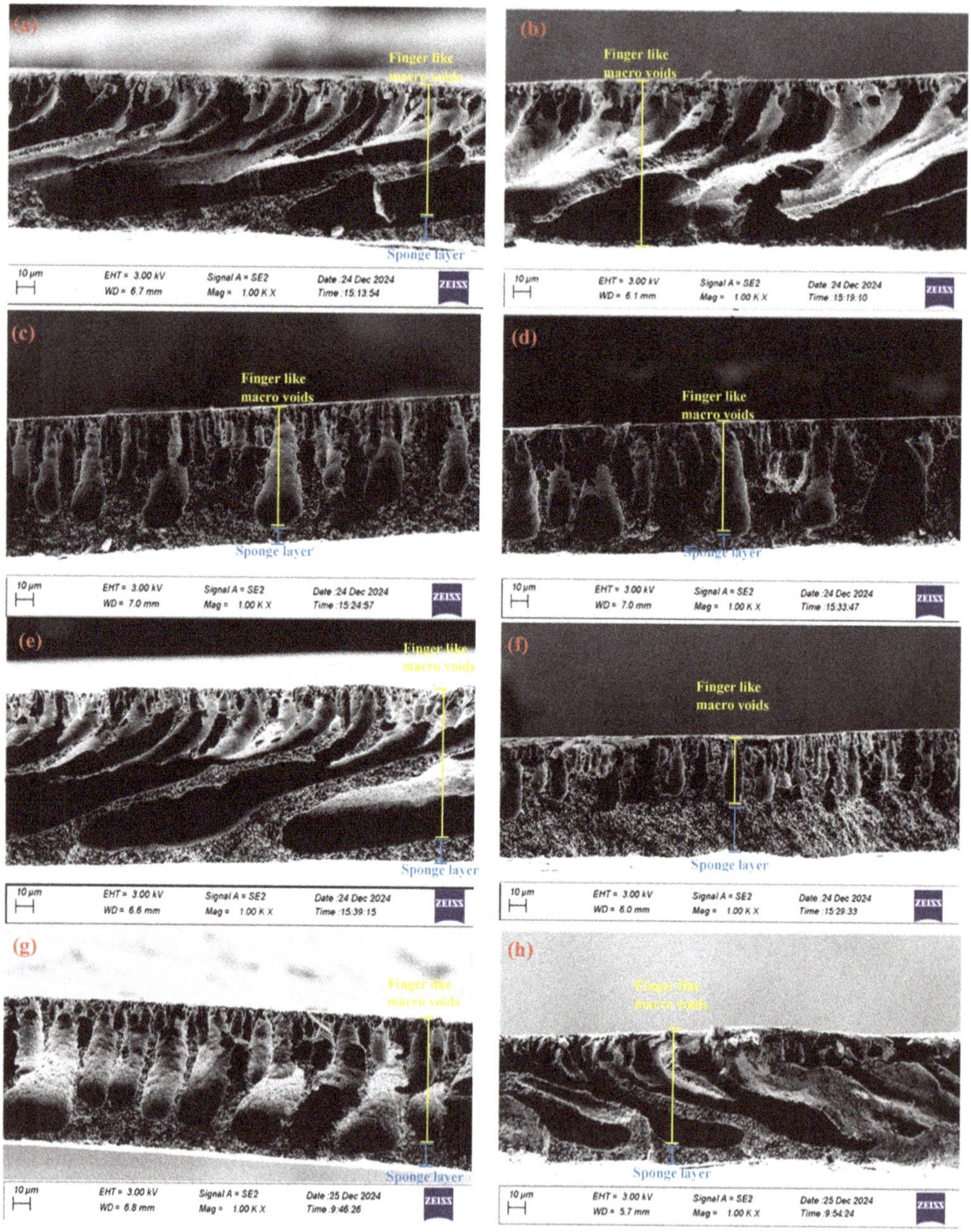

Figure 5. FE-SEM cross-section image of various concentrations of CNM/PAC/PVDF membranes (**a**) P0 (0 wt.%), (**b**) P1 (0.1 wt.%), (**c**) P2 (0.2 wt.%), (**d**) P3 (0.3 wt.%), (**e**) P4 (0.4 wt.%), (**f**) P5 (0.5 wt.%), (**g**) (1 wt.% FP4), (**h**) (2 wt.% FP4) membranes.

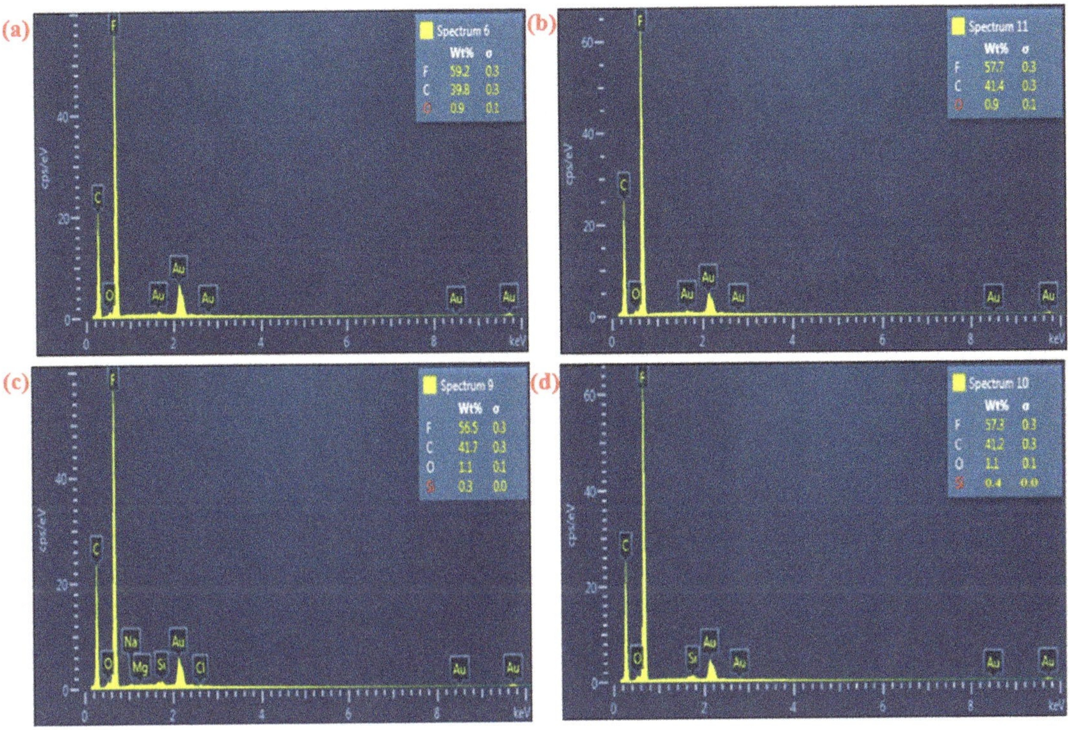

Figure 6. EDX image of (**a**) pristine PVDF (P0), (**b**) 0.4 wt.%CNM/PAC (P4) before and after chemical modification with 1–2 wt.% PFTES (**c**) (1 wt.% FP4), and (**d**) (2 wt.% FP4) membranes.

The mean pore size of all the prepared membranes is shown in Figure 7c. It can be seen that as the CNM/PAC increases, the pore size increases. The pore size increased from 0.32 μm to 0.52 μm as the CNM/PAC concentrations increased from 0.1% (P1) to 0.4 wt.% (P4) and then decreased to 0.47 after loading 0.5 wt.% CNM/PAC (P5). The addition of 0.5 wt.% CNM/PAC may be caused by increasing the viscosity of the casting solution, delaying the rate of exchange between the solvent and non-solvent, and reducing the pore size. In addition, Figure 7c displays the effect of CNM/PAC on the LEP. It can be seen from that figure the LEP decreased as the CNM/PAC went up from 0.1 to 0.4 wt.% due to the inverse relation with pore size. On the other hand, Figure 7d displays the effect of chemical modification on pore size and LEP. After modification with 1 and 2% PFTES, the mean pore size decreased slightly from 0.52 to 0.5 μm and 0.49 μm. This reduction in mean pore size may be due to pore shrinkage caused by the deposition of PFTES on the membrane surface. The decrease in pore size and porosity may cause a reduced permeate flux [19] while increasing the LEP. The increase in LEP post-modification indicates enhanced hydrophobicity and wetting resistance, which are crucial for maintaining membrane performance during long-term MD operation [46]. However, it can be seen that the 2% FP4 membrane shows only a slightly smaller pore size than the P4 membrane but significantly higher LEP than the P4 membrane. This observation was attributed to the slight reduction in pore size of the 2% FP4 membrane compared to the P4 membrane because of the effect of the fluoro-silane solution. At high concentrations of fluoro-silane, the pore size may be partially reduced by the silane solution (PFTES), resulting in a reduction in the pore size, whereas the enhancement in LEP was due to the inverse relation with the pore size and also attributed to the effect of PFTES on the hydrophobicity of the surface. When coating the membrane with 2%PFTES, the contact angle increased from 84° to 117° and the LEP was increased accordingly.

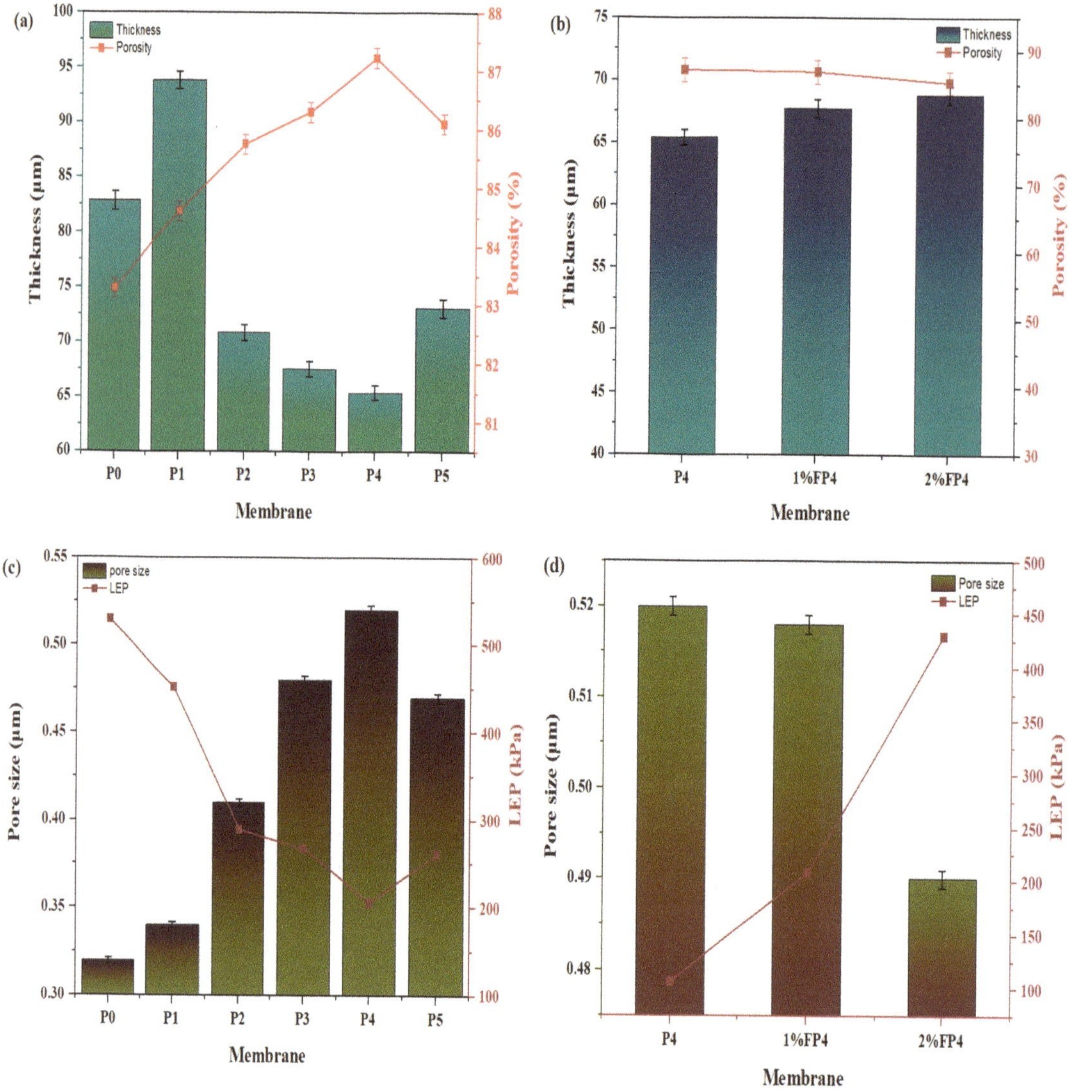

Figure 7. Physical characteristics of membrane before and after modification. (**a**,**b**) Thickness and porosity; (**c**,**d**) pore size.

5.5. Water Contact Angle

The contact angle plays an important role in evaluating the wettability of the membrane surface, which directly influences its hydrophobicity and efficiency in membrane distillation (MD) [47]. Figure 8 shows that the water contact angle for the pristine PVDF membrane (P0) was 65°. After the addition of 0.1% (P1), 0.2 wt.% (P2), 0.3 wt.% (P3), and 0.4 wt.% (P4) CNM/PAC, the contact angle increased to 73°, 78°, 79°, and 84°, respectively (Figure 8). The result indicates that embedding of CNM/PAC nanomaterials enhances the hydrophobicity of the membrane surface. The loaded nanomaterials increased roughness, leading to micro/nano-scale surface roughness, which in turn traps air pockets and increases the membrane's hydrophobic character [48,49]. This effect can be attributed to the surface modification achieved through nanomaterial embedding, which promotes the formation of air gaps that reduce liquid–solid contact, thereby increasing the contact angle [35]. However, as the CNM/PAC concentration was further increased to 0.5 wt%(P5), the contact angle decreased to 78°, suggesting that at higher concentrations of CNM/PAC, it may aggregate, potentially reducing the surface roughness and impairing the hydrophobicity [42]. Additionally, the water contact angle slightly decreased over time.

Also, Figure 8 displays the contact angle of the P4 membrane before and after chemical modification with 1 wt.% and 2 wt.% PFTES. It is evident that increasing the fluoro-silane concentration led to a significant increase in the water contact angle. The contact angle rose from 84° (P0) to 111° (1 wt.% FP4) and 117° (2 wt.% FP4). This suggests that the fluoro-silane coating effectively reduced the surface energy of the membrane, rendering it more hydrophobic. The increased contact angle can be explained by the deposition of fluoro-silane molecules, which create a uniform hydrophobic layer on the membrane surface [19]. Additionally, as shown in Figure 8, the contact angle of P4 and the PFTES-coated membranes (1 wt.% FP4 and 2 wt.% FP4) decreased slightly and remained hydrophobic over time, indicating the durability of the hydrophobicity under continuous exposure to water. This stability is critical for maintaining membrane performance during prolonged operation in MD systems.

5.6. Atomic Force Microscopy (AFM)

Since the surface contact angle affects the wetting, it is crucial to evaluate the surface roughness of any membrane to measure its effect on wetting [50–53]. Figure 9 displays the surface roughness of the PVDF membrane before and after modification with CNM/PAC and chemical modification. The pristine PVDF (P0) membrane had the smoothest surface roughness (Sa = 64.76 nm), as seen in Figure 9. After adding 0.1–0.4 wt.% of CNM/PAC, the roughness of the surface rose to 66.96, 81.53, 93.94, and 95.93 nm and then decreased after loading 0.5 wt.%(P5) (Figure 9). The aggregate of nanomaterials lowers both surface hydrophobicity and roughness [20]. Also, Figure 9 demonstrates the utilization of AFM to test the roughness of the chemically modified P4 membrane. As the concentration of PFTES solution increased, the chemically modified membrane showed a higher surface roughness compared to the original CNM/PAC membrane. The average surface roughness for the P4 (0.4 wt.%) CNM/PAC), (1 wt.% FP4), and (2 wt.% FP4) membranes increased from 95.93 to 156.3 and 175 nm, respectively, as shown in Table 2. In addition to hydrophobicity, the AFM analyzer has been used to determine the porosity and pore size distribution of the prepared membranes. The surface porosity, however, behaves in the opposite way. For P0, P1, P2, P3, P4, and P5, the corresponding surface porosities were found to be 26.64, 26.75, 27.01,27.88, 32.01, and 19.99%. After adding 1 wt.% PFTES to P4 (1 wt.% FP4), the surface porosity increased to 64.83. However, after being treated with 2 wt.% PFTES (2 wt.% FP4) it reduced to 38.97%, but it remains higher than the origin P4 membranes. This increase in surface porosity was due to the large number of maximum pore sizes on the membrane surface. This indicates the increase in the surface porosity after dip-coating [41].

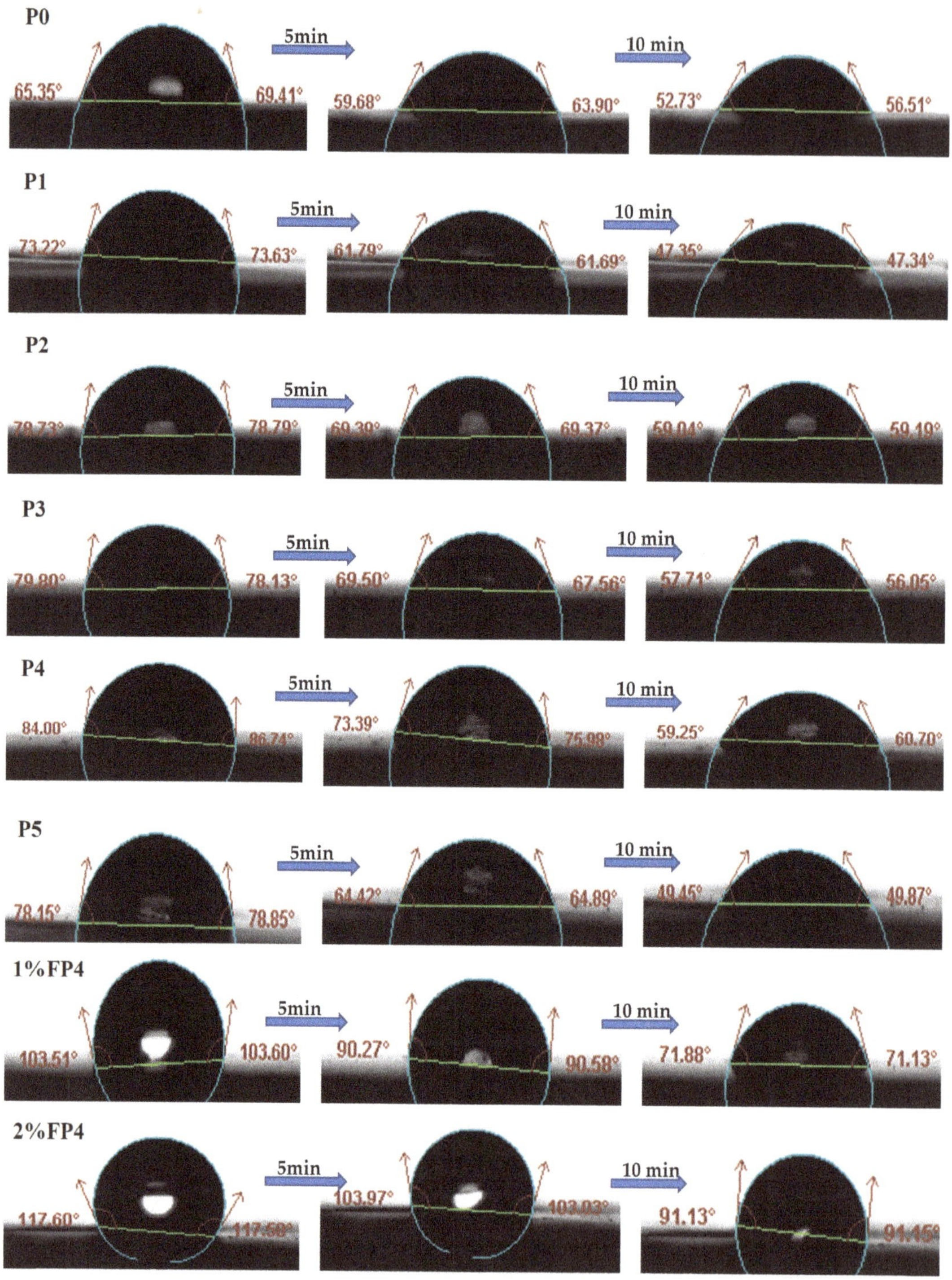

Figure 8. Contact angle with time for P0, P1, P2, P3, P4, P5, and chemical modification of (P4) with 1–2 wt.% PFTES (1 wt.% FP4), (2 wt.% FP4) membranes.

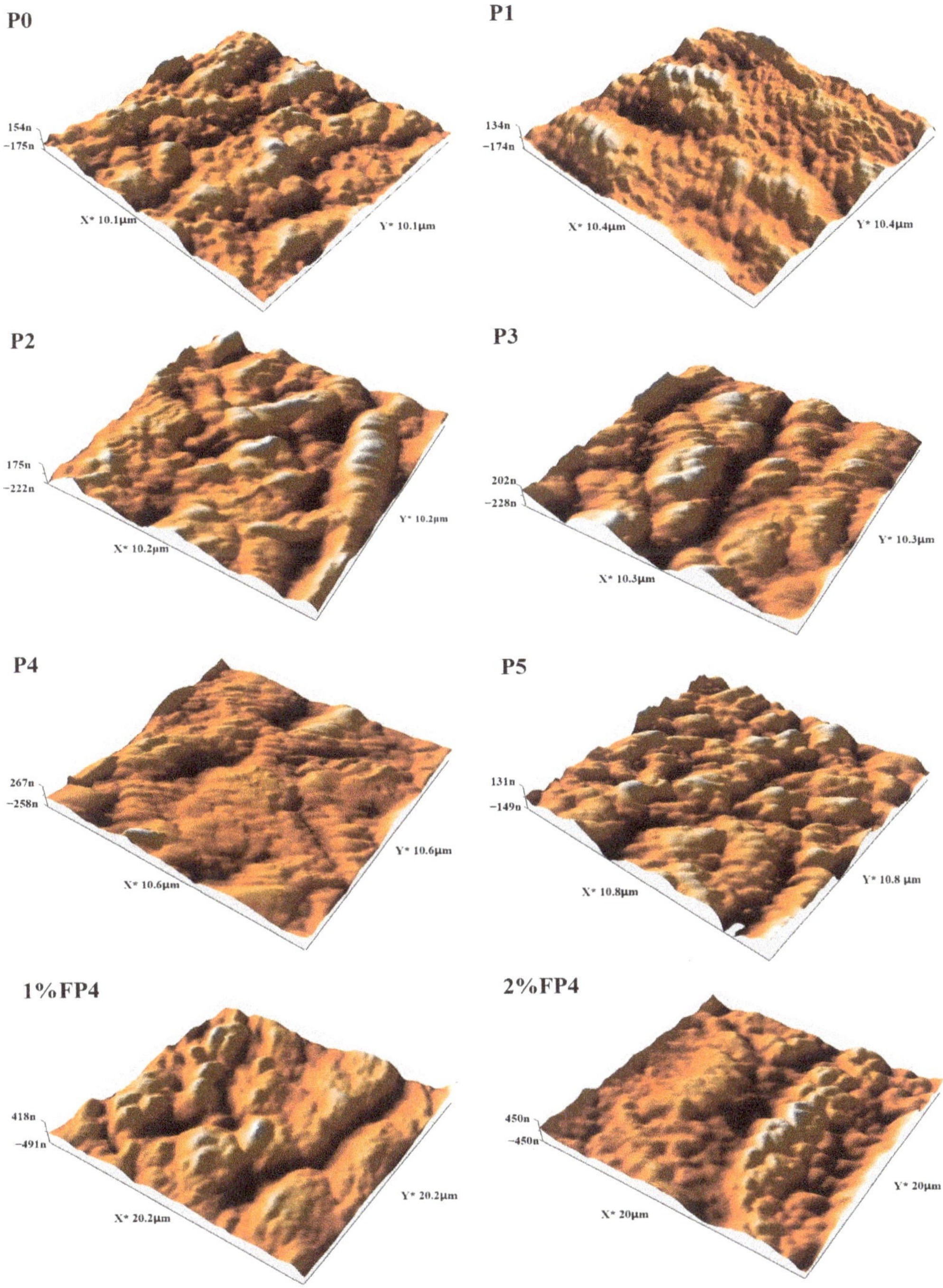

Figure 9. AFM images for P0, P1, P2, P3, P4, P5, and chemical modification of (P4) with 1–2 wt.% PFTES (1% FP4), (2% FP4) membranes.

The surface pore size indicates an increase as the concentration of CNM increases. The surface pore size increased from 0.044 µm to 0.087 µm and 0.131 µm for P0, P1, and

P4, respectively, then decreased after loading 0.5 wt.% CNM to 0.107 μm. After chemical modification with 1 wt.% PFTES, the pore size of the P4 membrane increased from 0.131 to 0.232 μm and then decreased to 0.142 μm with the saline solution increased up to 2%.

Table 2. Roughness characteristic of all the prepared membranes.

Membrane Code	Average Surface Roughness (Ra) nm	Root Mean Square (Rq) nm	Maximum Height (Rz) nm
P0	64.76	78.68	455.6
P1	66.96	83.28	499.9
P2	81.53	100.2	641.6
p3	93.94	115.2	706.3
p4	95.93	115.1	600
P5	56.17	69.56	466.2
1% FP4	156.3	196.3	1294
2% FP4	175	212.7	1122

5.7. Mechanical Characteristic of Membrane

The mechanical characteristics of membranes play an important role in finding their ability to withstand high pressure differences over a long period of VMD [50]. Figure 10a demonstrates the stress–strain properties of membranes with varying CNM/PAC loadings. It is clear that increasing the concentration of CNM/PAC generally resulted in an increase in tensile strength. The pristine PVDF membrane (P0) showed the lowest tensile strength at 3.8 MPa. However, the tensile strength of the modified membranes increased gradually as the CNM/PAC was increased. The tensile stress increased from 3.8 MPa to 4.1 MPa, 4.5 MPa, and 5.025 MPa for P1, P2, and P3, respectively. On the other hand, as a CNM/PAC concentration increased to 0.4 wt.% (P4) and 0.5 wt.% (P5), the tensile strength decreased to 3.312 MPa and 3.845 MPa, respectively. This decline may be attributed to poor dispersion and excessive aggregation of CNM/PAC within the polymer matrix. The aggregation of nanomaterials leads to stress concentration points, which reduce the overall mechanical strength of the membrane [42]. Thus, while CNM/PAC incorporation influences membrane properties, high loading levels must be carefully controlled to avoid compromising mechanical integrity through structural defects and material agglomeration.

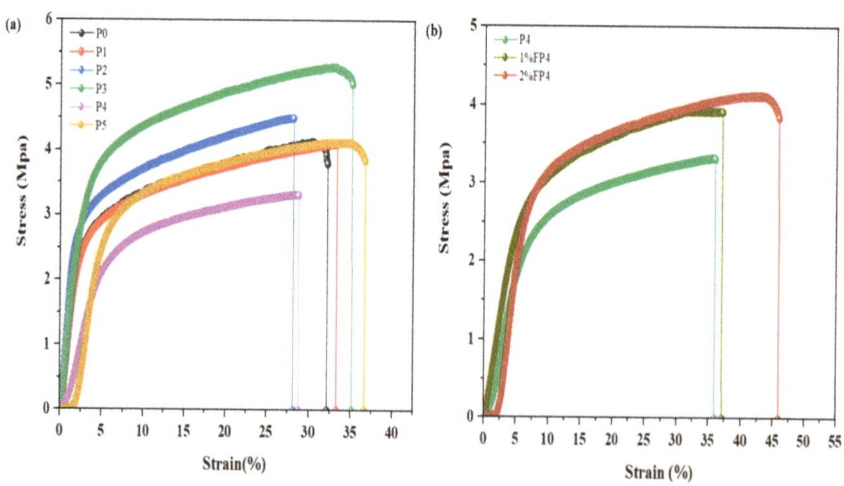

Figure 10. Stress–strain characteristics of (**a**) P0, P1, P2, P3, P4, P5 membrane; (**b**) P4 before and after chemical modification with 1–2 wt.% PFTES (1% FP4, 2% FP4) membranes.

Figure 10b displays the tensile stress–strain characteristics of the (P4) membrane after chemical modification with 1 wt.% and 2 wt.% PFTES. The chemically modified membrane of P4 enhanced its mechanical properties upon modification with PFTES. The origin P4 membrane exhibits a tensile strength of 3.8312 MPa, while the chemically modified membrane with 1 and 2% PFTES displays an increase to 3.915 MPa and 4.11 MPa. This enhancement in mechanical properties suggests that the dual modification (incorporating CNM/PAC and PFTES) prevents a significant loss of mechanical strength. The improved tensile strength in 1% FP4 and 2% FP4 makes them more suitable for long-term vacuum membrane distillation (VMD) processes. The increase in tensile strength for the dual-modified membranes, especially when compared to the pristine and solely modified membranes, provides an advantage in terms of mechanical durability, which is critical for MD applications that involve high operating pressure difference and extended usage periods.

6. Membrane Performance

6.1. Pure Water and Salt Solution

The performance of all prepared membranes in vacuum membrane distillation (VMD) was examined for both distilled water and simulated seawater at 35,000 ppm NaCl under operating conditions of 65 °C feed temperature, 0.6 L/min feed flowrate, and 21 kPa(abs) vacuum pressure. Figure 11a shows the permeate flux through the pristine PVDF (P0) membrane before and after loading different concentrations of CNM/PAC P1 to P5. The results demonstrated that the permeate flux increased by about 11.9%, 43.4%, 94.1%, 99.6%, and 74.7% for P1, P2, P3, and P4, respectively, compared to the pristine PVDF membrane (P0). This enhancement in flux may be due to the addition of CNM/PAC, which improves membrane properties such as porosity and surface hydrophobicity and reduces thickness, offering a larger surface area for evaporation and lower resistance to mass transfer. The highest permeate flux was observed in the P4 membrane (0.4 wt.% CNM/PAC), achieving 23.81 L/m^2·h. The presence of CNM/PAC also enhances water vapor transfer by acting as an adsorbent, thereby improving both flux and membrane selectivity. This trend suggests that the optimal amount of CNM/PAC leads to enhanced membrane morphology, hydrophobicity, and porosity, which contributes to enhanced flux performance.

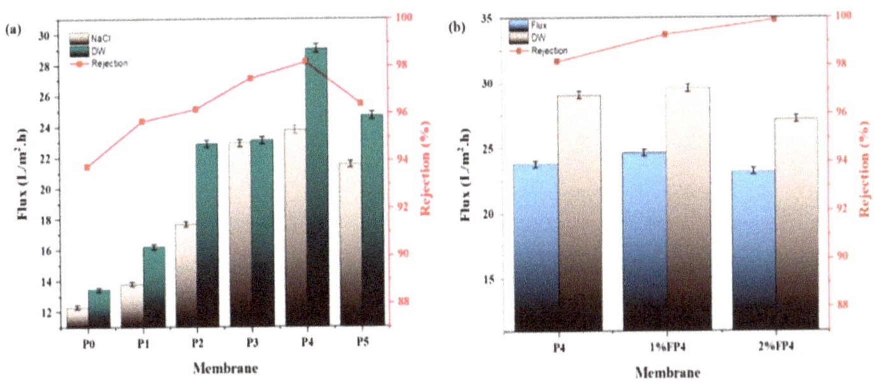

Figure 11. VMD performance for (**a**) P0, P1, P2, P3, P4 and P5 membranes; (**b**) P4 before and after chemical modification with 1–2%PFTES (1% FP4), (2% FP4).

Also, Figure 11a demonstrates the salt rejection performance of all the membranes. The pristine PVDF membrane (P0) exhibited a salt rejection rate of 93.8%, indicating the presence of pore wetting and low hydrophobicity. In contrast, the salt rejection rates of P4 exceeded 98.2%, signifying improved membrane hydrophobicity and resistance to pore wetting, which are crucial for effective desalination. Figure 11b displays the effect of varying silane solutions on the performance of the membrane. The figure shows an increase in permeate flux to 24.64 $L/m^2 \cdot h$ upon coating with 1% silane solution as a result of expanding the surface pore size caused by the evaporation of the hexane-silane solution and increased surface porosity. However, this increase was decreased to 23.2 $L/m^2 \cdot h$ when the concentration of silane solution increased to 2 wt% PFTES, as the surface pore size shrank, with the stability of salt rejection above 99.9%.

It is worth mentioning here that the membrane coated with 1 wt.% PFTES had a higher permeate flux and lower rejection than the membranes coated with the 2 wt.% PFTES. This was due to the moderate effect of a coated layer of 1 wt.% PFTES on the physical properties of the membrane surface, such as surface porosity and pore size, caused by the evaporation of the hexane-silane solution, resulting in higher permeate flux and low rejection of salt solution compared to using a high concentration of PFTES (2%) as a coated layer, as shown in Figure 4h.

6.2. *Effect of Operating Conditions*

The temperature of the feed solution has a direct impact on evaporation rates; it is an important parameter affecting the performance of the membrane in the MD process. This study investigates the effect of feed temperature at 45–65 °C on permeate flux for salt solution at 35 g/L while keeping other parameters constant, such as feed flow rate and vacuum pressure of 0.6 L/min and 21 kPa (abs), respectively. Figure 12a illustrates that an increase in feed temperature enhanced the permeate flux; this is due to increased evaporation of the feed solution at elevated temperatures, which raises the driving force between the two sides of the membrane [51–53]. As the temperature increased from 45 °C to 65 °C, the permeate flux for the P4 membrane rose from 8.6 to 23.82 $L/m^2 \cdot h$. CNM/PAC-loaded membranes showed considerably higher permeate flux. After being chemically modified with 2 wt.% PFTES, the 2% FP4 membranes' permeate flux rose from 7.53 to 23.2 $L/m^2 \cdot h$. The reason for this rise in flux is that the permeate side vacuum pressure remained constant, but the partial vapor pressure increased at higher feed temperatures, increasing the driving force for vapor transport. Also, Figure 12a highlights the effect of feed rejection. The results indicate that higher feed temperatures negatively impacted salt rejection due to increased pore wetting. As feed temperature rises, the surface tension decreases, making it easier for the pores to wet, which reduces the membrane's rejection capacity [54].

Feed flow is a critical parameter in the VMD process due to its impact on temperature gradients across the membrane and its ability to reduce both temperature and concentration polarization phenomena at the membrane surface [55]. Figure 12b shows the variation in permeate flux under different feed flow rates (0.4, 0.5, and 0.6 L/min). The data reveal that the permeate flux increased across the membranes. The permeate flux of P4 increased from 8.9 to 23.82 $L/m^2 \cdot h$. After PFTES chemical modification, the flux for P4 further increased from 7.82 to 23.2 $L/m^2 \cdot h$. The observed increase in permeate flux with higher feed flow rates can be attributed to improved Reynolds numbers, which reduce mass and heat transfer resistance by minimizing concentration and temperature polarization effects (i.e., reducing the thickness of the boundary layer) on the feed side, thus enhancing flux performance [54,56,57]. Also, Figure 12b shows the effect of feed flow rate on salt rejection. It was observed that water conductivity increased at higher feed flow rates, likely due to the increased internal pressure that may surpass the liquid entry pressure (LEP) of the membrane, thereby enhancing wettability and reducing salt rejection [58].

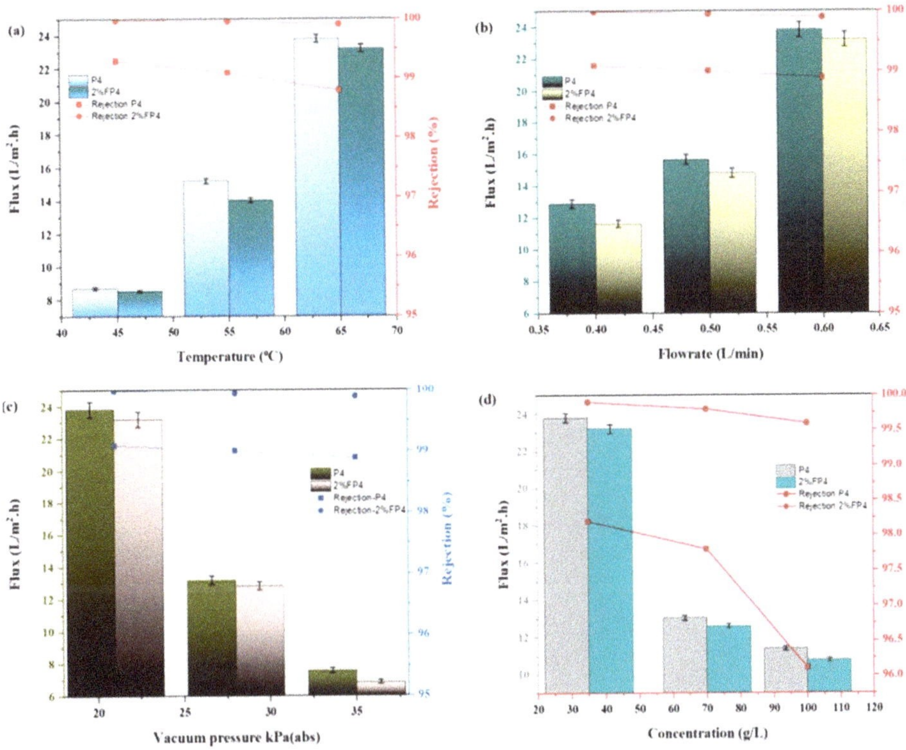

Figure 12. VMD performance for P4 before and after chemical modification (a) effect of feed temperature; (b) effect of feed flow rate; (c) effect of vacuum pressure; (d) effect of feed concentration.

Vacuum pressure is a critical parameter that significantly influences both permeate flux and the overall operation of the VMD process [59]. Figure 12c illustrates the effect of vacuum pressure on the permeate flux for a 35 g/L salt solution using PVDF membranes, both before and after chemical modification at various vacuum pressures from 21 to 35 kPa (abs), while other conditions remained constant. In general, the efficiency of the VMD process increased linearly as vacuum pressure decreased. Specifically, the permeate flux of P4 increased from 7.56 to 23.8 $L/m^2{\cdot}h$. Following chemical modification, the permeate flux of 2% FP4 further improved, increasing from 6.8 to 23.2 $L/m^2{\cdot}h$ as the vacuum pressure was reduced from 35 to 21 kPa (abs). This rise in permeate flux is mostly due to the improved driving force (transmembrane vapor pressure gradient) between the feed and permeate sides of the membrane, which becomes more pronounced as vacuum pressure decreases, minimizing thermal losses [56,60,61]. Moreover, Figure 12c highlights the effect of vacuum pressure on salt rejection. A higher rejection percentage was observed with decreasing vacuum pressure, likely due to the intensified driving force at lower pressures, which reduces the likelihood of pore wetting and improves separation efficiency [54,62].

Figure 12d shows the effect of varying feed salt concentrations on the permeate flux in VMD for the membranes P4, and 2% FP4. Increasing salt concentration negatively affected permeate flux across all membranes. For instance, the permeate flux of P4 dropped from 23.81 to 11.4 $L/m^2{\cdot}h$ whereas the chemically modified membrane with 2 wt.% PFTES (2% FP4) experienced a decline from 23.2 to 10.78 $L/m^2{\cdot}h$ as salt concentration increased from 35 to 100 g/L. This reduction in permeate flux with increasing salt concentration is attributable to a decrease in water activity and partial vapor pressure. Additionally, a higher salt concentration exacerbates concentration polarization, which adds resistance to mass transfer and reduces the vapor pressure differential across the membrane [63].

6.3. Leaching Nanomaterials and Longtime of Operation

The leaching of nanomaterials is one of the challenges that affect membrane performance. Figure 13a displays the leaching test of P0 and CNM/PAC-PVDF(P4) before and after chemical modification with 2 wt.% PFTES (2% FP4). The P4 membrane exhibited an initial leaching ratio of 0.54%, which eventually stabilized at 0.61%. The stability of CNM/PAC within the membrane matrix is crucial for maintaining optimal performance in vacuum membrane distillation (VMD) applications. The chemically modified membrane (2% FP4) demonstrated a higher initial leaching rate of 0.43%, likely due to the release of residual fluoro-silane. However, the leaching rate gradually decreased and stabilized at 0.36%, suggesting improved stability over time. Studies suggest that fluoro-silane, when used as a cross-linking agent, enhances the mechanical properties of the membrane, reducing long-term leaching [64]. Figure 13b illustrates the operational stability of the P0, P4, and 2% FP4 membranes over 96 h. The permeate flux for P0 decreased by approximately 44.7%, while P4 experienced a more moderate decline of 15.5%. In contrast, the chemically modified 2% FP4 membrane exhibited superior stability, with only an 8.4% reduction in flux. This slight decrease indicates that the modified membrane maintained its structural integrity and performance during prolonged operation, preserving its morphological properties.

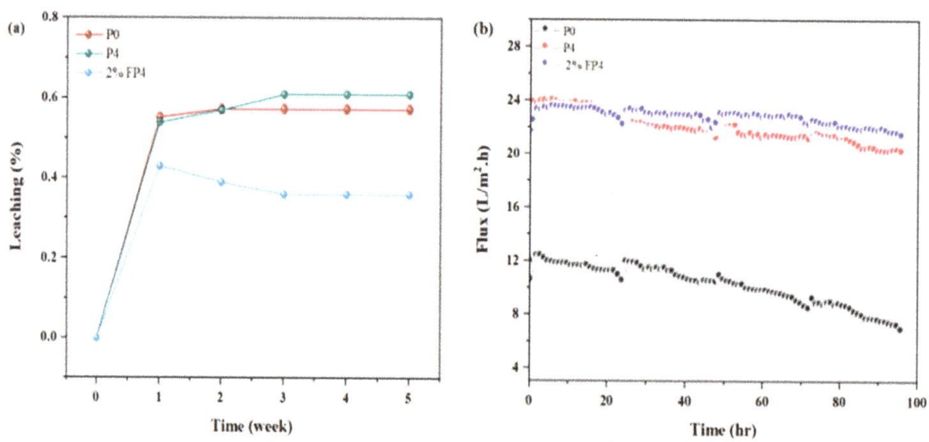

Figure 13. (**a**) Leaching ratio for P0, P4, and 2% FP4 membranes; (**b**) long-term operation test.

The evaluation of the chemical stability of the membrane surface prepared by the current work was assessed according to the results of the long-term operation of the membranes shown in Figure 13b. It can be seen that the permeation flow rate for long periods of time of P4 and 2% FP4 membranes was approximately stable for 4 days of long-term operation. This observation indicated that the surface of the membrane was chemically stable.

Table 3 shows a comparison of PVDF membranes coated with different types of silane solution and used for water desalination by membrane distillation listed in the literature. The comparison presents many properties of the membranes, such as contact angle, operating time, flux, and rejection. The CNM/PAC/PVDF membrane coated with 2% PFTES demonstrates high performance in terms of permeate flux and high removal.

Table 3. Some surveys from the literature for fluoro-silane-coated PVDF membranes in MD.

MD Conf.	Polymer	Fluorosilane Type	Contact Angle (°)	Operating Time (h)	Flux (kg/m^2·h)	Rejection (%)	Ref.
DCMD	PVDF-HFP	perfluorodecyldimethylchlorosilane (FAS-17)	138	/	19	99.9	[65]
DCMD	PVDF	Perfluorodecyltriethoxysilane (PFDES)	162	/	/	99.9	[66]
AGMD	PVDF	Heneicosafluorododecyltrichlorosilane (FAS-21)	>150	/	19.09	99.9	[41]
DCMD	PVDF	methyl trichloroalkyl silane (MTCS) perfluorooctane trichlorosilane silanes (PTCS)	112 116	/	/	/	[19]
VMD	PVDF	Perfluorodecyltriethoxysilane (PFTES)	117	96	23.2	99.9	This work

7. Conclusions

In this study, we enhanced the hydrophobicity and the salt rejection of the CNM/PAC/PVDF membrane by chemical modification with PFTES at various concentrations. Embedded CNM/PAC nanomaterials improved porosity, pore size, mechanical properties, and hydrophobicity and reduced thickness, which in turn enhanced the permeate flux and salt rejection. The permeate flux for P4 was 23.81 L/m^2·h, which was 48.2% higher than the pristine PVDF membrane (P0), and the salt rejection was 98.2%. Further chemical modification with 1% PFTES enhanced the permeate flux to 24.6 L/m^2·h and salt rejection to 99.3% due to the increase in surface pore size and porosity, while chemical modification with 2 wt.% PFTES reduced the permeate flux to 23.2 L/m^2·h and enhanced the salt rejection and mechanical properties to 99.9% and 4.1 MPa. Also, it can be concluded from the FTIR analysis of the PVDF membrane modified by different amounts of PFTES that the Si-O peak in FTIR corresponded to the covalent bonds between the PFTES and the PVDF membrane surface, which indicated that the surface of the membrane was saturated with PFTES. Furthermore, the long-term stability test revealed a lower degradation (7.76%) after chemical modification with 2 wt.% PFTES with the stability of salt rejection above 99.9%. In order to confirm the stability of chemical modification, leaching tests were conducted over five weeks, and it was found that the 2% FP4 membrane showed a stable leaching rate of 0.36, which confirms the minimal release of nanomaterials after chemical modification. These results highlight the potential of different concentrations of PFTES for membrane performance and salt rejection to improve desalination efficiency over VMD and maintain durability over a long period of operation.

Author Contributions: Conceptualization, Q.F.A. and S.S.I.; methodology, Q.F.A., S.R.K., A.C. and S.S.I.; software, A.F., D.U.L., Q.F.A., S.S.I., S.R.K. and A.C.; formal analysis, A.C. and S.S.I.; investigation, Q.F.A., S.S.I. and A.C.; writing—original draft preparation, Q.F.A., S.R.K. and S.S.I.; writing—review and editing, Q.F.A., S.S.I. and A.C.; supervision, Q.F.A. and S.S.I. All authors have read and agreed to the published version of the manuscript.

Funding: This research received no external funding.

Institutional Review Board Statement: Not applicable.

Data Availability Statement: Dataset available on request from the authors.

Conflicts of Interest: The authors declare no conflict of interest.

References

1. Patel, R.V.; Chaubey, S.; Yadav, A.; Shahi, V.K. Chemical grafting of hydrophobic functional groups on polyvinylidene fluoride side chain for vacuum membrane distillation applications. *J. Environ. Chem. Eng.* **2024**, *12*, 112746. [CrossRef]
2. Asadolahi, M.; Fashandi, H.; Zamani, H. Disclosing the pivotal role of Poly(vinylidene fluoride) omniphobic membrane surface morphology in performance of photothermal vacuum membrane distillation. *Chem. Eng. J.* **2024**, *496*, 153970. [CrossRef]
3. Zhang, N.; Zhang, J.; Gao, C.; Yuan, S.; Wang, Z. Emerging advanced membranes for removal of volatile organic compounds during membrane distillation. *Desalination* **2025**, *597*, 118372. [CrossRef]
4. Sun, B.; Wu, M.; Zhen, H.; Jia, Y.; Li, P.; Yuan, Z.; Li, X.; He, G.; Jiang, X. Tailored PVDF membrane with coordinated interfacial nano/micro-structure for enhanced membrane distillation. *Desalination* **2024**, *573*, 117177. [CrossRef]
5. Julian, H.; Nurgirisia, N.; Qiu, G.; Ting, Y.-P.; Wenten, I.G. Membrane distillation for wastewater treatment: Current trends, challenges and prospects of dense membrane distillation. *J. Water Process Eng.* **2022**, *46*, 102615. [CrossRef]
6. Chin, J.Y.; Ahmad, A.L.; Low, S.C. Anti-Wetting Membrane Distillation to Treat High Salinity Wastewater: Review. *J. Membr. Sci. Res.* **2020**, *6*, 401–415. [CrossRef]
7. Asadolahi, M.; Fashandi, H.; Ghodsi, A. One-step development of photothermal omniphobic membrane for vacuum membrane distillation towards sustainable water desalination using solar energy. *Chem. Eng. J.* **2024**, *500*, 157428. [CrossRef]
8. Samadi, A.; Ni, T.; Fontananova, E.; Tang, G.; Shon, H.; Zhao, S. Engineering antiwetting hydrophobic surfaces for membrane distillation: A review. *Desalination* **2023**, *563*, 116722. [CrossRef]
9. Pangarkar, B.L.; Deshmukh, S.K.; Sapkal, V.S.; Sapkal, R.S. Review of membrane distillation process for water purification. *Desalin. Water Treat.* **2016**, *57*, 2959–2981. [CrossRef]
10. Ahmadi, H.; Ziapour, B.M.; Ghaebi, H.; Nematollahzadeh, A. Optimization of vacuum membrane distillation and advanced design of compact solar water heaters with heat recovery. *J. Water Process Eng.* **2024**, *67*, 106212. [CrossRef]
11. Rashid, K.T.; Rahman, S.B.A.; Alsalhy, Q.F. Optimum Operating Parameters for Hollow Fiber Membranes in Direct Contact Membrane Distillation. *Arab. J. Sci. Eng.* **2016**, *41*, 2647–2658. [CrossRef]
12. Abid, M.B.; Wahab, R.A.; Salam, M.A.; Gzara, L.; Moujdin, I.A. Desalination technologies, membrane distillation, and electrospinning, an overview. *Heliyon* **2023**, *9*, e12810. [CrossRef]
13. Francis, L.; Ahmed, F.E.; Hilal, N. Advances in Membrane Distillation Module Configurations. *Membranes* **2022**, *12*, 81. [CrossRef]
14. Alsebaeai, M.K.; Ahmad, A.L. Membrane distillation: Progress in the improvement of dedicated membranes for enhanced hydrophobicity and desalination performance. *J. Ind. Eng. Chem.* **2020**, *86*, 13–34. [CrossRef]
15. Suga, Y.; Takagi, R.; Matsuyama, H. Effect of the characteristic properties of membrane on long-term stability in the vacuum membrane distillation process. *Membranes* **2021**, *11*, 252. [CrossRef] [PubMed]
16. An, Y.; Li, F.; Wang, H.; Zhang, X.; Di, Y.; Wang, L.; Yan, Z.; Wang, W.; Lu, J.; Liu, M.; et al. Preparation and properties of cellulose triacetate (CTA)-PVDF composite membrane with hydrophobic/hydrophilic properties for membrane distillation. *Appl. Surf. Sci.* **2025**, *684*, 161891. [CrossRef]
17. Zheng, R.; Chen, Y.; Wang, J.; Song, J.; Li, X.-M.; He, T. Preparation of omniphobic PVDF membrane with hierarchical structure for treating saline oily wastewater using direct contact membrane distillation. *J. Memb. Sci.* **2018**, *555*, 197–205. [CrossRef]
18. Schnittger, J.; McCutcheon, J.; Hoyer, T.; Weyd, M.; Fischer, G.; Puhlfürß, P.; Halisch, M.; Voigt, I.; Lerch, A. Hydrophobic ceramic membranes in MD processes—Impact of material selection and layer characteristics. *J. Membr. Sci.* **2021**, *618*, 118678. [CrossRef]
19. Jin, Q.; Zhang, X.; Li, F.; Zhao, X. Hydrophobic modification of a PVDF hollow fiber membrane by plasma activation and silane grafting for membrane distillation. *Water Sci. Technol.* **2023**, *87*, 2806–2819. [CrossRef]
20. Swaminathan, M.; Swaminathan, G. An Insight into Enhanced Membrane Hydrophobicity Using Silane Functionalization for Effective Membrane Distillation. *Sep. Purif. Rev.* **2024**, *31*, 1–21. [CrossRef]
21. Van Tran, T.T.; Nguyen, C.H.; Lin, W.-C.; Juang, R.-S. Improved stability of a supported liquid membrane process via hydrophobic modification of PVDF support by plasma activation and chemical vapor deposition. *Sep. Purif. Technol.* **2021**, *277*, 119615. [CrossRef]
22. Arkles, B. Hydrophobicity, hydrophilicity and silanes. *Paint. Coat. Ind.* **2006**, *22*, 114–135.
23. Arkles, B. Tailoring Surfaces with Silanes. *Chemtech* **1977**, *7*, 766–778.
24. Aljumaily, M.M.; Alsaadi, M.A.; Das, R.; Abd Hamid, S.B.; Hashim, N.A.; AlOmar, M.K.; Alayan, H.M.; Novikov, M.; Alsalhy, Q.F.; Hashim, M.A. Optimization of the synthesis of superhydrophobic carbon nanomaterials by chemical vapor deposition. *Sci. Rep.* **2018**, *8*, 2778. [CrossRef]
25. Almajras, Q.A.; Hassan, A.K.; Al-Juboori, R.A.; Alsalhy, Q.F. Green and sustainable biosynthesis of hybrid iron/palladium nanoparticles functionalized PES membranes for dye removal. *Desalin. Water Treat.* **2025**, *321*, 100973. [CrossRef]
26. Tai, B.; Ma, X. Superhydrophobic Surface Modification of Carbon Nanotubes for Polymer Membrane in Direct Contact Membrane Distillation. Master's Thesis, Texas A&M University, College Station, TX, USA, 2019. Available online: https://oaktrust.library.tamu.edu/server/api/core/bitstreams/beb4b8af-dc27-4f48-8838-dbfc91de714c/content (accessed on 19 March 2025).

27. Mpala, T.J.; Richards, H.; Etale, A.; Mahlangu, O.T.; Nthunya, L.N. Carbon nanotubes and silver nanoparticles modification of PVDF membranes for improved seawater desalination in direct contact membrane distillation. *Front. Membr. Sci. Technol.* **2023**, *2*, 1165678. [CrossRef]
28. Hussein Al-Timimi, D.A.; Alsalhy, Q.F.; AbdulRazak, A.A. Polyethersulfone/amine grafted silica nanoparticles mixed matrix membrane: A comparative study for mebeverine hydrochloride wastewater treatment. *Alex. Eng. J.* **2023**, *66*, 167–190. [CrossRef]
29. Ibraheem, B.M.; AlAani, S.; Alsalhy, Q.F.; Nemeth, Z.; Hernadi, K. Improving surface characteristics and operational parameters of polyvinyl chloride ultrafiltration membranes: How to manipulate the performance of thin film composite forward osmosis membranes? *Desalination* **2023**, *565*, 116896. [CrossRef]
30. Francis, L.; Hilal, N. Electrosprayed CNTs on Electrospun PVDF-Co-HFP Membrane for Robust Membrane Distillation. *Nanomaterials* **2022**, *12*, 4331. [CrossRef]
31. Saffarini, R.; Arafat, H.; Thomas, R. Influence of Pore Structure on Membrane Wettability in Membrane. In Proceedings of the Sixth Jordan International Chemical Engineering Conference, Amman, Jordan, 12–14 March 2012; pp. 1–5.
32. Racz, G.; Kerker, S.; Schmitz, O.; Schnabel, B.; Kovacs, Z.; Vatai, G.; Ebrahimi, M.; Czermak, P. Experimental determination of liquid entry pressure (LEP) in vacuum membrane distillation for oily wastewaters. *Membr. Water Treat.* **2015**, *6*, 237–249. [CrossRef]
33. Yang, C.; Peng, X.; Zhao, Y.; Wang, X.; Cheng, L.; Wang, F.; Li, Y.; Li, P. Experimental study on VMD and its performance comparison with AGMD for treating copper-containing solution. *Chem. Eng. Sci.* **2019**, *207*, 876–891. [CrossRef]
34. Rao, L.; You, X.; Chen, B.; Shen, L.; Xu, Y.; Zhang, M.; Hong, H.; Li, R.; Lin, H. A novel composite membrane for simultaneous separation and catalytic degradation of oil/water emulsion with high performance. *Chemosphere* **2022**, *288*, 132490. [CrossRef]
35. Aljumaily, M.M.; Alsaadi, M.A.; Hashim, N.A.; Alsalhy, Q.F.; Das, R.; Mjalli, F.S.; Awanis Hashim, N.; Alsalhy, Q.F.; Das, R.; Mjalli, F.S. Embedded high-hydrophobic CNMs prepared by CVD technique with PVDF-co-HFP membrane for application in water desalination by DCMD. *Desalin. Water Treat.* **2019**, *142*, 37–48. [CrossRef]
36. Tian, X.; Jiang, X. Poly(vinylidene fluoride-co-hexafluoropropene) (PVDF-HFP) membranes for ethyl acetate removal from water. *J. Hazard. Mater.* **2008**, *153*, 128–135. [CrossRef] [PubMed]
37. Silva, T.L.S.; Morales-Torres, S.; Figueiredo, J.L.; Silva, A.M.T. Multi-walled carbon nanotubes/PVDF blended membranes with sponge-and finger-like pores for direct contact membrane distillation. *Desalination* **2015**, *357*, 233–245.
38. Avilés, F.; Cauich-Rodríguez, J.V.; Moo-Tah, L.; May-Pat, A.; Vargas-Coronado, R. Evaluation of mild acid oxidation treatments for MWCNT functionalization. *Carbon* **2009**, *47*, 2970–2975. [CrossRef]
39. Wu, T.; Zhou, B.; Zhu, T.; Shi, J.; Xu, Z.; Hu, C.; Wang, J. Facile and low-cost approach towards a PVDF ultrafiltration membrane with enhanced hydrophilicity and antifouling performance via graphene oxide/water-bath coagulation. *RSC Adv.* **2014**, *5*, 7880–7889. [CrossRef]
40. Khayet, M.; Cojocaru, C.; García-Payo, M.C. Experimental design and optimization of asymmetric flat-sheet membranes prepared for direct contact membrane distillation. *J. Memb. Sci.* **2010**, *351*, 234–245. [CrossRef]
41. Baig, U.; Azeem, M.A.; Lawal, D.U.; Al Abdulgader, H.; Baroud, T.N. Facile and scalable fabrication of superhydrophobic PVDF membrane for the desalination of highly saline water using air-gap membrane distillation system. *J. Environ. Chem. Eng.* **2022**, *10*, 109025. [CrossRef]
42. Zhou, R.; Rana, D.; Matsuura, T.; Lan, C.Q. Effects of multi-walled carbon nanotubes (MWCNTs) and integrated MWCNTs/SiO_2 nano-additives on PVDF polymeric membranes for vacuum membrane distillation. *Sep. Purif. Technol.* **2019**, *217*, 154–163. [CrossRef]
43. El-Zanati, E.; Taha, E.; Ettouney, R.; El-Rifai, M. Functionalized CNT/PVDF Blended Flat Sheet Membrane for Water Desalination via Vacuum Membrane Distillation, Fabrication and Characterization. *Water Energy Food Environ. J. Int. J.* **2022**, *3*, 53.
44. Jiménez-Robles, R.; Moreno-Torralbo, B.M.; Badia, J.D.; Martínez-Soria, V.; Izquierdo, M. Flat PVDF Membrane with Enhanced Hydrophobicity through Alkali Activation and Organofluorosilanisation for Dissolved Methane Recovery. *Membranes* **2022**, *12*, 426. [CrossRef]
45. Jiménez-Robles, R.; Izquierdo, M.; Martínez-Soria, V.; Martí, L.; Monleón, A.; Badia, J.D. Stability of Superhydrophobicity and Structure of PVDF Membranes Treated by Vacuum Oxygen Plasma and Organofluorosilanisation. *Membranes* **2023**, *13*, 314. [CrossRef]
46. Chimanlal, I.; Nthunya, L.N.; Mahlangu, O.T.; Kirkebæk, B.; Ali, A.; Quist-Jensen, C.A.; Richards, H. Nanoparticle-Enhanced PVDF Flat-Sheet Membranes for Seawater Desalination in Direct Contact Membrane Distillation. *Membranes* **2023**, *13*, 317. [CrossRef] [PubMed]
47. Abdelrasoul, A. (Ed.) *Advances in Membrane Technologies*; IntechOpen: London, UK, 2020.
48. Madaeni, S.S.; Zinadini, S.; Vatanpour, V. Preparation of superhydrophobic nanofiltration membrane by embedding multiwalled carbon nanotube and polydimethylsiloxane in pores of microfiltration membrane. *Sep. Purif. Technol.* **2013**, *111*, 98–107. [CrossRef]
49. Aljumaily, M.M.; Alayan, H.M.; Mohammed, A.A.; Alsaadi, M.A.; Alsalhy, Q.F.; Figoli, A.; Criscuoli, A. The influence of coating super-hydrophobic carbon nanomaterials on the performance of membrane distillation. *Appl. Water Sci.* **2022**, *12*, 28. [CrossRef]

50. Aljumaily, M.M.; Alsaadi, M.A.; Hashim, N.A.; Alsalhy, Q.F.; Mjalli, F.S.; Atieh, M.A.; Al-Harrasi, A. PVDF-co-HFP/superhydrophobic acetylene-based nanocarbon hybrid membrane for seawater desalination via DCMD. *Chem. Eng. Res. Des.* **2018**, *138*, 248–259. [CrossRef]
51. Devi, S.; Ray, P.; Singh, K.; Singh, P.S. Preparation and characterization of highly micro-porous PVDF membranes for desalination of saline water through vacuum membrane distillation. *Desalination* **2014**, *346*, 9–18. [CrossRef]
52. Pagliero, M.; Comite, A.; Soda, O.; Costa, C. Effect of support on PVDF membranes for distillation process. *J. Memb. Sci.* **2021**, *635*, 119528. [CrossRef]
53. Liu, J.; Albdoor, A.K.; Lin, W.; Hai, F.I.; Ma, Z. Membrane fouling in direct contact membrane distillation for liquid desiccant regeneration: Effects of feed temperature and flow velocity. *J. Memb. Sci.* **2022**, *642*, 119936. [CrossRef]
54. Chen, L.; Xu, P.; Wang, H. Interplay of the Factors Affecting Water Flux and Salt Rejection in Membrane Distillation: A State-of-the-Art Critical Review. *Water* **2020**, *12*, 2841. [CrossRef]
55. Xu, J.; Singh, Y.B.; Amy, G.L.; Ghaffour, N. Effect of operating parameters and membrane characteristics on air gap membrane distillation performance for the treatment of highly saline water. *J. Memb. Sci.* **2016**, *512*, 73–82. [CrossRef]
56. Iqbal, F.; Asif, M.; Jafry, A.T.; Bibi, W. Experimental Investigation of Vacuum Membrane Distillation (VMD) Performance Based on Operational Parameters for Clean Water Production. *MATEC Web Conf.* **2024**, *398*, 01028. [CrossRef]
57. Pangarkar, B.L.; Sane, M.G.; Parjane, S.B.; Abhang, R.M.; Guddad, M. The heat and mass transfer phenomena in vacuum membrane distillation for desalination. *World Acad. Sci. Eng. Technol.* **2010**, *37*, 977–982.
58. Anvari, A.; Azimi Yancheshme, A.; Kekre, K.M.; Ronen, A. State-of-the-art methods for overcoming temperature polarization in membrane distillation process: A review. *J. Memb. Sci.* **2020**, *616*, 118413. [CrossRef]
59. Abu-Zeid, M.A.E.R.; Zhang, Y.; Dong, H.; Zhang, L.; Chen, H.L.; Hou, L. A comprehensive review of vacuum membrane distillation technique. *Desalination* **2015**, *356*, 1–14. [CrossRef]
60. Hussain Mana, T.; Alam, J.; Shukla, A.K.; Alkhudhiri, A.; Mohammed, A.N.; Alhoshan, M. Performance investigation of poly(vinylidene fluoride-cohexafluoropropylene) membranes containing SiO_2 nanoparticles in a newly designed single vacuum membrane distillation system. *Water Environ. Res.* **2024**, *96*, e10980. [CrossRef]
61. Madupathi, M.M.; Srishti, S.; Fatima, S.; Sridhar, S. Sea and brackish water desalination through a novel PVDF-PTFE composite hydrophobic membrane by vacuum membrane distillation. *Discov. Chem. Eng.* **2024**, *4*, 7. [CrossRef]
62. Kim, H.; Yun, T.; Hong, S.; Lee, S. Experimental and theoretical investigation of a high performance PTFE membrane for vacuum-membrane distillation. *J. Memb. Sci.* **2021**, *617*, 118524. [CrossRef]
63. Giraldo-Mejía, H.; Quintero, Y.M.; Mery, F.; Rodriguez, F.; Curcio, E.; Estay, H.; Garcia, A. Plasma-grafting surface modifications to enhance membrane hydrophobicity for brine membrane distillation. *Desalination* **2023**, *567*, 116942. [CrossRef]
64. Ji, J.; Li, H.; Wang, W.; Li, J.; Zhang, W.; Li, K.; Yang, T.; Jin, W.; Tang, Y.; Li, W.; et al. Silane-crosslinked polybenzimidazole with different hydroxyl content for high-temperature proton exchange membrane. *J. Memb. Sci.* **2024**, *694*, 122423. [CrossRef]
65. An, X.; Liu, Z.; Hu, Y. Amphiphobic surface modification of electrospun nanofibrous membranes for anti-wetting performance in membrane distillation. *Desalination* **2018**, *432*, 23–31. [CrossRef]
66. Li, B.; Yun, Y.; Wang, M.; Li, C.; Yang, W.; Li, J.; Liu, G. Superhydrophobic polymer membrane coated by mineralized β-FeOOH nanorods for direct contact membrane distillation. *Desalination* **2021**, *500*, 114889. [CrossRef]

MDPI AG
Grosspeteranlage 5
4052 Basel
Switzerland
Tel.: +41 61 683 77 34

Membranes Editorial Office
E-mail: membranes@mdpi.com
www.mdpi.com/journal/membranes

www.ingramcontent.com/pod-product-compliance
Lightning Source LLC
Chambersburg PA
CBHW040752250926
51850CB00007BB/1477
* 9 7 8 3 7 2 5 8 7 2 3 2 9 *